Claiming Land, Claiming Water

STUDIES IN EARLY AMERICAN ECONOMY AND SOCIETY
FROM THE LIBRARY COMPANY OF PHILADELPHIA

Jessica Choppin Roney, *Series Editor*

Sponsored by the Program in Early American Economy and Society (PEAES) at the Library Company of Philadelphia, the series supports advanced research in the history of American economy and society to approximately 1850. We define our remit broadly to encompass topics that relate to political economy, agriculture, commerce, business, entrepreneurship, banking, money, transportation, labor, servitude, slavery, technology, the environment, cartography, oceanic and maritime circulations, and continental expansion, invasion, and resistance.

Claiming Land, Claiming Water

Borders and the People Who Crossed Them in the Early Modern Atlantic

EDITED BY
Rachel B. Herrmann and
Jessica Choppin Roney

PENN

UNIVERSITY OF PENNSYLVANIA PRESS

PHILADELPHIA

This work was supported by a Networking Scheme Grant from the UK Arts and Humanities Research Council [AH/R002320/1]

Published by
University of Pennsylvania Press
Philadelphia, Pennsylvania 19104-4112 USA
www.pennpress.org

EU Authorized Representative: Easy Access System Europe - Mustamäe tee 50, 10621 Tallinn, Estonia, gpsr.requests@easproject.com

Printed in the United States of America on acid-free paper
10 9 8 7 6 5 4 3 2 1

Cataloging-in-Publication record for this book is available from the Library of Congress.

Hardcover ISBN 978-1-5128-2925-9
eBook: 978-1-5128-2927-3
OA electronic: 978-1-5128-2926-6

CONTENTS

INTRODUCTION

Borders, Places, and Movement

RACHEL B. HERRMANN AND JESSICA CHOPPIN RONEY

Often historians fall into a double bind of reading historical maps as conveying facts rather than arguments, and the fallacy of imposing modern geographical constructs backwards. The twenty-first-century borders of nation-states, which had neither reality nor meaning in the seventeenth, eighteenth, or nineteenth centuries, continue to inflect the organization of current thinking about this period. *Claiming Land, Claiming Water* asserts that modern scholars bear the responsibility to acknowledge the pitfalls of such practices and to question how the disciplinary origins of history and cartography shape our own assumptions, questions, methods, and archives.

The authors in *Claiming Land, Claiming Water* investigate how and why some people imagined and made claims to bounded space—and how and why other people confounded or challenged those claims—through a formative period of intense change in North America and the Atlantic world (c. 1630–1860). The early modern period was characterized by the rise of oceanic exploration, trade, migration, transcontinental invasion, continental (re)exploration, colonization, empire-building, and the nimble responses of Indigenous actors to the possibilities and perils that newcomers presented.

In this era farmers, fishers, hunters, religious leaders, colonial projectors, traders, sailors, soldiers, diplomats, and cartographers all had their own viewpoint on how space was—or should be—organized. Shaped by their own material and ideational worlds, they experienced land- and waterscapes in distinctive ways and asserted claims accordingly. In the process, as people moved to new places, they imposed onto space their own distinctive ideologies and imaginings.

This volume's authors demonstrate that these ideologies of space were deeply informed by culture and experience, and yet subject to transformation. Whether traveling to new worlds or experiencing an old world made anew by colonizers, pathogens, and objects, no society could reproduce exactly old social, political, and economic structures. Claims upon space and the power to enforce those claims, particularly if the claims were meant to be exclusive, required evolving political, civil, juridical, military, fiscal, and cartographic technologies.[1] The development of those technologies was neither linear nor uncontested, nor was there broad agreement about the sources of authority that legitimated one or another system.

To describe actors moving in and making claims upon early modern space, historians themselves invoke spatial paradigms, none of which have been more influential than one based on water and one on land: Atlantic history and North American hemispheric or continental history. This volume bridges between the two.

The Atlantic world paradigm centers the ocean as the vital link that connected the peoples and histories of Europe, Central and West Africa, and the Americas beginning in the fifteenth century. Characterized by "multiracial, multiethnic, multinational, and multi-imperial" experiences and connections, the Atlantic world's parameters were characterized by the rise of transoceanic exploration and movement, invasion and resettlement, colonization, voluntary and forced migration, commercial agriculture, long-distance trade, empire-building, and the strategic adaptations, reactions, and responses of Indigenous people.[2] Distinct but related to imperial history, Atlantic history insists on multipolarity, polyvocality, and the contingent interplay of many factors at once. While some historians have criticized the Atlantic framework for drawing the scholarly gaze away from the power of Native nations, others have demonstrated that Indigenous American maritime power was formidable not only across the "Red Atlantic," but also in the Pacific Ocean.[3]

In contrast to the Atlantic paradigm, continental historians center the North American continent itself. The impetus of looking into the interior stretches back at least to the nineteenth-century scholarship of Francis Parkman, but received new energy in the late nineteenth century, with Frederick Jackson Turner's thesis that the (white) American character was deeply intertwined with the frontier.[4] Later work reframed the spatial metaphor from a border line to a zone or multiple zones of contact, trade, and contest between Indigenous nations and newer arrivals from Europe and Africa.[5] Over time, much of the center of gravity of early modern continental histories has turned to heavy emphasis on Indigenous peoples because, as Kathleen DuVal has trenchantly reminded us, into the nineteenth century most of North America remained "Native ground."[6] These places, such as the Mississippi River, Mobile Bay, and its environs, were places that Elizabeth Ellis characterizes as borderlands with "multiple and layered land rights." The Petites Nations did not aspire to exercise "uncontested power" and understood that their reciprocal relationships with other people in the Gulf South were what controlled space.[7] Continental history focuses on the political, economic, military, and kinship networks that constituted authority in North America and the contested, brutal process by which the continent was eventually made over from, in Michael Witgen's words, "Native homelands into American homesteads."[8] That process had not concluded by the time period that this volume considers.

Historians of the Atlantic world and of the North American continent share a common goal, which is to recover the multivalent and competitive geopolitical world that predated the dominance of territorial empires or nation-states. In both cases scholars tell rich, interconnected histories that show how entangled were the political, economic, demographic, and cultural processes; how richly varied the cast of characters; and how uncertain and contingent the events that shaped the early modern world. Yet, in doing so, and despite common aims, historians of the North American continent and Atlantic world have pulled in different directions. Thus, while both sets of historians have produced scholarship theorizing space, borders, and zones of interaction, these different historical interpretations have been siloed for too long, making it impossible comprehensively to theorize the making and unmaking of borders on water and land. Our volume brings specialists from both subfields together and asks them to attend to the implications of their methodology in conversation with one another.

The most current scholarship has encouraged authors to move away from dichotomies pitting continents against oceans and to think more carefully about where land and water literally met to thereby revise categories, methodologies, and paradigms. Where some scholars adopt Stephen Hornsby's perspective that the geography of the British Atlantic can best be understood as two contrasting spatial systems—one fluid and maritime, from the Atlantic to the Indian Ocean, and the other a system of combined nodes, staple territories, and agricultural frontiers along eastern North America—others recognize that neither land nor water was neatly "contained."[9] Simone Müller and David Stradling warn against privileging "a container perspective," by which they mean the geomorphological formations that contained water, such as a river's banks or the coastlines of an ocean. Such a perspective runs the risk of creating artificial boundaries between land and water or producing a study whose scale is too large. Müller and Stradling recommend moving beyond analyses of containers and watersheds to consider water's physical characteristics and how people imagined and understood them.[10]

We urge in this volume that historians benefit when we consider water alongside land and land alongside water. Here, we draw on Tiffany Lethabo King's theorization of shoals, ecotones characterized by both land and water and that drifted and moved over time. Their blended nature and their movement made them dangerous for navigators. King writes that before the 1860s and the advent of bathymetry—which allowed people to calculate the probability curves that estimated latitude and longitude to locate shoals in geographic space—coral beds, rock formations, and sandbars were difficult to map and to avoid. Their danger forced people to pause before proceeding. We draw on King to encourage scholars of borders analytically "to slow down . . . to move from the automaton to a more alert driver and navigator." This "shoaling effect," this deliberate slowing down and careful attention to terrestrial features alongside aqueous ones, offers new interpretive avenues and the potential to reframe our paradigms about space, control, and sovereignty in the early modern period.[11]

By centering attempts to make and unmake borders not just on land but also on water, this book participates in a shift within the last decade and a half to theorize the use of waterways in the early modern period. This scholarship about watery spaces—including border seas, coasts, saltwater frontiers, the Red Atlantic, the Transimperial Greater Caribbean, river valleys, and undercurrents—has established how central water-based trade, travel, and transportation was to the development of early modern North America. European imperial officials strove—unsuccessfully—to dominate maritime zones, ports, rivers, and the Great Lakes. Indigenous nations and people of African descent challenged Europeans' plans, retained control of water routes (especially rivers) and used them for their own purposes, and took advantage of aqueous mobility to further their own ends.[12]

Because our authors address the questions of *what* makes and unmakes a border and *how* people used or destabilized those constructs, they necessarily see motion and human agency in what some might misread as teleological boundaries of nation-states or as static images. This volume therefore models methods for using and interpreting historical maps. The authors take maps not as objective depictions of reality but as historically contingent material artifacts expressing a range of desires. The maps discussed are objects with arguments and were co-created by Indigenous informers, colonial observers, European mapmakers, and imperial officials—many of whom had conflicting motives and goals.[13] According to this volume's authors, borders were processes before sometimes (but not always) becoming lines. The volume thus models how to use and theorize maps in practice.

In their origins both the disciplines of history and cartography served power, and more specifically, the project of empire and the rise of the modern nation-state. The very origin of the field of academic history as we know it emerged in the late nineteenth century, "the ideological handmaid of the nation state."[14] Early historians relied on maps that the geographer J. B. Harley contended were themselves "inextricably linked to the rise of the nation-state in the modern world." Far from neutral images, Harley urged, maps must be read as "'thick' texts or as a socially constructed form of knowledge."[15] Historians who have taken this approach include Christian Koot, who has demonstrated how locally constructed knowledge could be transformed and repurposed in metropolitan centers to create engraved maps that might serve as "apparatus of empire." Max Edelson in turn has shown how such maps misled metropolitan policymakers into believing they had mastery over geographies that turned out to be beyond their grasp.[16] We continue to interpret maps as constructed forms of knowledge by thinking in this volume about archives.

Established and more recent scholarship insists on recognizing non-European ways of seeing and depicting space, a project that necessitates expanding our concept of archives. Gregory A. Waselkov reminds us that "drawing maps was within the competence of every adult southeastern Indian of the colonial period"; it was not a specialized skill confined to a limited number of surveyors and mapmakers as in Europe. With different priorities than Europeans, Indigenous mapmakers produced maps that symbolically communicated political relationships as "graphic depictions of the balance of power." Native Americans had mental maps, in addition to physical ones, even if they declined to share them with colonizers or if colonizers declined to acknowledge them as co-authors. Lisa Brooks and Christine DeLucia engage in recovering Indigenous landscapes, establishing that Native Americans made place-worlds through kin relationships, and that local and tribal archives, oral traditions, and memories can and should inform how scholars interpret these places today.[17]

The essays in this volume explore how colonists, Native Americans, and African-descended people saw and ordered space: how they claimed it, asserted power over it, and forged relationships across and beyond it. Collectively, the essays show people shaping political economies of space on land and water, which is to say, creating the social, political, legal, economic, and intellectual conditions that structured power within imagined, constructed, and often contested basins of geographic space. Human assertions and technologies inscribed certain areas with particular meaning, but our authors show that in the early modern era most zones remained porous and open to rival uses by people within and without. In the end, claiming land and claiming water was an endless work in progress.

This volume is organized into three parts, though as with any rigorous study of land and water, contributors speak across and beyond their containers. The authors in Part I, "Ways to Think About Borders," describe European officials, and free settlers' schemes, hopes, and assumptions about the relationship between borders, power, and reality throughout North America and on the Atlantic coast. Edward Gray studies the peculiar constitutional history of the founding of Maryland to argue that it was the first English continental colony to be legally enclosed by formal boundaries. Gray reminds us that "sovereignty . . . was unlikely to be territorial in any modern, spatial sense. It would remain, throughout the colonial era, a matter of persons and produce."[18] In Christian Koot's chapter, John Povey's addition by hand of watercolor borders to a 1676 map of New England offers readers insight into the process of imperial reimagining for that region as the Lords of Trade in England sought to rein in the territorial ambitions of the Massachusetts Bay Colony. Borders on maps, Koot argues, might be used not only as implements of authority or legitimacy, but also and alternatively as "things to think with." When Povey inked new colored lines, he helped to imagine a spatially defined entity called "New England," and so to bring it into being.[19] Bitterly as English colonists sometimes grieved to their monarch about the boundaries between English colonies, still more laden were contests between rival empires. Alex Zukas uses the early eighteenth-century map "war" between cartographers Herman Moll and Guillaume Delisle, working on behalf of the British and French empires respectively, to explore Europeans' competing and self-serving justifications for empire, which can be contrasted visually in the kinds of maps they produced. Moll's and Delisle's maps stressed those continental features that buttressed their empire's claims and thus functioned as "weaponized" "symbolic resources to support the empire-building."[20] These first three chapters, in common with much of the archives of empire, tell us much about the aspirations and preoccupations of colonizers. By contrast,

John Morton shows that in reality locals attained clarity over disputed territory earlier than metropolitan authorities. This assertion builds from unusual sources, which he calls "data maps," detailed charts of local settlements that itinerant missionaries assembled in the early nineteenth century. While US and British metropolitan authorities fussed at one another for decades about the precise location of the border that separated Maine from New Brunswick, Morton shows that by 1804 "the boundary was a fixed reality" to the people who lived there. On the Maine–New Brunswick border, "itinerant ministers and their parishioners fixed the line in place," an example of a border drawn not by metropolitan officials or religious antagonists (e.g., Catholic vs. Protestant), but by co-religionists concerned about the material support of their respective churches.[21]

Part II of the volume, "Creating Place," focuses on how people intellectually understood and in practice shaped the spaces and places they occupied, from the plantation landscape of the Caribbean, to the riverine networks of ambitious merchants, to the battlefield terrain that military men had rapidly to sketch. Authors explore what Paul Musselwhite calls in this volume "spatial technology," insisting that paradigms about the occupation, possession, control, and depiction of spaces humans inhabit are historically contingent, constructed, and subject to change.[22] Musselwhite challenges the inevitability of the English plantation complex and private property ownership, tracing the uneven shift in English definitions of what constituted a "plantation" from earlier ideas that cast it as a public, civic project to later ideas that redefined it as a private space of economic profit that expanded state power and paved the way for the growth of capitalism. Kim Gruenwald explores the compelling interest of merchants to harness the capacity of moving water to transport goods and people to distant markets. These interests led merchants to argue for their own unfettered access to the Ohio and Mississippi Rivers even when the Lenape (Delaware), Miami, and Shawnee nations hunted south of the Ohio and sought to protect their villages, farms, and claims to the north and west of it. Where officials from rival Atlantic empires wanted rivers to act as inflexible, nonporous boundaries under their control, merchants saw rivers as conduits and arteries of trade. Gruenwald suggests that the region often denominated "the trans-Appalachian west" might be better called the "Riverine West" to recognize that it was the swift-flowing downstream water, more than fertile land beyond the Appalachians, that first enticed empires and white settlers alike.[23] Finally, Karen Rann reminds us that the perception and strategic mapping of space included not only distance but also height. She explores the creation of contour lines, a new element that had contentiously to be worked out in carto-technology and introduced into carto-literacy in response to the difficulties that British military men faced when striving to map mountainous, contested territory. She reminds us that in the early modern period the production of maps required "bodies moving through terrain."[24] Finding and agreeing on a single, reproducible carto-technology that could more accurately communicate that experience and a terrain's myriad features—gradient, height, soil, vegetation, and so on—was a contentious work of time and considerable effort well into the nineteenth century.

Part III, "Movement," the final section of the volume, centers the people navigating through and across borders, attempting to invest such lines with meaning or subvert the meanings imposed by others. Charles Prior uses John Mitchell's 1755 map of North America to show how this European source centers Haudenosaunee (Iroquois) territoriality, history, power, and epistemologies. As Anglophone officials tried to create an asymmetrical colonial world, they found themselves instead required to work in a larger, polyvalent international world shaped by Indigenous sovereignties. In the North American northeast, Prior argues, "the international was characterized by embedded diplomatic customs and rituals that were intrinsic to the political culture of the Haudenosaunee. . . . drawing Europeans into the orbit of the Confederacy along multiple paths, routes and circuits of exchange and power."[25] Chad McCutchen and Jennifer Monroe McCutchen shift the focus southeast and further forward in time, finding similarly that Indigenous power, particularly of the Creek (Muscogee) Confederacy, served to check US, British, and Spanish ambitions and ensure instead "the persistence of community-level authority and coalition-based politics," even to the frustration of the centralizing ambitions of a Creek leader like Alexander McGillivray.[26] The McCutchens argue that Creek strategies, in blending colonial experience with new sociopolitical ideologies and in substituting diplomatic and commercial relationships for territorial hegemony, symbolize the transitional phase of geopolitics in the Gulf region after the US War for Independence. Rachel Herrmann extends the focus on commercial relationships between Spaniards, Britons, Creeks, and Seminoles in the Gulf South with a chapter examining the Spanish policy of *comercio libre*. She argues that Spanish inabilities to acquire geographic knowledge of rivers and Creek geographic control of those vital networks enabled Muscogees to continue serving Creek women's interests through the sale of dressed deerskins. In semi-aqueous Creek homelands characterized by local town politics and Spanish forts and trading houses on the coast, the Spanish allowed relaxations to *comercio libre*'s tariffs and rules about foreign traders.[27] By contrast Harvey Amani Whitfield and Sarah Chute study people uprooted to an entirely new geography seeking autonomy but often finding the opposite. "The history of Black people crossing the early American-Canadian border," they remind us, "is usually framed as a crossing from American enslavement to Canadian freedom." In fact, "the aquatic highway of the Atlantic" connected Black people in the Maritimes to both spaces of freedom and (re)enslavement.[28] By following three case studies of two men and one woman striving for freedom, they demonstrate how elusive such freedom could be in the Maritimes and how a combination of political jurisdiction, water borders, personal networks, and pure chance shaped each person's quest for freedom. We end with Samuel Truett's chapter on border-crossing strangers and villagers, whose lives were entangled in terrestrial and aqueous spaces. Truett follows the life of John Denton Hall, "a child of the Atlantic World," to model one way to "circumnavigate" the different histories considered in this volume and to disrupt our tendencies to write toward the formation of aggregated cultures, empires, or nations.[29]

Collectively, the essays in this volume speak to how space informed practice and how people reshaped space. This volume works to bring into one frame land and water; the power dynamics around imagining, enforcing, and subverting borders; and the dexterity of people seeking to assert their own rights over particular spaces. It offers methodological suggestions, contributing to a praxis of how people in the early modern Atlantic and North America claimed land and water. Weaving through all the spatial paradigms runs a productive tension between stasis and change; endurance and innovation; solidity and fluidity.

Many of the essays in this volume originated from an expanding network supported by funding from the United Kingdom's Arts and Humanities Research Council that has aimed to bring together scholars working on several different types of boundaries: landed, watery, continental, Native American, and hemispheric borders in the early modern Atlantic world. This book emerges from four linked international conferences we organized in the United States and United Kingdom that supported the research of historians, geographers, and literary scholars in considering these questions, collecting together diverse perspectives on early modern borders.[30]

In the decade since we embarked on the first grant application for this project, much has changed for us in our own subject positions respectively in Bristol, England, and Philadelphia, Pennsylvania, and for our authors.

Our thinking, writing, and editing took place in the context of Brexit; a US presidency characterized by the building of a border wall and the banning of people from predominantly Muslim countries; continuing debate about the US-Mexico border; the invasion of Ukraine by Russia; the political and economic reconfiguration of Europe; and a deadly pandemic that crossed national, state, city, and household lines to infect communities worldwide. As the book went to press the US president began forecasting radically new global priorities, provoking diplomatic and trade realignments in North America, Europe, and beyond. Suddenly there was once again disagreement over what to call the Gulf of Mexico, and long-established control of spaces from Gaza to Greenland seem to at least some policymakers to be up for grabs. In short, now more than ever it seems important to hone the analytic tools—and the long scholarly vision—to situate claims to water and land as constructs of power. And to remember that however powerful borders seemed, people, networks, and nations always subverted and contested them.

Notes

1. See for example, William Cronon, *Changes in the Land: Indians, Colonists, and the Ecology of New England* (Hill and Wang, 1983); James Merrell, *The Indians' New World: Catawbas and Their Neighbors from European Contact Through the Era of Removal* (University of North Carolina Press, 1989); Timothy Silver, *A New Face on the Countryside: Indians, Colonists, and Slaves in South Atlantic Forests, 1500–1800* (Cambridge University Press, 1990); Anthony Pagden, *Lords of All the World: Ideologies of Empire in Spain, Britain and France, c. 1500–c. 1800* (Yale University Press, 1995); Patricia Seed, *Ceremonies of Possession in Europe's Conquest of the New World, 1492–1640* (Cambridge University Press, 1995); Andrew Fitzmaurice, *Humanism and America: An Intellectual History of English Colonisation, 1500–1625* (Cambridge University Press, 2003); Lisa Brooks, *The Common Pot: The Recovery of Native Space in the Northeast* (University of Minnesota Press, 2008); Christine M. DeLucia, *Memory Lands: King Philip's War and the Place of Violence in the Northeast* (Yale University Press, 2018).

2. Philip Morgan and Jack P. Greene have suggested Atlantic history should be "thought of . . . primarily as a *perspective*," rather than a discrete discipline. Philip D. Morgan and Jack P. Greene, "Introduction: The Present State of Atlantic History," in *Atlantic History: A Critical Appraisal*, ed. Jack P. Greene and Philip D. Morgan (Oxford University Press, 2009): 3–33, esp. 4 and 24.

3. For the critique, see for example Julianna Barr, "The Red Continent and the Cant of the Coastline," *William and Mary Quarterly* 69, no. 3 (July 2012): 521–26; Allan Greer, "Commons and Enclosure in the Colonization of North America," *American Historical Review* 117, no. 2 (April 2012): 365–86. On Indigenous maritime power, see Jace Weaver, *The Red Atlantic: American Indigenes and the Making of the Modern World, 1000–1927* (University of North Carolina Press, 2014); Andrew Lipman, *The Saltwater Frontier: Indians and the Contest for the American*

Coast (Yale University Press, 2015); Joshua L. Reid, *The Sea Is My Country: The Maritime World of the Makahs, an Indigenous Borderlands People* (Yale University Press, 2015); Matthew R. Bahar, *Storm of the Sea: Indians and Empires in the Atlantic's Age of Sail*, (Oxford University Press, 2019); James Hill, "'Bring Them What They Lack': Spanish-Creek Exchange and Alliance-Making in a Maritime Borderland, 1763–1784," *Early American Studies* 12, no. 1 (2014): 36–67.

4. Frederick Jackson Turner, "The Significance of the Frontier in American History," in *The Frontier in American History* (Henry Holt and Company, 1921), 1–38; William Cronon, "Revisiting the Vanishing Frontier: The Legacy of Frederick Jackson Turner," *Western Historical Quarterly* 18, no. 2 (1987): 157–76; Andrew R. L. Cayton and Fredrika J. Teute, "Introduction: On the Connection of Frontiers," in *Contact Points: American Frontiers from the Mohawk Valley to the Mississippi, 1750–1830*, ed. Andrew R. L. Cayton and Fredrika J. Teute (University of North Carolina Press, 1998): 1–15.

5. Herbert Bolton, *The Spanish Borderlands: A Chronicle of Old Florida and the Southwest* (Yale University Press, 1921); Jeremy Adelman and Stephen Aron, "From Borderlands to Borders: Empires, Nation-States, and the Peoples in Between in North American History," *American Historical Review* 104, no. 3 (Jun. 1999): 814–41.

6. Kathleen DuVal, *The Native Ground: Indians and Colonists in the Heart of the Continent* (University of Pennsylvania Press, 2006).

7. Elizabeth N. Ellis, *The Great Power of Small Nations: Indigenous Diplomacy in the Gulf South* (University of Pennsylvania Press, 2023), 6, 25.

8. Michael Witgen, "A Nation of Settlers: The Early American Republic and the Colonization of the Northwest Territory," *William and Mary Quarterly* 76, no. 3 (2019), 391–98 (quote 393); Pekka Hämäläinen, "The Shapes of Power: Indians, Europeans, and North American Worlds from the Seventeenth to the Nineteenth Century," in *Contested Spaces of Early America*, ed. Juliana Barr and Edward Countryman (University of Pennsylvania Press, 2014), 31–68.

9. Stephen J. Hornsby, "Geographies of the British Atlantic World," in H. V. Bowen, Elizabeth Mancke, and John G. Reid, *Britain's Oceanic Empire: Atlantic and Indian Worlds, c. 1550–1850* (Cambridge University Press, 2012), 15–44, esp. 43.

10. Simone M. Müller and David Stradling, "Water as the Ultimate Sink: Linking Fresh and Saltwater History," *International Review of Environmental History* 5, no. 1 (2019): 23–41, esp. 27–29, 30 (containers).

11. Tiffany Lethabo King, *The Black Shoals: Offshore Formations of Black and Native Studies* (Duke University Press, 2019), xv.

12. See note 3 above. See also S. Max Edelson, *The New Map of Empire: How Britain Imagined America Before Independence* (Harvard University Press, 2017), 24; Christian Koot, *A Biography of a Map in Motion: Augustine Herrman's Chesapeake* (New York University Press, 2018); April Lee Hatfield, *Atlantic Virginia: Intercolonial Relations in the Seventeenth Century*, (University of Pennsylvania Press, 2004); Susan Sleeper-Smith, *Indigenous Prosperity and American Conquest: Indian Women of the Ohio River Valley, 1690–1792* (University of North Carolina Press, 2018); John Nelson, *Muddy Ground: Native Peoples, Chicago's Portage, and the Transformation of a Continent* (University of North Carolina Press, 2023); Renaud Morieux, "Anglo-French Fishing Disputes and Maritime Boundaries in the North Atlantic, 1700–1850," in *Governing the Sea in the Early Modern Era: Essays in Honor of Robert C. Ritchie*, ed. Peter C. Mancall and Carole Shammas (Huntington Library, Art Collections, and Botanical Gardens, 2015), 41–75; David A. Chang, *The World and All the Things Upon It: Native Hawaiian Geographies of Exploration* (University of Minnesota Press, 2016); Kevin Dawson, *Undercurrents of Power: Aquatic Culture in the African*

Diaspora (University of Pennsylvania Press, 2018), 21; Ernesto Bassi, *An Aqueous Territory: Sailor Geographies and New Granada's Transimperial Greater Caribbean World* (Duke University Press, 2016); Ada Ferrer, *Freedom's Mirror: Cuba and Haiti in the Age of Revolution* (Cambridge University Press, 2014); Jane Landers, *Atlantic Creoles in the Age of Revolution* (Harvard University Press, 2010); Michael J. Chiarappa, "Working the Delaware Estuary: African American Cultural Landscapes and the Contours of Environmental Experience," *Buildings & Landscapes* 25, no. 1 (2018): 64–91; Julius S. Scott, *The Common Wind: Afro-American Currents in the Age of the Haitian Revolution* (Verso, 2018); Matthew J. Smith, *Liberty, Fraternity, Exile: Haiti and Jamaica After Emancipation* (University of North Carolina Press, 2014); Rebecca J. Scott and Jean M. Hébrard, *Freedom Papers: An Atlantic Odyssey in the Age of Emancipation* (Harvard University Press, 2012); Douglas Hamilton and John McAleer, eds., *Islands and the British Empire in the Age of Sail* (Oxford University Press, 2021); Tessa Murphy, *The Creole Archipelago: Race and Borders in the Colonial Caribbean*, (University of Pennsylvania Press, 2021).

13. Reuben Rose-Redwood, Natchee Blu Barnd, Annita Hetoevėhotohke'e Lucchesi, Sharon Dias, and Wil Patrick, "Decolonizing the Map: Recentering Indigenous Mappings," *Cartographica* 55, no. 3 (Fall 2020): 151–62; Martin Brückner, *The Social Life of Maps in America, 1750–1860* (University of North Carolina Press, 2017), 3, 10–12; Koot, *A Biography of a Map in Motion*, 9, 11, 176; Nicholas Gliserman, "Introduction: Unpacking the Meaning of Maps, Power, and Boundaries," in *The Power of Maps and the Politics of Borders: Papers from the Conference Held at the American Philosophical Society, October 2019: Transactions of the American Philosophical Society* 110, no. 4 (American Philosophical Society Press, 2021), 1–10; Reed, "Thinking Multidimensionally," 57–70; Juliana Barr, "Borders and Borderlands," in *Why You Can't Teach United States History Without American Indians*, ed. Susan Sleeper-Smith, Juliana Barr, Jean M. O'Brien, Nancy Shoemaker, and Scott Manning Stevens (Chapel Hill: University of North Carolina Press, 2015), 9–25, esp. 16–17.

14. Allan Greer, "National, Transnational, and Hypernational Historiographies: New France Meets Early American History," *The Canadian Historical Review* 91, no. 4 (2010): 695–724 (quote 696); Priya Satia, *Time's Monster: How History Makes History* (Harvard University Press, 2020).

15. J. B. Harley, *The New Nature of Maps: Essays in the History of Cartography*, ed. Paul Laxton (Johns Hopkins University Press, 2001), 59, 52.

16. Koot, *A Biography of a Map in Motion*, 176; Edelson, *The New Map of Empire*, 332–33. Harley's call aligned with anthropologist Clifford Geertz's practice of "thick description," the classic articulation of which remains Clifford Geertz, "Deep Play: Notes on the Balinese Cockfight," *Daedelus* 101 (1972): 1–37.

17. Gregory A. Waselkov, "Indian Maps of the Colonial Southeast," in *Indians in the Colonial Southeast*, revised and expanded edition (University of Nebraska Press, 2006), 435, 445; Louis de Vorsey Jr., "American Indians and the Early Mapping of the Southeast," in William P. Cumming, *The Southeast in Early Maps*, 3rd ed., revised by Louis de Vorsey Jr. (University of North Carolina Press, 1998), 65–98; Christine DeLucia, *Memory Lands: King Philip's War and the Place of Violence in the Northeast* (Yale University Press, 2018); Lisa M. Brooks, *Our Beloved Kin: A New History of King Philip's War* (Yale University Press, 2018); Christine Delucia, "Recovering Material Archives in the Native Northeast: Converging Approaches to Traces, Indigeneity, and Settler Colonialism," *Early American Literature* 55, no. 2 (2020): 355–94, esp. 356, 366, 368. On expanding the concept of archives, see Marisa J. Fuentes, *Dispossessed Lives: Enslaved Women, Violence, and the Archive* (University of Pennsylvania Press, 2016).

18. Edward G. Gray in this volume, "Toward a Prehistory of Territory: Thomas Hobbes, the Maryland Palatinate, and the Colonial Boundary Problem," 33.

19. Christian J. Koot in this volume, "Things to Think With: The Use of Borders on a Seventeenth-Century Map of New England," 38, 39.

20. Alex Zukas in this volume, "Lines on a Map: Crafting and Contesting Borders in Guillaume Delisle's and Herman Moll's Early Eighteenth-Century Maps of North America," 62.

21. John Morton in this volume, "Data Maps of Downeast Maine: Missionary Records from the Early Republic Borderlands," 91, 94.

22. Paul Musselwhite in this volume, "Depicting and Defining the Plantation in the Early English Caribbean, 1625–1675," 118.

23. Kim M. Gruenwald in this volume, "When a River Is a Border: Rivalries and Commercial Networks in the Riverine West," 159–60.

24. Karen Rann in this volume, "Military Lines: How the Introduction of Contours Affected Maps and Movement," 181.

25. Charles Prior in this volume, "Indian Centers, Colonial Peripheries: Locating the International in Early America," 207.

26. Chad McCutchen and Jennifer Monroe McCutchen in this volume, "'Playing the Old Game of Double': Navigating Creek and Spanish Geopolitics in the Post-Revolutionary Gulf South," 219.

27. Rachel B. Herrmann in this volume, "*Comercio Libre*: Revisiting a Concept on Trade and Borders in Creek Homelands," 242, 244–45.

28. Harvey Amani Whitfield and Sarah Chute in this volume, "Possibilities and Peril: Exploring the Transnational Experiences of Black People in the Maritimes, 1783–1792," 264, 265.

29. Samuel Truett in this volume, "Amphibious Tales: Villagers and Strangers in a Border-Crossing World," 290, 293.

30. "On Edge: New Frontiers in Atlantic History," University of Southampton, Southampton, England, June 2016 (sponsored by an Adventures in Research Award from the University of Southampton, 2015); and "Geographies of Power on Land and Water: Space, People, and Borders," American Philosophical Society, Philadelphia, PA, July 2018; "Zones and Lines: New Conversations on Borders," Cardiff University, Cardiff, Wales, May 2019; and "Lines on a Map: Crafting and Contesting Borders in the Early Modern Atlantic and Beyond," British Library, London, England, December 2019 (Grant ref: AH/R002320/1, a Networking Scheme Grant from the Arts and Humanities Research Council, 2017–19).

PART I

Ways to Think About Borders

CHAPTER 1

Toward a Prehistory of Territory

Thomas Hobbes, the Maryland Palatinate, and the Colonial Boundary Problem

EDWARD G. GRAY

Thomas Hobbes has no known connection to the Maryland colony. At the time King Charles I's seal was applied to the colony's charter, in 1632, Hobbes was probably living at Hardwick Hall, a grand Derbyshire house owned by his patrons, the Earls of Devonshire. In 1630, Hobbes had begun serving as advisor to the widow of the second earl and to the earl's cousin, the Earl of Newcastle. In 1634, the year Maryland's first colonists settled St. Mary's City on a small eponymously named tributary of the Potomac River, Hobbes accompanied the third Earl of Devonshire on a European tour.

Insofar as there is any historical connection between seventeenth-century England's greatest political philosopher and the obscure colonial enterprise of a prominent Jacobean Catholic family, it most likely lies with the much better known, better capitalized, and altogether more catastrophic colonial enterprise known as Virginia. Both Hobbes and the eventual founder of Maryland, George Calvert, member of James I's Privy Council, owned shares in the quasi-private entity that controlled that other Chesapeake colony. Hobbes most likely received his share from his employer and former student, William, Lord Cavendish, the second Earl of Devonshire. Cavendish probably made the grant to his trusted former tutor as a means of strengthening his own position on the Virginia Company Court (as the company's board was known). At Cavendish's behest, between 1622 and 1624, Hobbes also seems to have attended no fewer than thirty-seven Virginia Company meetings. In this same period, Calvert served as a secretary of state and in the period Hobbes was active, Calvert's responsibilities included corresponding with the Virginia Company Court. Hobbes surely knew of Calvert; Calvert surely knew of the associate and secretary of one of the Virginia Company's most influential shareholders. Beyond this circumstantial intersection of the as yet little-known philosopher and the prominent courtier, the connection of Thomas Hobbes to Maryland is pure conceit.[1]

But it is, this chapter suggests, an illuminating conceit. Hobbes is obviously much less well known as a colonial bureaucrat than a political philosopher. And the work for which he is most known happened years after his service to Lord Cavendish—his masterpiece, *Leviathan*, appeared in 1651. That work has much to say about the relationship between colonies and the sovereign, but very little to say about any given colony and nothing to say about Maryland.[2] And yet, it contains insights very much relevant to the Maryland colony and, ultimately, all of England's mainland colonies. That so capacious and multifaceted a treatise should say something about contemporaneous historical developments is not, one must acknowledge, an especially bold proposition. Historians of English America have long found in Hobbes's writings a useful interpretive frame through which to view their subject matter.

Hobbes's famously dark characterization of the hideous, barbaric, and disordered condition beyond the sovereign state, often misunderstood as Hobbes's characterization of human nature, has been especially appealing to historians of England's Chesapeake colonies. As places where humans faced "continuall feare, and danger of violent death," and where "the life of man [is] solitary, poore, nasty, brutish, and short," the plantations of the Chesapeake have seemed the embodiment of Hobbes's characterization of the "natural condition of mankind," an ungoverned state of permanent war.[3] Whether historians' characterization of a Hobbesian Chesapeake accords with Hobbes's actual understanding of humans' natural state is debatable. What is less debatable, despite American historians' tendency to imply otherwise, is the fact that Hobbes's larger philosophical project, especially as articulated in *Leviathan*, goes far beyond the terrifying characterization of human beings living in the absence of stable governing authority.

Although any brief characterization of Hobbes's purpose in *Leviathan* is certain to miss the full complexity and subtlety of that masterpiece, for the purposes of what follows, we can say that Hobbes sought a rational, empirical justification for formally incorporated government—the state, or what Hobbes usually referred to as the "common-wealth." The ecclesiastical implications of

Hobbes's argument have justifiably drawn historians' attention. Contemporaries widely and correctly saw the book as an attack on political doctrine upheld by both the Church of England and its Puritan critics. The notion at the center of Hobbes's analysis, that sovereignty was effectively indivisible, whether inhering in the person of the monarch or the abstraction of the state, left no place for an ecclesiastical sanction on the English sovereign—itself a not terribly controversial position. Hobbes went further, however, in dismissing the Anglican view that ordination conferred on clergy spiritual powers independent of the sovereign. Clergy were, according to *Leviathan*, little different from any other servants of the sovereign. The argument repelled Anglicans, but proved appealing both to ultra-royalists and Puritan Parliamentarians, the latter seeing in Hobbes's conception of unitary sovereignty justification for rule by a unitary, godly Parliament.[4] As a means of contextualizing Hobbes's most celebrated work, it thus makes a good deal of sense to interpret *Leviathan* and its reception through England's seventeenth-century confessional strife. Because the Maryland colony was a product of that same strife—albeit a very different phase in its eruptions—one might thus expect an analysis of Hobbes and Maryland to emphasize English religious politics.

The conceit advanced here depends on a distinctly more arcane intersection of *Leviathan* and the Maryland colony. That intersection concerns the term territory, a term not formally associated with Maryland, but frequently used by Hobbes. There is a certain anachronism in introducing this term and its most familiar connotations in a chapter that deals with the seventeenth century. For the most part, contemporary scholars have understood territory and territoriality—the normative convergence of mapped space with governing authority—as contingent on variables associated with the modern nation-state. As one geographer has defined it, territory applies when an individual or group assert "influence, or control of people, phenomena, and relationships, by delimiting and asserting control over a geographic area." What implies modernity in this definition is the correspondence of jurisdiction—"control"—to clearly articulated, lawful spatial limits, or borders. The latter make possible another attribute of the modern nation-state, the rights-bearing citizen-subject. States' mechanisms for securing citizens' rights are, in large measure, a function of clear insider-outsider status, and along with modern identity documentation, the national border represents a crucial instrument for achieving such totalizing forms of identity. In a global history of the bordered nation-state, Charles Maier writes that territory is space "empowered by borders." For those borders "allow people to be governed or taxed or imbued with loyalty by virtue of their shared spatial location, not their race or their kinship ties or their faith or their professional affiliation."[5]

Hobbes has a very different notion of territory. He assumes no clear, lineal borders because the Hobbesian common-wealth rests on a well-known premodern form of political affiliation. Rather than the kinds of horizontal, fraternal ties associated with citizenship, the Hobbesian state is bound by personal and contractual ties, the vertical bonds between subjects and sovereign. Although monarchical authority had clear spatial dimensions for Hobbes, those dimensions had less to do with the status of subjects within the realm than the specific legal claims of sovereigns to persons, land, and other resources.[6] To grasp the point, we might consider *Leviathan*'s treatment of the common-wealth's political economy. "The Nutrition of a Common-wealth," as Hobbes characterizes the state's economic life-blood, "consisteth, in the *Plenty*, and *Distribution of Materials* conducing to Life." Those materials, the "matter" needed for the common-wealth's survival, are "commonly called Commodities" and are of two kinds: the native and the foreign. The native is "that which is to be had within the Territory of the common-wealth: forraign, that which is imported from without." What defines territory in this context, in other words, is not lineal borders or fraternal bonds of citizens, but the material yield of a kingdom. In an overwhelmingly agricultural age, this notion of territory obviously has a spatial dimension. Commodities come primarily from land, and if that land lies within the sovereign's domain, those commodities are "native." Insofar as territory is bounded for Hobbes, it is thus bounded in the way titled property is bounded. Title confers exclusionary rights to the produce of property, but indicates very little about the relationship between individual landholders and the sovereign. Bounded territory, according to this usage, thus does not denote territoriality—or the sovereign authority to confer rights-bearing subjecthood or citizenship.[7]

This is not to say that Hobbes has no conception of a spatial-limit to sovereignty, or even a kind of insider-outsider status. Hobbes notes, for example, a distinction between ordinary subjects and ambassadors or messengers. The distinction arises when the latter travel on state business. Hobbes does not explain exactly when a servant of the sovereign becomes an ambassador—there is no literal or metaphoric border crossing that turns servants of the realm into ambassadors. The same problem presents itself in the case of the criminal exile: Such a person is no longer a subject precisely because she or he is banished from the common-wealth, but banishment implies some kind of territorial limit—a point at which a subject becomes an exile. Again,

Hobbes does not explain exactly where—as a spatial matter—exile begins. In a Europe still populated by city-states and kingdoms with vaguely defined frontiers and marchlands, what we get from Hobbes is an idea of territory subordinate to Hobbes's contractual idea of sovereignty: The latter is limited in space, and the status of the subject implies spatial limits, but as a means of locating those spatial limits, actual international borders do not enter Hobbes's analysis.[8]

In addition to trade and travel, there is a third context in which Hobbes's analysis confronts the problem of sovereign territorial limits. Travel and trade suggest borders insofar as they involve transit between distinct sovereign domains, themselves defined by control of persons and produce. But Hobbes's common-wealth does not exist in an international order defined solely by the presence of other common-wealths. It exists in a mosaic of adjoining territories and while some of those are governed, in theory, at least, some are not. These barbarous realms, enveloped in perpetual war, point to another spatial limit, one in which clearly articulated subject–sovereign relations give way to marchlands characterized by the absence of government. In the words of Annabel S. Brett, Hobbes's common-wealth gives way to an "uneasy frontier," one that affords passage between the civil and governed city and its obverse, the realm of nature where human beings are perpetually at war. Again, Hobbes offers little specificity about the conceptual nature of this frontier—he leaves his readers nothing that could, for instance, be represented on a map. Instead, he offers the distinction between barbarous humans in nature and the common-wealth as an "ontological ground on which [the common-wealth's] structure of laws and rights is erected."[9]

To summarize: Territory for Thomas Hobbes suggests a spatial dimension of sovereignty determined by various legal instruments that afford sovereigns control of subjects, lands, and the commodities they generate. International borders exist in *Leviathan*, but they exist as ontological abstractions—points at which envoys become ambassadors or subjects become banished exiles. Hobbes offers little about the exact nature of these points of transition. Insofar as the sovereign state is bounded, it is vaguely bounded by frontiers and marchlands. These are as likely to be zones in which one common-wealth's jurisdiction gives way to another's, as they are frontiers where governance gives way to the state of perpetual war.

Hobbes's understanding of territory is generally consistent with the geopolitical order inhabited by seventeenth-century states and empires. Mountains and rivers often provided convenient territorial limits but artificial lineal boundaries were largely unknown. Insofar as there were clearly articulated, artificial limits on sovereignty, these tended to exist in the form of regulatory architecture—most notably, the walls and gates surrounding cities. What was true of the Old World also tended to be true of England's first mainland American colonies. Beyond the palisades surrounding initial outposts, these were bounded not by any clearly articulated, artificial limits but by vague assertions of proprietary and jurisdictional control. Hobbes's state of perpetual war may resonate with the nastiness, the brutishness, and the shortness of life in the English Chesapeake. But a far more resonant intersection of Hobbes's political thought and English colonial reality lies in the territorial character of England's first mainland North American colonies. These entities were defined in space above all by the goods and people over which they exerted putative legal control. Insofar as English authorities understood the spatial limits of England's first colonies, they thus understood those limits in a way Hobbes would have recognized: a vague dissipation of jurisdiction rather than any clearly articulated territorial limits.

There is an important exception to this pattern of premodern territoriality in mainland English America. That exception emerged from a political problem of the kind that inspired Hobbes's *Leviathan*: Under what circumstances does the common-wealth achieve for its citizen-subjects peace and security? The Chancery lawyers who drafted the Maryland charter in 1632 established a geopolitical entity with clearly articulated, lineal boundaries—the first in English North America. These territorial limits were born of political concerns—an emerging Anglo-Dutch contest and the fractious religious politics of the period of Charles I's Personal Rule. That those political concerns—which obviously existed before Maryland's founding and would exist after—yielded unique territorial form cannot be understood without first recognizing that, much like modern territorial limits, Maryland's boundaries were of a fundamentally constitutional nature: They provided a crucial limit on governing authority. That such a limit was at all necessary was a function of Maryland's peculiar constitutional order.

* * *

As an instrument of colonial government, at least in the mainland colonies, Maryland's constitution followed the charter of Avalon, a colony in Newfoundland established by George Calvert on lands he purchased in 1620 from a bankrupted former Oxford friend, Sir William Vaughan. Because Calvert would convert to Catholicism shortly before the actual colony was established, Avalon was England's first colony founded by Catholics. In fact, this had very little impact on the colony's charter, which Calvert received in 1623, before his conversion was public. For the most part, the Avalon colony would reflect the general religious attitudes of the court of James Stuart. A colony established by Catholics, as far as the Crown was concerned, was far from a Catholic colony.

Of much greater importance for the future Maryland colony was Avalon's form of government. Avalon was to be a feudal palatinate, the latter a term derived from the Latin adjective *palatinus*. According to one authority, the term signified "a very close relation to the person of the sovereign." In practice, what this meant was not so much political proximity as political authority: The lord or earl-palatine would effectively possess powers over his feudal grant undifferentiated from those of the sovereign.[10] He would have the authority to enforce law; he would appoint and oversee the courts; he would levy taxes; he would dictate ecclesiastical affairs; he would make law, in consultation with his subjects; he would enfeoff land, and he would otherwise govern and defend his fiefdom. This palatinate model represented a sharp departure from earlier colonial charters. As commercial trading ventures, dependent on private capital, Maryland's mainland predecessor colonies had been granted to limited-liability joint stock companies. With unitary ownership, governance of the Avalon colony would also be effectively unitary—it would reside with the person of a vice-regal Sir George Calvert, Lord Baltimore. There would be no feuding board of directors or intriguing shareholders to disrupt the work of governance.

The unitary propriety of a palatinate served aims consistent with ancient practice: rewarding a favorite with hereditary title to some portion of the realm. That a palatinate carried much greater constitutional autonomy than baronies or other common feudal conveyances had less to do with the political significance of the bequest than with political geography. James I's Privy Council began granting proprietary charters with a palatinate clause not because of the perceived value of the grant—to grant palatine rights, in addition to normal feudal rights *in capite*, whereby land tenure is exchanged for direct service to the sovereign, did not reflect a judgment as to the value of the title. Instead, it reflected a geopolitical strategy. England's palatinate counties, Cheshire, Lancashire, and the oldest of them, Durham, were all in the north of the country, far from the seat of monarchical authority. The Bishop of Durham, the ruler of that county-palatinate, had much broader powers to assemble armies and organize military action than did other feudal lords, and this proved vital in mobilizing defenses against Scottish invaders. In deference to geographic reality, in other words, the palatinate originated as a means of maintaining control in regions far beyond the reach of ordinary institutions of governance. In exchange for policing English borders, the Bishops of Durham could expect general immunity from royal writs and other proclamations; lawmaking was left to the bishops and their courts. Similarly, the bishops related to the king through diplomatic channels and the palatinate had no representation in Parliament. As one historian of the Durham Palatinate has observed, until reforms began in the sixteenth century, "the position of the palatinate of Durham in the legal machinery of the [English] kingdom . . . may well be compared to the status of a dependent foreign country."[11]

In the 1620s, the pressing need was to preserve order in England's expanding colonial periphery—a mandate made all the more urgent by the slaughter that enveloped Virginia in 1622. But the Avalon colony would never have the chance to establish the palatinate as a viable mechanism to do for the colonial periphery what it had been intended to do for the medieval English periphery. After less than a decade, the frigid colony had been able to attract only about a hundred English and Irish settlers, few of whom found much more than deprivation and illness.

For George Calvert, the fact that Avalon attracted only a few Catholics—mostly wealthy associates of the Calverts—was justification enough to consider the venture a failure. In 1629, Lord Baltimore and his family left Ferryland, the Calvert family seat at Avalon, and relocated to Virginia. With the collapse of the Virginia Company in 1624, local officials now served at the pleasure of King Charles I, who had succeeded his father and Calvert's patron after James I's death in March 1625. Although happy for such an esteemed courtier as Lord Baltimore to bring paying tenants and capital to bolster the growing tobacco trade, Virginians, who now included a powerful Puritan faction, were less happy about the presence of an influential Catholic. Calvert might have allayed the Virginians' concerns, but he refused to take requisite oaths of allegiance and supremacy.[12]

With little hope for a cross-confessional political alliance in England's Chesapeake colony, Calvert abandoned plans for a Virginian Avalon and returned to England. He would now spend the final two years of his life on the design for a new Avalon, one that avoided both the climatic constraints of Newfoundland and the religious ones of the king's existing colonies. Because of his loyalty to the king's late father, Calvert remained a royal favorite and worthy agent of the Stuart realm's American expansion. With the help of friends on the Privy Council, Calvert petitioned to establish a colony south of the James River, and the petition was nearly approved before word reached former directors of the Virginia Company. Although the Crown had revoked the company's charter, former shareholders expected to retain control of the company's land claims. Because it meant parceling up portions of Virginia's original grant, Virginia Company partisans argued that Calvert's proposed colony was a violation of their property rights.[13]

It is a measure of how arcane and improvised the legal apparatus of English colony building was (and of how poorly understood the revocation of colonial or any other charters remains) that such a claim had any validity at all. The Crown had revoked the Virginia Company's charter; but the former holders of that charter argued that the property rights it conferred were effectively unchanged. In revoking the charter the only thing the king had done, according to this logic, was affirm his authority to govern his dominions. Investors still controlled the land, at least as stipulated by the colony's founding documents, and a grant to Calvert on lands so close to the company's original settlements clearly violated land titles of former Virginia Company proprietors. The idea that the Crown was prepared to disburse former Virginia Company lands to another colonizer was all the more alarming when that colonizer was a Catholic who had already refused to subscribe to Virginia's standing legal order.

Rather than settle the question of whether ownership by a joint-stock company carried the sorts of property rights associated with feudal land conveyance, whatever the status of an originating patent, the Crown accepted the opposition argument and proposed to Calvert a new location, this one in Virginia's northern reaches. Here, it seemed, there was less danger of intruding on the former Virginia Company claims, but a northern location would also accomplish a pressing new strategic end, namely establishing a buffer between Virginia and an avaricious and expansive New Netherland, now extending its trading activities south into the Delaware River Valley.[14]

* * *

George Calvert died on April 15, 1632, at the age of fifty-two, two months before the Maryland charter received its final royal seal. The charter would be granted to Calvert's eldest son, Cecil, the second Baron Baltimore, and it would look very much like that awarded George for an earlier Calvert-family venture. Like Avalon, Maryland was to be a palatinate. Regarding civil and ecclesiastical affairs in the new colony, Lord Baltimore and his heirs would enjoy the same "Rights, Jurisdictions, Privileges, Prerogatives, Royalties, Liberties, Immunities, and Royal Rights, and temporal Franchises whatsoever . . . to be had, exercised, used, and enjoyed, as any Bishop of Durham, within the Bishopric or County Palatine of Durham, in our Kingdom of England, ever heretofore." Cecil Calvert would have sweeping powers not only over the government of his new colony but also, much like the Bishops of Durham, over the ecclesiastical affairs of his colony. In fact, he would have far more control—at least in theory—than the seventeenth-century Bishops of Durham had. After 1536, King Henry VIII had constrained the judicial autonomy of the Durham bishops, and by the end of the century royal magistrates had assumed control of the palatinate's courts. In Maryland, Lord Baltimore and his heirs would be empowered "by judges by them delegated, to award Process, hold Pleas, and determine in those Courts, Preaetorian Judicatories (equity courts), and Tribunals, in all Actions, Suits, Causes, and Matters whatsoever, as well Criminal as Personal, Real and Mixed, and Praetorian."[15]

Cecil Calvert had more power than the seventeenth-century Bishops of Durham in another respect as well. Theirs was an elective office. The Maryland charter left no doubt that this palatinate would carry conventional feudal tenure: "Whereas our well beloved and right trusty Subject Caecilius Calvert, Baron of Baltimore . . . Son and Heir of George Calvert, Knight, late Baron of Baltimore . . . treading in the steps of his father, being animated with a laudable, and pious Zeal for extending the Christian Religion, and also the Territories of our Empire . . . with some certain Privileges, and Jurisdiction, appertaining unto the wholesome Government, and State of his Colony and Region aforesaid, may by our Royal Highness be given, granted and confirmed unto him, and his Heirs."[16]

Whatever the precise political origins of the palatinate colony, the political consequences of a chartered entity enjoying such extensive constitutional autonomy were far-reaching. For Virginians and their London advocates, the fact that the Calvert family had such sweeping powers only made the Maryland colony that much more threatening. The first Lord Baltimore had already proven reluctant to disavow any papal allegiance, and now the second Lord Baltimore was being granted a minor kingdom in lands nominally claimed by Virginia. Able to conduct his own diplomacy and warfare, Lord Baltimore could ally his colony with his father's old (and mostly alleged) Spanish-Catholic friends or perhaps even build direct ties to Rome. In a more basic sense, there was no constitutional clarity whatsoever about the palatinate's powers relative to other colonies. Should Virginia and Maryland come to dispute their respective jurisdictions, for example, the terms of the palatinate suggested that Maryland would negotiate as a quasi-independent state while Virginia, now a royal colony, would have to defer to royal authority. For Virginia's partisans, this presented a near-hopeless political predicament, given the king's widely assumed Catholic sympathies and his continued support for the Calvert family.

In defense of their putative claims, former directors of the Virginia Company railed against a Catholic Maryland. The well-connected supporters of the company knew full well that popular anti-Catholicism retained much political potency. In aligning themselves with Puritan opponents of the Stuarts' tolerance, they hoped to pressure the king and his ministers to restore the Virginia Company charter. A rechartered anti-Catholic Virginia Company, they seem to have believed, would be a price the king would pay to quiet critics of the Maryland venture. But the Virginia partisans were wrong. Charles and his ministers never restored Virginia's charter but neither did they ignore partisan hostility to Maryland.[17]

* * *

The true novelty of the Maryland charter lay not in the powers it afforded its proprietors but in the limits it imposed on those powers—limits in the form of lineal boundaries. The contrast with the early joint-stock colonial charters is instructive.[18]

According to the first Virginia Company charter (which effectively established two joint-stock entities, the London Company, which controlled a southern portion of Virginia, and the Plymouth Company, granted control of a northern portion that would become New England), the London Company was entitled to establish "a Plantation and Habitation in some fit and convenient Place" between thirty-four and forty-one degrees of latitude. The company's plantation could be situated anywhere along the coast between those lines of latitude or "on the islands thereunto adjacent, or within one hundred miles of the coast thereof." Any additional claims would be a function of the first settlement. Much like a medieval fortress town or a walled city-state, that initial settlement's territorial control would radiate outward, fifty miles north and south along the coasts and one hundred miles inland. Two subsequent charters expanded the company's legal claim to two hundred coastal miles north and south.

Virginia would encompass "all those lands, countries, or territories situate, lying, and being in that part of America called Virginia, from the point of land called Cape or Point Comfort [at the mouth of the James River] all along the sea coast to the southward two hundred miles, and all that space and circuit of land lying from the sea coast of the precinct aforesaid up into the land throughout from sea to sea west and northwest."[19] Neither the southern Virginia colony nor any northern counterpart would, in effect, be bounded grants. Any limits on the colonies' respective claims would be a function of initial settlement and subsequent conquest. From the first outposts, there would be a rough extension of some sort of jurisdiction for some two hundred coastal miles, including any territory "not now actually possessed by any Christian Prince or People."[20] But beyond these terms, the Crown made no effort to create anything that could be mistaken for firm, territorial boundaries. The original colonies would have influence over hinterlands and frontiers, but nothing that suggested a bounded, spatial estate.[21]

In its efforts to appease Virginia Company partisans and their Puritan allies, the lawyers of the Chancery took a very different tack with the Maryland charter. Lord Baltimore and his heirs would have title to a tract as clearly defined as seventeenth-century cartographic technology afforded. Maryland would include what is now known as the Delmarva Peninsula, excluding its southern tip, which would remain in Virginia, affording the latter control of the mouth of the Chesapeake Bay. That section of Eastern Shore Virginia would be separated from Maryland by a lineal boundary running from Watkins Point on the west, due east to the Atlantic coast. Maryland's northern boundary, and the one that would eventually be a point of bitter and prolonged contestation with William Penn and his heirs, would begin "unto that part of the Bay of Delaware . . . which lieth under the Fortieth degree of North Latitude . . . where New England is terminated" and would follow a direct westward line to a "true meridian" intersecting the "first Fountain of the River of Pattowmack." From that point, the colony's boundary would form a southward, lineal boundary and then travel along the Potomac's far bank to "a certain place . . . where [the river] disembogues into the aforesaid Bay of Chesapeake, and thence by the shortest Line unto the aforesaid Promontory of Place, called Watkin's Point."[22]

In the history of English colony building, the Maryland charter was momentous. It represents the first time the Crown authorized extensive lineal, territorial limits on a portion of its American claims. It also marks the beginning of a cartographic and political process that would ultimately make eastern North America the birthplace of one of settler colonialism's most familiar features: its reliance on arbitrary geometric boundaries, imposed by distant authorities in deference to political imperatives far removed from colonized lands. In some ways this pattern suggests growing sophistication: from boundaries defined by the organic expansion of fortress-like outposts, England's colonies came to have lineal boundaries, facilitated by improvements in cartographic knowledge. But progress would be the wrong frame through which to view Maryland's boundaries. Maryland was, at base a feudal conveyance. It represented a monarch's efforts to reward a favorite while serving some larger geopolitical aim. In this way, it differed little from the palatinates of the middle ages or really any other form of feudal estate. What made the Maryland palatinate distinct was the Crown's efforts to forestall territorial conflict by imposing clear jurisdictional limits.

* * *

Understanding his colony's jurisdiction as a function of carefully articulated territorial boundaries, the second Lord Baltimore interpreted his own powers accordingly. Territorial control, as articulated in the Maryland charter, was essentially absolute within the bounds of Maryland. Native peoples and Europeans who happened to reside in the Maryland grant would, accordingly, become subjects of their new Lord Baltimore. But, of course, political authority is not so readily fungible—as Hobbes recognized. Insofar as the early Chesapeake was a place of nastiness, brutishness, and shortness of life, it was so in no small measure precisely for this reason. A bounded palatinate in a geopolitical order populated by Native nations and European colonies was akin to a Protestant city-state among Catholic kingdoms, or a republic surrounded by monarchies. Without a common state system, a bounded palatinate among unbounded Native nations and European colonies was, as Hobbes would surely have recognized, destined for perpetual, internecine conflict. Within a very short time after the settlement of St. Mary's City, Maryland's first and, in some ways, most damaging jurisdictional crisis began. The crisis centered on the Kent Island fur-trading factory controlled by a Virginian named William Claiborne.

Claiborne was a Cambridge graduate with Puritan leanings who had migrated to Virginia in 1621. Like many of Virginia's early settlers, he was not the kind to take up axe and hoe and face the hard work of carving an agricultural settlement from the Virginia countryside. His parents were of middling status but with his Cambridge education, Claiborne clearly expected to achieve a higher social rank. He was initially employed as the Virginia Company's chief land surveyor, a lucrative post, but not lucrative enough to place him among the colony's emerging planter elite. For that, Claiborne would need political office, something that came to him after the 1622 Powhatan Rebellion. As member of a faction favoring royal governance for the colony, Claiborne was rewarded with an appointment to the Governor's Council, a position he retained after 1624, the year the Crown assumed control of the colony. Among the benefits of council membership was access to the colony's Indian trade. No single member of the Governor's Council exploited this commercial advantage more effectively than Claiborne.

In 1630, just after the first Lord Baltimore had visited Virginia, Claiborne found partners in London whose investments allowed him to expand his stake in the northern Chesapeake Indian trade. By the time Lord Baltimore's colonists had settled on the banks of the St. Mary's River, Claiborne had established his factory on Kent Island, a Chesapeake outcropping near present-day Annapolis. From this outpost, which Claiborne had acquired from the Susquehannock Indians, his agents were able to dominate the Indian trade of the northern Chesapeake.[23]

Given Claiborne's Puritan sympathies, it is not surprising that Lord Baltimore identified him as a serious security threat. It was, after all, unlikely that a Puritan would accede to Catholic rule. Similarly, Claiborne's business interests rested on a trading alliance with the Susquehannocks, the dominant Native group in Maryland's far north. As long as this was the case—that the region's most powerful Native group was allied with Virginians and that those Virginians had little reason to embrace the authority of a new Catholic proprietor, Cecil Calvert could claim no real territorial integrity for his colony. The far north remained, in effect, part of Virginia. The second Lord Baltimore complained about the matter to Claiborne's London partners, to little effect. Maryland's government at St. Mary's City, meanwhile, minuscule and largely powerless, did its best to ignore the Kent Island outpost.

The continued expansion of Dutch colonial activities south from Manhattan Island along the Delaware River (which the Dutch called the South River) brought an infusion of wampum, firearms, textiles, and other European trade goods to Maryland's far north. The Susquehannocks had attempted to capitalize on this new Dutch trade, sending emissaries to Manhattan in 1626 to negotiate a trading alliance. Fortunately for Lord Baltimore, the alliance was never secured. Dutch traders chose instead to trade primarily with the Lenape of the lower Delaware and the Eastern Iroquois, or Mohawks, initiating what would be a decade of mercantile rivalry among these Native groups. Instead of a powerful Indian nation allied with England's Dutch rivals, Maryland's northern neighbor was thus a powerful Indian nation dependent largely on English trade, although dependency is not an entirely accurate characterization of Susquehannock trading relations. Despite struggles with their neighbors, the Susquehannocks, from their settlements near present-day Lancaster, Pennsylvania, exerted immense power over the mid-Atlantic fur trade, mostly by controlling the movement of beaver pelts acquired from Huron allies in the Great Lakes. Advantageous as this trade was for the Susquehannocks, Dutch trade, particularly with the Lenape to the east, meant increased pressure on fur supplies throughout the Native northeast, a circumstance that only exacerbated Native trade rivalries. By the late 1620s, this competition for furs had ignited a decade-long war between the Susquehannocks and the Lenape.[24]

Although founded as a fur-trading outpost, like most such outposts, the Kent Island factory was actually a fairly diverse economic enterprise. Its residents engaged in a variety of agricultural pursuits, including tobacco production, dairy farming, and cattle ranching—the latter proving particularly well-served by the factory's island location. Natural and human predators were mostly nonexistent. By 1637, there were somewhere near 150 head of cattle on the island. As a political entity, Kent Island was unambiguously Virginian—the islanders sent a representative to the Virginia House of Burgesses. As prescribed in its original charter, the Virginia colony was less a contiguous territory than a sort of inland archipelago composed of distinct plantations, engaged in trade and agriculture under the auspices of a single governing authority. Now that archipelago stretched across contiguous territory claimed by the second Lord Baltimore. As James Rice has written, the Kent Island factory is thus best viewed as "a thriving colony within a colony."[25]

By the time the Maryland colony was established, in sum, Claiborne had built a thriving Virginia outpost in the upper Chesapeake, secured by a trading alliance with the region's dominant Native nation. Whatever legal dispensations Lord Baltimore enjoyed, whatever the intentions of the English Crown, the reality was that the most powerful European geopolitical entity between Virginia and New Netherlands was not the colony of Maryland, but the Kent Island factory of Claiborne and his English partners. The crucial qualifier here is "European." Powerful as Claiborne and his London partners may have been, the reality was that their power depended on Native allies. Perhaps, then, the best way to characterize the geopolitical entity that dominated the northern Chesapeake in the early 1630s is as a fur-trading network consisting of Susquehannock Indians and English factors associated with William Claiborne's Kent Island factory.

* * *

That Maryland's proprietor ever thought his colony could free itself of Claiborne was mostly owing to Claiborne's London business partners. After eight years, they saw modest returns, but at a steep political price. Their North American agent's refusal to recognize the second Lord Baltimore's authority had infuriated a close Stuart ally and emboldened Virginia's fractious Puritan faction. When approached by Lord Baltimore about Claiborne's expansive Kent Island presence, the London partners ultimately reversed course, and in 1638 Maryland authorities were able to evict Claiborne. But an end to the Kent Island fur-trade factory was far from an end to the Claiborne problem. Claiborne moved his traders to Palmer's Island, now known as Garrett Island, at the mouth of the Susquehanna River. Again, Lord Baltimore protested and ultimately Maryland's militia, reinforced by Potomac Indian allies, seized Claiborne's Susquehanna outpost. These actions obviously infuriated Claiborne but now the greater threat to Maryland were Claiborne's Native trading partners. Seeking access to Dutch trade, the Susquehannocks had forged a peace with their old Lenape enemies, effectively replacing one European trading alliance with another.[26]

Claiborne's absence, in sum, changed very little for Lord Baltimore. Maryland's northern borderlands were still controlled by Native groups and European traders overwhelmingly hostile to Lord Baltimore's great American experiment. The Maryland palatinate, however clearly defined in space and however unitary its putative palatinate powers, faced problems much like those of its neighbor colony to the south. If on a map (although no such map would exist for decades) Maryland suggested sharply bounded territory, in geopolitical reality it functioned as all of England's colonies did—a collection of far-flung legal claims and mercantile alliances with little territorial or jurisdictional clarity.

It is significant that Claiborne's Kent Island factory occupied an island and that its successes were in no small measure a function of its island circumstances. In some ways, we might say that Hobbes's ideal of territorial sovereignty is applicable above all to island polities. As long as the shoreline is contiguous with sovereignty, islands present a much simpler answer to the question of territory. From Thomas More's *Utopia* to Shakespeare's *The Tempest* and Defoe's *Robinson Crusoe*, the island stands as the ideal setting for the polity of fantasy, precisely because it solves the problem of territorial sovereignty. There is no question of boundaries or frontiers, leaving rulers unchallenged in their benignity or their tyranny. As Lauren Benton has written, "as naturally bounded spaces that could be thoroughly discovered and surveyed, islands seemed by their very nature to simplify the processes of taking possession and making dominion transparent."[27] They seemed, in other words, utopian precisely because they made possible the fantasy of total and complete contiguousness of authority and spatial limits. Land-bound colonies presented precisely the opposite problem. Territorial sovereignty would never be more than a political fiction for England's mainland colonies.

* * *

Hobbes, it seems, understood geopolitical reality much better than the Privy Counsellors and Chancery lawyers responsible for Maryland's charter. Sovereignty, even a layered, composite, colonial sovereignty, was unlikely to be territorial in any modern, spatial sense. It would remain, throughout the colonial era, a matter of persons and produce and although the Lords of Trade would employ lineal boundaries in limiting the jurisdiction of future colonial governments, those boundaries would accomplish little. As a political matter, regions adjoining colonial borders would look much more like transnational borderlands than nationalized and bordered lands. Nowhere would this be more true than along the borderlands separating Maryland from its future northern neighbor, Pennsylvania. Ultimately, the proprietors of the two colonies would turn to two English astronomers, Charles Mason and Jeremiah Dixon, to transform their common frontier from borderlands into bordered lands. But here too, as the struggles of a new American federal government would make clear, boundaries did little to resolve the jurisdictional tensions inherent in federated empire.

Hobbes would surely have anticipated the failure of Maryland's northern boundary. Unable to translate constitutional principle into geopolitical reality,

Maryland's northern borderlands descended into decades of ruthless conflict that often bore hallmarks of Hobbes's war of all against all. In considering the metaphoric virtues of Hobbes's state of nature, historians of early America might thus shift their gaze from the Chesapeake to the region's northern borderlands. Although violence pervaded the Chesapeake from very early in its colonial history, by the early eighteenth century the region had stabilized, its violence assuming a more institutionalized form. To the north, along the newly incorporated Maryland-Pennsylvania borderlands, that violence continued and, in fact, would intensify across most of the eighteenth century.

Notes

1. Although it is likely that Hobbes attended meetings of the Virginia Company, some qualification is necessary since his given name is never used and company records refer variously to a "Mr. Hobbs," "Hobs," and "Hobbes." Historians have found no plausible alternative to Cavendish's secretary and associate. Noel Malcolm, "Hobbes, Sandys, and the Virginia Company," *The Historical Journal* vol. 24 (1981), 297–321. On Calvert's early engagement with the Virginia Company court, see 1623 correspondence listed in "List of Records," *The Records of the Virginia Company of London: Volume I: The Court Book*, ed. Susan M. Kingsbury (Govt. Printing Office, 1906), 174, 175, 177.

2. For an insightful analysis of Hobbes's colonial concerns, see Alexander B. Haskell, *For God, King, and People: Forging Commonwealth Bonds in Renaissance Virginia* (UNC Press for the Omohundro Institute of Early American History and Culture, 2017), esp. 272–89.

3. *Leviathan*, I, xiii, 89. References to Richard Tuck, ed. (Cambridge Univ. Press, 1991). For Hobbesian characterizations of the Chesapeake, see for example, James Horn, *Adapting to a New World: English Society in the Seventeenth-Century Chesapeake* (Univ. of North Carolina Press for the Omohundro Institute, 1994), 334; Jack P. Greene, *Pursuits of Happiness: The Social Development of Early Modern British Colonies and the Formation of American Culture* (Univ. of North Carolina Press, 1988), 12; Jack N. Rakove, *James Madison and the Creation of the American Republic* (Scott, Foresman/Little, Brown, 1990), 1. The formulation has had particular appeal for textbook writers. See, for instance, David M. Kennedy, Lizabeth Cohen, and Mel Piehl, *The Brief American Pageant*, 9th ed. (Cengage Learning, 2017), 49.

4. See, for example, Johann P. Sommerville, *Thomas Hobbes: Political Ideas in Historical Context* (St. Martin's Press, 1992), esp. ch. 6; Richard Tuck, *Philosophy and Government, 1572–1651* (Cambridge Univ. Press, 1993), 325–35; Teresa M. Bejan, *Mere Civility: Disagreement and the Limits of Toleration* (Harvard Univ. Press, 2017), esp. ch. 3; and Haskell, *For God, King, and People*, esp. 287–88.

5. Robert David Sack, *Human Territoriality: Its Theory and History* (Cambridge Univ. Press, 1986), 19; Charles S. Maier, *Once Within Borders: Territories of Power, Wealth, and Belonging since 1500* (Harvard Univ. Press, 2016), 1–2. I am not suggesting that borders are the only mechanisms available to states for this purpose or that insider-outsider status is purely a function of international borders. As a historical matter, domestic, national law very clearly had a

more central role in defining subjects and citizens. See, for example, Barbara Young Welke, *Law and the Borders of Belonging in the Long Nineteenth Century United States* (Cambridge Univ. Press, 2010), and more generally, Josep M. Fradera, *The Imperial Nation: Citizens and Subjects in the British, French, Spanish, and American Empires* (Princeton Univ. Press, 2018).

6. The relationship between territory and subjecthood is explored in Michel Foucault's *Security, Territory, Population: Lectures at the College de France 1977–1978*, ed. Michelle Senellart, trans. Graham Burchell (Picador, 2007), esp. 12–16, 100–105. Foucault's interest is neither territory nor subjecthood per se, but what he terms governmentality, or the exertion of governance over the subject. Territorial sovereignty or "territorial monarchies" represent one of the preconditions for governmentality to the extent that a territorial monarch must take stock of the geographic limits of her or his powers. This implies some kind of data gathering, perhaps of a cartographical sort, and it lays the groundwork—in Foucault's interpretation—for a more thoroughgoing statistical understanding that allows the sovereign to identify and define subjects. What is most helpful in Foucault's analysis, for my purposes, is the recognition that sovereignty is more a legal than a political or governmental fact. The premodern head of state may be sovereign and exert bounded, territorial authority, but that in no way means that the sovereign governs in any sense comparable to governance in the modern liberal state. In Foucault's interpretation, the former controls things and persons in the service of the sovereign through the use of legally sanctioned violence; the latter controls things and persons in the service of abstractions such as safety, security, health, and well-being and does so through more tacit, constitutional, and normative disciplinary systems. Where the distinction lies for Foucault is in efficacy: The modern liberal order is much more effective than the territorial sovereign at achieving the aims of government. What is crucial for the present analysis is the notion that a form of territoriality (or sovereignty with specific, delineated cartographical limits) existed before the modern nation-state. My interest is not the relative power of these geopolitical forms, but the historical shifts they suggest. See also Saskia Sassen, *Territory, Authority, Rights: From Medieval to Global Assemblages* (Princeton Univ. Press, 2006), esp. Part I.

7. *Leviathan*, I, xxiv, 170–71.

8. *Leviathan*, I, xxiii, 169; I, xxviii, 218–19. See also, Arash Abizadeh, "Sovereign Jurisdiction, Territorial Rights, and Membership in Hobbes," in *The Oxford Handbook of Hobbes*, ed. A. P. Martinich and Kinch Hoekstra (Oxford Univ. Press, 2016), www.oxfordhandbooks.com. David Runciman has observed that in fact Hobbes's conception of the state rejects the very notion of boundaries, implying a kind of universal *imperium*. For Hobbes, "the state does not exist within the state, so it is not a world that can be mapped on the co-ordinates given by its laws—it has none, beyond the laws of nature, which ascribe the ownership of actions to natural persons alone. Instead, it must be described though whatever encounters take place between the sovereign representative and an audience who take his words and actions to be the state's own. This audience has only two possible constituencies: either the sovereign addresses his subjects, or he addresses the sovereigns of other states." *Pluralism and the Personality of the State* (Cambridge Univ. Press, 1997), 16–17.

9. Annabel S. Brett, *Changes of State: Nature and the Limits of the City in Early Modern Natural Law* (Princeton Univ. Press, 2011), 5, 7. Hobbes's *De cive* quoted in ibid., 5.

10. Gaillard Thomas Lapsley, *The County Palatine of Durham: A Study in Constitutional History* (Longmans, Green, 1900), 3.

11. Lapsley, *The County Palatine of Durham*, 257; Albert J. Martinez Jr., "The Palatinate Clause of the Maryland Charter, 1632–1776: From Independent Jurisdiction to Independence,"

American Journal of Legal History vol. 50:3 (July 2008–2010), 308–309; Francis A. Mullin, "The Palatinate of Durham," *Catholic Historical Review* vol. 21:2 (July 1935), 186; Viola Florence Barnes, "Land Tenure in English Colonial Charters of the Seventeenth Century," in *Essays in Colonial History Presented to Charles McLean Andrews by His Students* (1931; reprint ed., Books for Libraries Press, 1966), 4–40; and see Ken MacMillan, *Sovereignty and Possession in the English New World: The Legal Foundations of Empire, 1576–1640* (Cambridge Univ. Press, 2006), esp. 89–105.

12. Antoinette Sutto, *Loyal Protestants & Dangerous Papists: Maryland and the Politics of Religion in the English Atlantic, 1630–1690* (University of Virginia Press, 2015), 23–24. More generally on Maryland's creation and early history, Bernard Bailyn, *The Barbarous Years: The Peopling of British North America: The Conflict of Civilizations, 1600–1675* (Alfred A. Knopf, 2012), ch. 6.

13. Charles M. Andrews, *The Colonial Period of American History: The Settlements, Volume 2* (1936; pbk, Yale Univ. Press, 1964), 277–80, 282, 283–85.

14. Andrews, *The Colonial Period of American History: The Settlements, Volume 2*, 279.

15. "Charter of Maryland Granted to Lord Baltimore June 20, 1632," *Foundations of Colonial America: A Documentary History, Volume II, Part I, Middle Atlantic Colonies*, W. Keith Kavenagh, ed. (Chelsea House Publishers, 1974), 757–59 (hereafter *FCA*). It should be said that the charter also suggests that the Maryland Lords Palatine would make laws "with the Advice, Assent, and Approbation of the Free-Men of the same Province, or the greater Part of them, or of their Delegates or Deputies, whom we Will shall be called together . . . when, and as often as Need shall require." Scholars have seen this as a concession by Calvert to attract settlers by reassuring them that proprietary rule would not be absolute. An alternative perspective is suggested by the fact that similar consensual elements had long been instituted in the Durham palatinate. These included a parliamentary-like assembly, called by the bishops to offer judicial advice and consider fiscal matters. The bishops also relied on a Bishop's Council for advice in civil and ecclesiastical affairs. That the Maryland charter is so vague about the nature of freemen's legislative authority might even suggest that far from a concession to potential colonists, the representative element of the charter involved a weakening of earlier patterns of colonial representation. This ambiguity would further suggest that the justification for the Maryland palatinate had much more to do with a sense in Charles's court that some sort of proto-absolutism, or personal rule, would be needed on England's far periphery. For the earlier consensus perspective, David W. Jordan, *Foundations of Representative Government in Maryland, 1632–1715* (Cambridge Univ. Press, 1987), 3–4. On the Durham councils, Lapsley, *The County Palatine of Durham*, ch. 4.

16. "Charter of Maryland . . . " in *FCA, Vol. II, Part I*, 756–57.

17. Sutto, *Loyal Protestants & Dangerous Papists*, 33–37.

18. Although the Massachusetts Bay Company charter of 1629 does establish a colony "lyeing and being in Bredth, from Forty Degrees of Northerly Latitude from the Equinoctiall Lyne, to forty eight Degrees Of the saide Northerly Latitude inclusively." But these stipulations, granting rights between parallels of latitude, though more explicit as a guide for colony building (e.g., assuming control and occupation of lands within these bounds), bear a closer relation to the spatial limits of the original London and Plymouth companies than they do to the clearly and fully bounded entity of Maryland. https://avalon.law.yale.edu/17th_century/mass03.asp. The consort colonies of Massachusetts—Rhode Island, New Hampshire, and Connecticut—came into being under Massachusetts Bay Company auspices, so lacked royal charters. Perhaps the most emblematic of all charters in the period of Maryland's founding were those for England's

island colonies, Bermuda (Somers Island) and Providence Island. Islands, granted to joint-stock trading companies, avoided the sort of territorial conflict that would be chronic for England's mainland grants. See MacMillan, *Sovereignty and Possession in the English New World*, 93–94.

19. *FCA, Vol. III, Part 1*, 1698–99; 1716.

20. Ibid., 1698.

21. Although the palatinate's origins were geopolitical, as a basis for colonial rule, it had clear economic significance. As the control of English colonies shifted from trading companies to individual proprietors, the palatinate model facilitated enfeoffment, the creation of rent-generating manors. See MacMillan, *Sovereignty and Possession in the English New World*, 96–97; and Christopher Tomlins, *Freedom Bound: Law, Labor, and Civic Identity in Colonizing English America, 1580–1685* (Cambridge Univ. Press, 2010), 166–83.

22. *FCA, Vol. II, Part I*, 757.

23. "William Claiborne (1600–1679)," *Encyclopedia Virginia*. https://www.encyclopediavirginia.org/Claiborne_William_1600-1679#start_entry.

24. Francis Jennings, "Glory, Death, and Transfiguration: The Susquehannock Indians in the Seventeenth Century," *Proceedings of the American Philosophical Society* vol. 112:1 (Feb. 15, 1968), 15–53; Jean R. Soderlund, *Lenape Country: Delaware Valley Society Before William Penn* (Univ. of Pennsylvania Press, 2015), 49–53; and James Rice, *Nature & History in Potomac Country: From Hunter-Gatherers to the Age of Jefferson* (Johns Hopkins Univ. Press, 2009), ch. 5.

25. Lorena S. Walsh, *Motives of Honor, Pleasure, & Profit: Plantation Management in the Colonial Chesapeake, 1607–1763* (UNC Press for the Omohundro Institute, 2010), 77; Rice, *Nature & History in Potomac Country*, 94.

26. Sutto, *Loyal Protestants & Dangerous Papists*, 42–45; Rice, *Nature & History in the Potomac Country*, 100–101.

27. Lauren Benton, *A Search for Sovereignty: Law and Geography in European Empires, 1400–1900* (Cambridge Univ. Press, 2010), 164.

CHAPTER 2

Things to Think With

The Use of Borders on a Seventeenth-Century Map of New England

CHRISTIAN J. KOOT

Sometime in 1681 John Povey, who had just begun working as clerk to William Blathwayt, secretary of the Committee of Lords of Trade and Plantations, picked up a brush and painted the borders of seven colonies in watercolors on Robert Morden and William Berry's *A Map of New England New Yorke New Iersey Mary-Land & Virginia* (ca. 1676) (Figure 2.1). At the same time either Povey or another individual in the Plantation Office adjusted the toponyms of Connecticut, Massachusetts, and Plymouth to conform to the new painted borders.[1] One of the many maps the committee stored in a series of custom-built "presses for papers, books and Maps" at their Scotland Yard office in London, this map is one of the most distinctive English maps of seventeenth-century New England.[2] The Lords of Trade and Plantations were voracious map consumers in the late 1670s as they worked with Charles II and his ministers to understand, consolidate, and reform the expanding English empire.[3] During 1681 their focus was largely on New England as they mediated disputes about Massachusetts's borders, and this work likely provoked Povey to annotate the Morden and Berry map. While borders had become a common feature on engraved and printed maps by the middle of the seventeenth century, this map's watercolor borders are a rare trace of the usually ephemeral tactile process of colonial border-making at the highest level of government. That the Lords of Trade turned to printed maps as a key apparatus of empire is less novel. Scholars have long described how European architects of empire relied on maps and the borders they represented to claim distant lands, chart colonial developments, and ultimately to bring empire into existence.[4] What is so surprising about Povey's embellishment of *A Map of New England New Yorke New Iersey Mary-Land & Virginia* is that the borders he drew did not accurately represent the actual borders of the New England colonies. The northern border of the Massachusetts Bay Colony, for example, is shown far below the Merrimack River, its true place, and the western borders of Connecticut, Massachusetts, and John Mason's Province of New Hampshire are equally vague. What should we make of Povey's lines? They did not represent Massachusetts's claimed borders, they did not represent those the lords recognized, nor were the map's borders where they would eventually be placed. Instead Povey appears to have been marking the general location of the New England colonies' borders. This imprecision suggests a more complicated understanding of the geographic space of early modern empire than scholars, even those who consider maps as "weapons of imperialism," have previously assumed.[5] Historians typically explain early modern English cartographic inaccuracies by pointing to the confluence of powerful Indigenous nations that restricted access to territory, poor cartographic instruments, rugged terrain, and unskilled draftsmen and engravers. If boundaries could have been rendered precisely, these scholars' narratives run, they would have been.[6] The Povey borders, however, suggest a different interpretation. They were generalized rather than exact because that fuzziness opened the space for the Lords of Trade to reimagine the political organization of New England and advance the imperial goal of restraining Massachusetts. Imprecise borders were not mistakes but part of the plan.

Focusing on the key moment in the seventeenth century when English state-builders were working to consolidate control of the growing empire after the Restoration of Charles II, this essay traces the multiple, often conflicting ways imperial projectors formulated boundaries in cartographic images of English New England when they drew borders on maps. Most immediate to the colonial project, administrators and sovereigns used maps as legal claims to empire that worked alongside colonial charters and evidence of occupation to solidify sovereignty over colonial places. Englishmen also believed that maps created order in the colonial world by naming territories and geographic features and by representing the unfamiliar in familiar ways. As important as marking borders on maps was for claiming sovereignty and creating order, rulers, government bureaucrats, mapmakers, and colonists did not consider the political borders rendered on maps

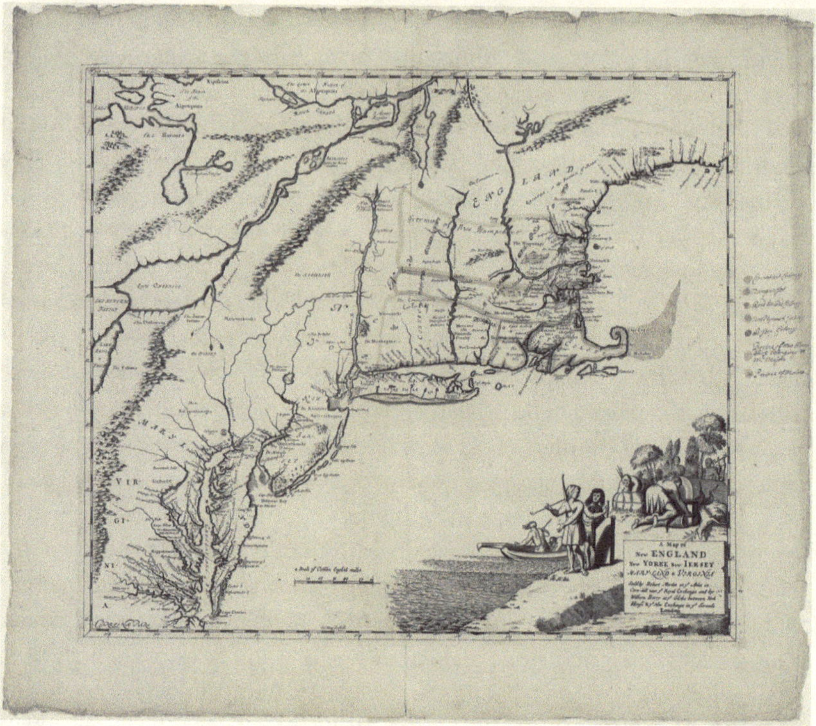

Figure 2.1. John Povey added the watercolored borders and captions to this map as clerk to William Blathwayt, Secretary of the Committee of Lords of Trade and Plantations, in 1681. Robert Morden and William Berry, *A Map of New England New Yorke New Iersey Mary-Land & Virginia*. London, ca. 1676. Blathwayt Atlas, Map #12. Courtesy of the John Carter Brown Library, Providence, R.I.

to be rigorous depictions of reality. This way of thinking about maps as scientific and rational instruments for state building whose authority rested on the accurate representation of space, the delimitation of boundaries, and the demarcation of those divisions on the landscape did not develop fully until the nineteenth century.[7]

To be sure, seventeenth-century mapmakers relied on precise geographic data to render landscapes and to denote borders when possible, and makers and viewers employed the language of accuracy and precision when evaluating maps' accomplishments. The steady accumulation of geographic knowledge in the seventeenth century and the increased precision with which cartographers could represent it enabled European empires to rationalize colonial space, to consolidate authority over colonial places and peoples, and to secure political boundaries.[8] But to say that maps had the power to make empire does not mean early modern Englishmen intended the borderlines drawn on maps to represent reality precisely. As John Povey's inexact lines on the Morden and Berry map suggest, English state-builders—including the high-ranking officials who made up the Committee of Trade and Plantations—did not yet believe that lines on maps were immutable boundaries. This was especially true in the Americas where competing empires and the presence of Indigenous nations meant the Westphalian principles of sovereign territorial statehood were contested and frontiers changed frequently.[9] In these places imprecise boundaries on maps accommodated ever-changing ambitions and opened the theoretical space that enabled opportunistic imperial state-builders to imagine (and thus create) new boundaries when needed. If we want to understand how conceptions of space shaped imperialism, we need to be more exact about how English authorities, maritime boosters, colonists, and—as others suggest—Indigenous Americans thought about the borders on maps in the seventeenth century and by extension the nature of borders themselves. Carefully examining the lines drawn on maps and the intentions behind them allows us to see imprecise and porous borders as intentional and even politically strategic.

Mapping New England

Let's return to John Povey and his map of New England. What exactly was Povey thinking about in 1681 as he colored the borders of New England provinces on Morden and Berry's map? First, his choice of watercolors to annotate the map was not unusual. Watercolors were commonly used for map-coloring because their transparency meant that they would not obscure the printed lines beneath. The medium was likely familiar to Povey. Adding watercolors to maps was a part of elite education and, as an important art manual first published in 1676 noted, was also "an excellent recreation for those gentry, and others, who delight in the knowledge of maps."[10] The map that Povey embellished was one of a host of new maps published in London during the 1660s and 1670s. As map consciousness expanded along with the English empire, Londoners sought new cartographic representations. Increased demand created new opportunities for mapmakers like Morden and Berry, who for the first time were able to compete with Continental suppliers. The copy of *A Map of New England New Yorke New Iersey Mary-Land & Virginia* that Povey colored is one of at least forty-eight maps the Lords of Trade collected in this period and is now part of what is known as the *Blathwyat Atlas*, held at the John Carter Brown Library in Providence, Rhode Island. The lords used this collection of both printed and manuscript maps at the Plantations Office in Whitehall during the 1670s and 1680s. Procured largely by Blathwayt from English as well as Continental suppliers, these maps—only a small part of that office's extensive cartographic holdings—enabled the Lords of Trade to imagine the expanding English empire, to chart its progress, and to manage its affairs.[11] Morden and Berry's map was purchased at this critical moment as Charles II and his ministers worked to consolidate and reform the expanding empire.[12]

There is much the Lords of Trade likely admired about *A Map of New England New Yorke New Iersey Mary-Land & Virginia*. The mapmakers present an unbroken line of English settlement running from Virginia in the south to New Hampshire in the north and stretching as far west as the Appalachian Mountains and Lake Ontario (see Figure 2.1). Though recording the presence of Indigenous nations on the land—the Mohegan, the Raritan, and the Susquehannock among others—the map's overriding effect is to indicate extensive English knowledge about the eastern seaboard. Buttressed by the confidence engendered by a large scale, graduated longitudinal and latitudinal lines, and the precise placement of numerous rivers and towns all bearing English names, the map is a declaration of English control of its North American empire.

It is still unknown exactly how and where Morden and Berry acquired their geographic information about New England, but their image marks a distinctive break from earlier maps of the region, many of which relied on Dutch surveys. The most likely source was a combination of a manuscript map of unknown origins supplemented by the addition of places newly important to English audiences for their association with King Philip's War (1675–78). The map, for example, includes references to "Mounthope," (the home of Pometacomet, or King Philip), "King Philip's Country," and "Spaw [Squaw] sachem" (the residence of Pometacomet's sister-in-law Weetamoo, the important Pocasset *saunkskwa* and diplomat). The inclusion of these places and the map's 1676 publication date suggest that Morden and Berry hoped to appeal to Londoners reading about the conflict in their newssheets. These references to King Philip's War also reinforced the map's accuracy as they testified to the freshness of the map's information. The section depicting the Chesapeake likewise breaks from earlier Dutch precedents and is based on the recently published *Virginia and Maryland As it is Planted and Inhabited this present Year 1670*, which had been drawn by Augustine Herrman of Maryland and engraved by William Faithorne, which John Seller began to sell in London in 1674. The only portion of the Morden and Berry map not based on recent English surveys is the section depicting New Jersey, which follows the well-known Dutch Jansson-Visscher map series supplemented with new information about English land grants. Published just two years after the final return of New Netherland to the English, *A Map of New England New Yorke New Iersey Mary-Land & Virginia* embodies the consolidated North American empire Charles II had pursued for a decade and a half.[13]

In its depiction of North America as an English space the Morden and Berry map was a key component of English efforts to defend Charles II's claim to authority over his colonial possessions against other empires. By the seventeenth century maps had become essential tools in articulating legal *imperium*—the sovereign claim to authority over a place—and *dominium*—the authority to govern that place. It was not enough to simply discover a territory. Rather, to receive international recognition of sovereignty empires also had to demonstrate long and continuous occupation and rule. Maps were effective markers of what is often referred to as "occupation and prescription" because they clearly demonstrated that land was known, measured, and occupied; that the land was held by a specific sovereign monarch whose presence was marked by symbols and nomenclature upon the map; and, when placed in a series of maps, that control was continual. The specificity of geographic information presented on a map—the location of rivers, mountains, and towns—helped to reinforce its power as an object of possession because these details indicated command and control of the geographic space.[14] *A Map of New England New Yorke New Iersey Mary-Land & Virginia* served these functions by presenting the most up-to-date cartographic knowledge of North America.

Reinforcing the map's cartographic claiming of the land is the map's cartouche, a vignette of eight Native people in the lower right-hand corner. Comprising four distinct but overlapping scenes, the figures—pictured bare-chested and wearing animal-skin skirts—are shown fishing, hunting, butchering deer, preparing hides, and packing generic bales for shipment. Drawn from stock representations of Native peoples common in European maps and prints—the fishing scene, for example, is based on Theodor de Bry's engravings of John White's Roanoke watercolors—the cartouche works with the map to make North America comprehensible for English viewers. Among the most familiar depictions of Amerindians, De Bry's images emphasized Native Americans' exoticness, but the figures' frequent duplication across a variety of media meant that period viewers were so versant with these representations that they no longer stood for the real Algonquin-speaking people they initially depicted but instead became abstracted into the type of the "Indian." Pictured here laboring to produce goods for European trade, the Amerindians' inclusion on the map served to underline supposed English mastery of colonial peoples and enabled Englishmen to compare their supposed civility with Native Americans' supposed savageness.[15] Together the familiar cartographic and iconographic modes of representation employed on *A Map of New England New Yorke New Iersey Mary-Land & Virginia* allowed viewers to imagine themselves in control of English North America.

So far this reading accords with the understanding of the map's role as a political tool, but if the ca. 1676 Morden and Berry map was already so potent an instrument of empire, why did Povey feel the need to edit it? The answer to this question lies in the timing of his act and the conversations underway in the Plantation Office in the 1670s and 1680s. John Povey became clerk there in 1680 and the seven New England colonies ceased to exist as depicted here when James II created the Dominion of New England in 1686. During these years the Lords of Trade engaged in a long-running and often acrimonious discussion about the Massachusetts Bay Colony and its place in the empire. Many in London were frustrated by the colony's leaders' constant infringement upon the territory of its English neighbors, their rejection of English political authority, and Massachusetts residents' growing intercolonial (and often illicit) trade, which critics argued was a drag on England's own commerce. The solution to all these problems, many in England began to argue, was the dissolution of the Massachusetts Bay Colony's charter and the imposition of royal authority on the colony. As the Lords of Trade put it in 1681, perhaps the only way to bring "New England" to a "perfect settlement" was to dispatch a "general governor" for the region to be maintained not at local expense but at the imperial government's so as to make him responsive to imperial, not colonial, concerns. Though no clear decision had yet been reached to withdraw Massachusetts's charter when Povey added the borders to the Morden and Berry map, English administrators were beginning to reconceptualize the political geography of New England. Ultimately that work was completed when James II, with the support of the Lords of Trade and their secretary William Blathwayt, established the Dominion of

New England in 1686. This new province was designed to consolidate metropolitan control of New England by bringing the region under a unified administration headed by a new royal governor.[16]

The Borders of New England

Drawn during this period of administrative reform and reconsideration of the Massachusetts Bay Colony's charter, Povey's imprecise borders can best be understood as part of Charles II and the Lords of Trade's efforts to assert authority over the empire and more directly as a tool to restrain the leaders of the Bay Colony's territorial ambitions. As numerous scholars have demonstrated, the fight over the Massachusetts charter was wrapped up in broader imperial, legal, religious, and political concerns, but it also was a product of differences in geography.[17] A key part of tensions with the colony were varying interpretations of the colony's geographic bounds as described in the Massachusetts Bay Company's 1629 charter and in subsequent charters for Rhode Island, Connecticut, and New York as well as preexisting land claims in what is now New Hampshire and Maine. The details of these debates are complex and involve both Massachusetts's northern and southern bounds, which were fixed in the charter as running from a point three miles south of the southernmost branch of the Charles River in the south and a point three miles north of the Merrimack River in the north.[18] The difficulty in settling these borders was twofold. First, James I and the Council of New England (a company convened to oversee settlement in New England) had set down the geographic bounds of the New England colonies in textual charters and land grants that hinged on the interpretation of poorly understood geographic features. This decision meant that colonial companies, proprietors, and officials on the ground struggled to agree about where exactly these lines were. Massachusetts' leaders exacerbated these conflicts in every case by interpreting their charter in the broadest ways possible so as to justify their land grabs from neighboring colonies.[19]

With the Restoration of Charles II in 1660 colonial governments weary of futile intercolonial negotiations brought their concerns to England. The result was a series of lawsuits in English courts and appeals made to the Crown and Lords of Trade to restrain the ambitions of the leaders of Massachusetts Bay Colony. In 1662 two years of lobbying by agents for Connecticut, for example, resulted in a new charter for that colony that settled its northern border with Massachusetts in Connecticut's favor.[20] The General Court of Massachusetts's hesitancy in proclaiming the new king in 1660 and Massachusetts's Puritan leadership put the colony in a difficult position in London. The arrival of a royal commission to investigate the Bay Colony's abuse of its authority, including its expansive territorial ambitions in 1664, brought these issues to a head. While the commission's primary responsibilities were not geographic, much of their work was designed to rein in Massachusetts's leaders' exuberant land grabs including along its border with Rhode Island (the so-called Narragansett Country or King's Province) and its westward boundary (its charter extended this to the Pacific Ocean). In addition, the commission investigated the situation in New Hampshire and Maine where Massachusetts authorities had ignored the land claims of Robert Mason and Ferdinando Gorges, both of whose grandfathers had been granted land between the Merrimack and Kennebec rivers a half-century earlier. Here the commissioners reestablished royal control of the Maine lands, though they left the decision about New Hampshire to the Crown. These measures, however, did not fully resolve the border disputes, and the work of the commissioners stalled once they returned to England. Many of the most damning petitions and papers they collected against Massachusetts were (providentially to Massachusetts Puritans) lost, and Charles II's administration was grappling with a variety of crises.[21]

The General Court of Massachusetts took advantage of waning pressure from the Crown and the commissioners' departure to reassert their authority in Maine and New Hampshire. These actions combined with continued petitioning from Mason and Gorges (they presented sixteen petitions in rapid succession) brought the issues of Massachusetts's expansive border policy back to the attention of the Lords of Trade in the mid-1670s. When combined with Massachusetts's authorities' continued flouting of royal authority, violations of English trade law, and provoking of King Philip's War the colony's actions were enough for English officials to intervene in Massachusetts policy again. This time the lords sent agent Edward Randolph (cousin to Robert Mason) to Massachusetts to investigate and gave him the power to demand that the colony send agents to England to answer the king directly.[22] Again the Massachusetts Bay Colony's borders were not the central issue, but they were important enough that one of Randolph's first actions after arriving in Boston on June 10, 1676, was to deliver Gorges and Mason's petitions to the governor and council. To counter their critics the Massachusetts General Court dispatched maps of New England that defended their aggressive border claims with agents William Stoughton and Peter Bulkeley to make their case directly in London. The Lords of Trade repeatedly held hearings to discuss Stoughton and Bulkeley's interpretation of Massachusetts's boundaries before the Lords Chief Justices ruled that the Bay Colony had "no right to governe" the lands claimed by Gorges in 1677, and the Lords of Trade restored Mason's lands and established the Province of New Hampshire in 1679. This action did not end matters, however, as the quarreling continued despite the lords' actions. Massachusetts's leaders tried to settle the conflict in Maine by buying much of the disputed Maine territories from Gorges's heirs, but the New Hampshire lands remained in Mason's hands. Meantime, the issue of the bounds of the Narragansett Country (or King's Province), which had first been settled by a royal boundary commission in 1665, arose again to occupy the lords' attention when agents from Rhode Island, Massachusetts, and Connecticut resumed their efforts to gain control of the region. It is clear, therefore, that in 1681 when Povey likely annotated the Morden and Berry map the matter of the borders of the New England colonies remained unsettled, and these matters continued to crowd the Lords of Trade's agenda.[23]

For a map so focused on borders, one of the most curious aspects of this map is that Povey completely ignored New York. Recently wrested from the Dutch, this colony's borders, especially those it shared with Connecticut, were also in dispute in the 1660s and 1670s. This omission highlights how imperial politics shaped Povey's work. When established in 1662 Connecticut's charter, which had been issued under Lord Clarendon's hand, awarded the colony a western border that terminated at the Pacific, putting it in direct conflict with Dutch New Netherland, which claimed a border with Connecticut some twenty miles east of the Hudson River. England's conquest of New Netherland two years later made this dispute less pressing, but the grant the Duke of York received complicated matters further as it placed New York's new eastern border at the west bank of the Connecticut River. Local negotiations between Richard Nicolls, governor of New York, and Connecticut governor John Winthrop Jr. settled the matter temporarily. However, because the agreement was not ratified by the Crown and because it was based on disputed geographic features, this border too remained an almost constant feature of dispute on both sides of the Atlantic. In the 1670s these issues flared after the brief reoccupation of New York by the Dutch in 1673. The return of English rule necessitated a new charter for New York in 1674 that firmly placed New York's eastern border at the Connecticut River. New York's new governor, Edmund Andros, attempted to enforce this border, but the duke refused to raise the issue in London because he believed that he would be in a more advantageous position if he waited until the completion of the investigations of the New England colonies, which had just begun. Before he could press his advantage, however, the Popish Plot and the Exclusion Crisis intervened to sideline the Duke of York as James was forced out of London by the king. Between 1679 and 1682 he was only briefly in London, spending most of his time first in Brussels and then in Edinburgh. The political campaign against James, which had its origins among radical Whigs in the Parliament, tainted his reputation in court, and his physical absence left him unable to use his influence to secure the expansive borders that his grant gave New York. This political sidetracking of New York's strongest voice likely explains why New York is missing from the 1681 annotations on the Morden and Berry map and why Povey, despite the Board of Trade's Tory sympathies under Blathwayt's leadership, didn't mark its boundaries.[24]

As a whole Povey's lines constituted a thoughtful and direct response to Massachusetts's violation of its neighbors' borders. Given the extensive attention that border disputes within and between the New England colonies generated locally and in London, the Lords of Trade and John Povey's lack of fidelity to the charter borders in his map in 1681 cannot be understood as accidental or sloppy. Instead the new borders Povey drew on the Morden and Berry map were a clear assertion of power—by the Lords of Trade, not Massachusetts's rulers—to set colonial boundaries. The generalized watercolor, rather than ink, borders helped to detach each of the New England colonies' boundaries from those that their charters articulated. And in their capricious placement Povey and the lords suggested disdain for the colony's punctiliousness. The use of the sobriquet "Boston Colony" for Massachusetts on the hand-lettered key to the painted borders on the one hand could reflect the inflated way that many in that city saw their importance to the wider colony. On the other hand, given the lords' frustration with Massachusetts in the 1670s the term more likely subtly mocks the territorial ambitions of what was still a small city.[25] Povey's hand-colored borders are not the only editing that was performed on *A Map of New England New Yorke New Iersey Mary-Land & Virginia* at the Plantation Office. Someone also scrawled out the parts of the toponyms of "Connecticut Colony," "Massachusetts Colony," and "Plymouth Colony" that extend beyond the bounds of those colonies and reinked them in new locations while adding "The Province of New Hampshire" as well (see Figure 2.1). The map's watercolor lines and inked corrections showed active engagement with the framing of New England policy in action.

As Povey painted in new borders and altered the map's toponyms he was participating in the ongoing process of creating borders. Once unmoored from the claims and counterclaims of colonial governments and proprietors the borders of the New England colonies could be redrawn and reframed. The Povey annotated map, in other words, helped accomplish the ideological work of reimagining New England's political geography. His alterations indicate that the lords would not be bound by past charters and colonial claims; it would be they and the Crown who determined the empire's shape. Amending the map in these two ways, even if it benefited metropolitan policy, was risky because the action called into question the power of maps themselves. If printed maps could be so casually altered, did they have the authority they claimed?[26] The Lords of Trade's reimagining of the political geography of New England was a potentially destabilizing act in that it undermined the authority of the map's political assertions and by extension the king and empire it proclaimed as sovereign. In this case, though, as the lords labored to remake the Stuart empire, the risky act of redrawing New England's borders afforded them an irresistible opportunity to reassert metropolitan authority. The lords judged gaining greater control over New England worth the risk.

Imagining New England

The effort to reframe New England's geography in the 1670s and 1680s extended beyond the Plantation Office. At the same time that the lords were discussing the future of the Massachusetts charter, a rush of maps went on sale in London's shops that represented the colonies of New England in innovative ways. The name "New England" was first used to identify the English possessions between New Netherland and New France on John Smith's map *New England, The most remarqueable parts* (1616). Soon it had become the standard term used for the region on European maps. But before the late 1670s it was unusual for these maps to mark out New England as a distinct political space. Rather, most maps, even if employing "New England" in the title or using the term as a toponym on the map itself, displayed the political borders and names of individual colonies as well. This convention is clear in the most famous map of seventeenth-century New England, the so-called Hubbard *A Map of New-England* (1677) (see Figure 2.2). Long celebrated as the first map to be printed in British North America, this woodcut map accompanied William Hubbard's history of King Philip's War, *A Narrative of the Troubles with the Indians* (Boston, 1677). Re-cut and printed in London to accompany the version of Hubbard's pamphlet printed there, *The Present*

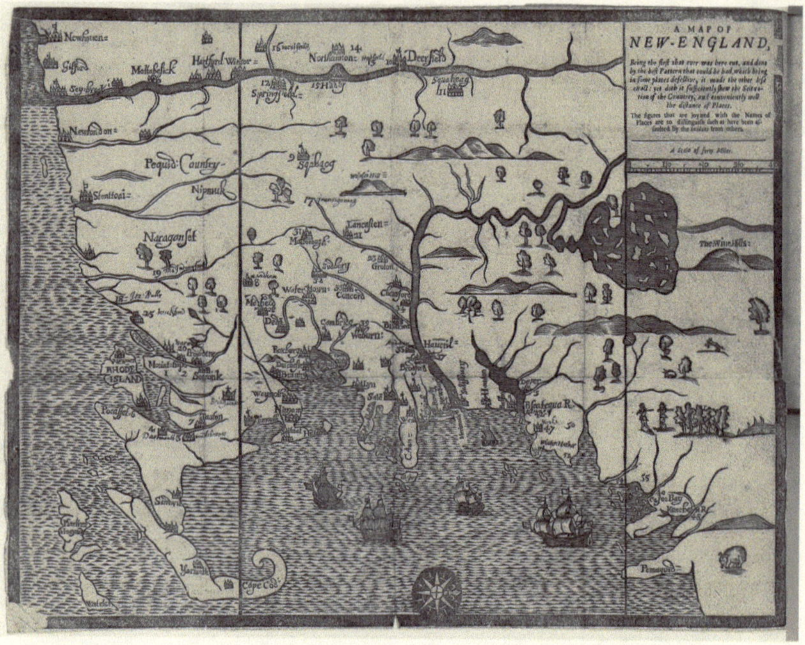

Figure 2.2. Drawn by William Hubbard, this map is the first to be produced in North America. Though titled as a "Map of New-England" its framing emphasizes Massachusetts's borders as standing for those of New England and indicates that in locals' minds no such place as "New England" yet existed. [John Foster], *A Map of New-England, Being the first that ever was here cut*. London, 1677. Courtesy of the John Carter Brown Library, Providence, R.I.

State of New-England (1677), the map circulated widely on both sides of the Atlantic and fundamentally shaped English understanding of the region's geography.[27] This image is labeled "*A Map of New-England*" but the map's framing of land and use of borders belies that title. Indeed, it is a map of Massachusetts. Among the thickest and darkest marks on the map are the two parallel lines that represent the northern and southern bounds of the Massachusetts Bay Colony according to its charter. This rendering of Massachusetts follows a manuscript map created by Bostonian William Reed in 1665 in response to a demand from Charles II that Massachusetts furnish "a mapp or draught . . . made with all exactness possible" so that the Crown could "well understand" the colony's "bounds . . . in discoursing of them."[28] Though the Reed map was lost at sea, a copy of it ultimately reached the Lords of Trade where it sparked anger because it placed Massachusetts's border so far north as to claim the coastal settlements of New Hampshire and Maine.[29] The Hubbard map's use of the distinctive Reed borders isolates Massachusetts from the rest of its neighbors and advances the Bay Colony's highly localized and politicized understanding of its geography. Residents' cartographic claim for an expansive Massachusetts works against the map's title, which presents a map of "New England." The darkly pictured borders focus the viewer's attention on the map's central panel, an effect enhanced by the map's orientation—with west to the top—and the darkly printed Connecticut River. In its representation of borders, its framing, and depiction of waterways this map makes clear that there was as yet no such political unit as "New England."[30]

Just a few years later, however, as the Lords of Trade debated the future of New England and two years before Povey remade Morden and Berry's *A Map of New England New Yorke New Iersey Mary-Land & Virginia*, several leading mapmakers with close ties to the Plantation Office and the Crown began to reimagine the region's political geography and to present New England as a politically distinct geographic unit. In 1679 Morden and Berry published *A Map of ye English Empire in ye Continent of America* by Richard Daniels. Replicating the framing of the Jansson-Visscher series, this map is most notable for its equation of New England with Virginia, Maryland, Carolina, New York, and New Jersey in its cartouche. Placing New England—made up by a group of colonies—alongside individual colonies, Daniels encourages viewers to understand New England as a single colonial unit. The map itself includes toponyms naming the individual New England colonies, but they are represented in a type far smaller than that used to denote New England and Virginia (though New Jersey and New York are rendered in the same size font as the New England colonies). The version of this map held at the John Carter Brown Library goes even further to erase New England's individual political units.[31] This copy includes hand-colored borders rendered in different colors as on the Povey map. These borders do not follow the boundaries of individual New England colonies, but rather combine New England and New York as one unit (as would the Dominion of New England) even as they represent Virginia and Maryland singularly. The next year John Seller, one of London's most prominent map publisher and sellers, brought out a map that even more assertively created a distinctive New England. Hydrographer to the King since 1671 and publisher of perhaps the most important cartographic achievement of the period, *The English Pilot*, Seller was tied closely to the Stuart court.[32] In 1680 Seller published a new atlas entitled *Atlas Terrestris: or a collection of choice mapps of all the Empires, Monarchies,*

Kingdoms, which included a map of eastern North America. *A chart of the sea coasts of New-England, New-Jarsey, Virginia, Maryland and Carolina* is derived from a number of different Seller maps and pictures the coastline in impressive detail.[33] Most striking about the map, however, are the decorative coats of arms that ornament each colony. Heraldic symbols were common on seventeenth-century maps where they served to denote English proprietary control of territory. What is most notable about their use on this map is that like the 1679 Morden and Berry map New England is placed alongside Carolina, Virginia, Maryland, New Jersey, and New York as a discrete political unit. The placement of the arms of Great Britain within New England reinforces Crown control over the region in the most direct way possible. Produced at a moment when Massachusetts's charter was being questioned because of the colony's challenges to royal authority, Seller's imprinting of the chief symbol of the monarch's sovereignty onto Massachusetts indicates that the colony's legitimacy rests not with its dissenting past but with the monarch. The distinctive local geographic imagination of Massachusetts that is so prominent on the Hubbard map has been subsumed by a metropolitan reimagining that erased that colony and replaced it with an imperially controlled New England.[34]

English policymakers' shift toward imagining New England as a single political unit is apparent on a wide range of maps of the region published in the last two decades of the seventeenth century. Though some English mapmakers continued to issue maps indicating the bounds of individual colonies, most cartographic images published in England in these years followed the shift toward representing New England as a single political unit even after the collapse of the Dominion of New England in 1689.[35] This trend is clear in the maps that represented the New England coastline that were included in John Thornton and William Fisher's newly published *The English Pilot, The Fourth Book (West-India)* (London, 1689). *The English Pilot* was the standard guide to sea navigation in the English empire and was reprinted with only occasional updates to its content until the 1790s.[36] New England is depicted multiple times in a series of overlapping maps in *The English Pilot*; each time it is shown as a discrete political entity alongside its neighboring colonies.[37] Philip Lea's *North America divided into its III principall parts* (London, ca. 1685) shares the Thornton map's depiction of New England.[38] This map captures the colonies of the three European empires that claimed North America from the Caribbean to the Arctic and from the Atlantic coast to the Pacific and makes little attempt to distinguish the individual colonies of New England. Instead, the maker imprinted the label "New England" over them even as it more clearly lists each other English colony, including the closely packed West Indian territories. Povey's 1681 hand-drawn borders, then, appear to be an early iteration of a larger cartographic reimagination of New England's political geography. As the lords debated revising Massachusetts's charter in the 1670s and 1680s, Povey's hand-inked borders created the metaphorical space to reframe the political organization of New England.

Pinning Down Borders

During the second half of the eighteenth century a new form of scientific cartography began to emerge in Britain and much of western Europe that had more confidence in the accuracy of maps to capture the reality of geographic space. This new thinking intersected with the British government's development of more centralized, systematic, and uniform ways of managing the empire during and after the Seven Years' War. The result was a series of British maps produced at mid-century that sought a more accurate representation of colonial borders to "impose order on a disordered colonial world," especially in terms of clearly demarcating political authority.[39] The very title of Emanuel Bowen's *An accurate map of North America* (London, 1755) stresses this new emphasis on precision. The map's text reinforces this message by explaining away any errors in the map as being a result of earlier cartographic and political failures. "The Limits of His Majestys several Provinces are here laid down as they at present exercise their Jurisdiction," the map reads, but the borders between "Massachusets [*sic*] Province with New York, New York with new Jersey, Connecticut with New York, and Pensilvania [*sic*] with Maryland" not to mention "the Boundary of North & South Carolina . . . or of South Carolina & Gorgia [*sic*]" were not as accurate as they could be because they had "not yet [been] finally determined."[40] Scholars have tended to echo Bowen's complaint and have argued that the failure of seventeenth-century English maps to render borders accurately was a product not of ideology but of technology. And even when borders were mapped government was too weak to enforce them. Imprecise borders, many historians contend, were a product not of ideology but of technology.[41]

What I hope to have shown here is that these borders remained imprecise in English perception and on English maps by design. As things to think with rather than definite, these borders could encompass and accommodate new imaginings of empire and allow actors at multiple levels—local, imperial, bureaucratic, royal, cartographer, and Native informant—to articulate borders. This fictionality of borders was not a bug or failure in the system but rather was vital for building an empire. The Lords of Trade's copy of *A Map of New England New Yorke New Iersey Mary-Land & Virginia* with Povey's watercolor borders shows one example of how English state-builders used this fictionality to the empire's advantage in the late 1670s and 1680s by exploiting it to help enforce their political control over New England.

Notes

1. The attribution (originally made by Jeannette D. Black) to John Povey is based on a comparison between the handwritten legend for the colors on the bottom right of the map. This writing matches other materials in Povey's hand. Povey had to have made these notations between 1680 when he began work as clerk to Blathwayt and 1686 when Massachusetts was subsumed into the Dominion of New England. As will be show below, 1681 is the most likely date because that is when the lords were most concerned with Massachusetts's border. Black argues that the edited toponyms were made in Blathwayt's hand, but there is not enough text to confirm this. See Jeannette D. Black, *The Blathwayt Atlas*, vol. 2, *Commentary* (Brown University Press, 1975), 86–87. A high-resolution image is available at: https://jcb.lunaimaging.com/luna/servlet/detail/JCBMAPS~1~1~1135~102010001:A-Map-of-New-England-New-Yorke-New.

2. Accounts of the Lords of Trade, 25 March 1677, Add. MS 9767, BL, quoted in Black, *The Blathwayt Atlas*, vol. 2, *Commentary*, 12.

3. Daniel A. Baugh, "Maritime Strength and Atlantic Commerce: The Uses of 'a Grand Marine Empire,'" in *An Imperial State at War: Britain from 1689 to 1815*, ed. Lawrence Stone (Routledge, 1994), 185–93; Owen Stanwood, *The Empire Reformed: English America in the Age of the Glorious Revolution* (University of Pennsylvania Press, 2011), 25–53.

4. There is a large and growing literature on the use of maps in English state-building. Begin with Richard L. Kagan and Benjamin Schmidt, "Maps and the Early Modern State: Official Cartography," 661–69; Peter Barber, "Mapmaking in England, ca. 1470–1650," 1598–1609; Robert C. D. Baldwin, "Colonial Cartography Under the Tudor and Early Stuart Monarchies, ca. 1480–ca. 1640," 1757–72, all in *The History of Cartography*, vol. 3, *Cartography in the European Renaissance*, ed. David Woodward (University of Chicago Press, 2007); Ken MacMillan, "Sovereignty 'More Plainly Described': Early English Maps of North America, 1580–1625," *Journal of British Studies* 42, no. 4 (2003): 423–31, 435–46. For the importance of maps in conceptualizing empire and claiming space see William J. Bouwsma, *The Waning of the Renaissance, 1550–1640* (Yale University Press, 2000), 16–17, 67–69, 112–13, 116–23, 143–55, 189–90; J. B. Harley, "Maps, Knowledge, and Power," in *The Iconography of Landscape: Essays on the Symbolic Representation, Design, and Use of Past Environments*, ed. Dennis Cosgrove and Stephen Daniels (Cambridge University Press, 1988), 278–84.

5. Harley, "Maps, Knowledge, and Power," 282. Also, see works cited above.

6. Patricia Seed, *Ceremonies of Possession: Europe's Conquest of the New World, 1492–1640* (Cambridge University Press, 1995), 140–48; Christopher Tomlins, "The Legal Cartography of Colonization, the Legal Polyphony of Settlement: English Intrusions on the American Mainland in the Seventeenth Century," *Law & Social Inquiry* 26, no. 2 (2001), 324–25; Benjamin Schmidt, "Mapping an Empire: Cartographic and Colonial Rivalry in Seventeenth-Century Dutch and English North America," *William and Mary Quarterly* 54, no. 3 (1997), 563. See also Jeremy M. Black, "The Historical Atlas: Teaching Tool or Coffee-Table Book?" *History Teachers* 25, no. 4 (1992): 495–96; Alan Greer, *Property and Dispossession: Natives, Empires and Land in Early Modern North America* (Cambridge University Press, 2018), 21, 271–90.

7. Benedict Anderson, *Imagined Communities: Reflections on the Origin and Spread of Nationalism*, rev. ed. (Verso, 1991), 163–64, 170–78; D. Graham Burnett, *Masters of All They Surveyed: Exploration, Geography, and a British El Dorado* (University of Chicago Press, 2000), esp. 204–12; Peter Sahlins, *Boundaries: The Making of France and Spain in the Pyrenees* (University of California Press, 1989), 2–5; James C. Scott, *Seeing Like a State: How Certain Schemes to Improve the Human Condition Have Failed* (Yale University Press, 1999), 2–3, 7, 24, 38, 44–47. Scholars have tended to read these arguments about political borders on maps back onto earlier periods even as they argue for the power of map symbolism as opposed to geographic accuracy. See, for example, Jordan Branch, *The Cartographic State: Maps, Territory, and the Origins of Sovereignty* (Cambridge University Press, 2013), esp. 100–19; D. W. Meinig, *The Shaping of America: A Geographical Perspective on 500 Years of History*, vol. 1, *Atlantic America, 1492–1800* (Yale University Press, 1986), 231–35; Schmidt, "Mapping an Empire," esp. 550–51. More recently Nathan Braccio, too, conflates map use with the search for "precise boundaries" even as he profitably uncovers a richer noncartographic variety of describing geography in early New England than scholars have previously assumed. Braccio, "Map Scarcity in Early Colonial New England," *Early American Studies* 19, no. 2 (2021), esp. 457–62, quotation, 460. Exceptions that consider a different map sensibility in the seventeenth century include Matthew H. Edney, "New England Mapped: The Creation of a Colonial Territory," in *La cartografia europea tra primo Rinascimento e fine dell'Illuminismo: Atti del Convegno internazionale "The Making of European Cartography,"* ed. Diogo Ramada Curto, Angelo Cattaneo, and André Ferrand Almeida (Leo S. Olshki, 2003), 160–66; MacMillan, "Sovereignty 'More Plainly Described,'" esp. 413–14. The one group of scholars that consistently has allowed for divergent geographic consciousness among early modern Americans is those studying Indigenous mapping. Juliana Barr and Edward Countryman, "Maps and Space, Paths to Connect, and Lines to Divide," in *Contested Spaces of Early America*, ed. Juliana Barr and Edward Countryman (University of Pennsylvania Press, 2014), 4–25; Braccio, "Map Scarcity in Early Colonial New England," 476–81; Gavin Hollis, "The Wrong Side of the Map? The Cartographic Encounters of John Lederer," in *Early American Cartographies*, ed. Martin Brückner (Published for the Omohundro Institute of Early American Culture by the University of North Carolina Press, 2011); Robert Paulett, *An Empire of Small Places: Mapping the Southeastern Anglo-Indian Trade, 1732–1795* (University of Georgia Press, 2012), 12–48; Gregory A. Waselkov, "Indian Maps of the Colonial Southeast," in *Powhatan's Mantle: Indians in the Colonial Southeast*, ed. Peter H. Wood, Gregory A. Waselkov, and M. Thomas Hatley (University of Nebraska Press, 1989), 292–343.

8. For two examples of this use of maps in two different places in the British empire see Matthew Edney, *Mapping an Empire: The Geographical Construction of British India, 1765–1843* (University of Chicago Press, 1999); MacMillan, "Sovereignty 'More Plainly Described.'"

9. On the theory of territorial borders (which has largely developed to explain the period after 1700) see John Agnew, "The Territorial Trap: The Geographical Assumptions of International Relations Theory," *Review of International Political Economy* 1, no. 1 (Spring 1994), 53, 60–62, 69–70; Burnett, *Masters of All They Surveyed*, esp. 204–12; Daniel-Erasmus Khan, "Territory and Boundaries," in *The Oxford Handbook of the History of International Law* (Oxford University Press, 2012), 231–34, 237–40; J. R. V. Prescott, *Political Frontiers and Boundaries* (Unwin Hyman, 1987), 1–13.

10. Smith's manual was just one of a host of similar manuals that instructed readers in using watercolors. By that date professional colormen offered watercolors for sale in London. John Smith, *Art of Painting in Oyl* (Samuel Crouch, 1705 ed.), 93, cited in David Woodward, "Techniques of Map Engraving, Printing, and Coloring in the European Renaissance," in *The History of Cartography*, vol. 3, *Cartography in the European Renaissance*, 603–6, quotation 605; Jackson Zyontz, "The Watercolor Manual: Legitimizing Watercolor Through Text," in Philip Earenfight et al., *A British Sentiment: Landscape Drawings and Watercolors 1750–1950 from the Collection of John Harbold* (The Trout Gallery Dickinson College, 2017), 12–16.

11. Black, *The Blathwayt Atlas*, 2:3–14.

12. Daniel A. Baugh, "Maritime Strength and Atlantic Commerce: The Uses of 'a Grand Marine Empire,'" in *An Imperial State at War: Britain from 1689 to 1815*, ed. Lawrence Stone (Routledge, 1994), 185–93; Owen Stanwood, *The Empire Reformed: English America in the Age of the Glorious Revolution* (University of Pennsylvania Press, 2011), 25–53.

13. Black, *The Blathwayt Atlas*, 2:82–87; Richard Boulind, "William Hack and the *Description of New England*," in *Sibley's Heir: A Volume in Memory of Clifford Kenyon Shipton* (The Colonial Society of Massachusetts and the University Press of Virginia, 1982), 119–20. For King Philip's War and Weetamoo see Lisa Brooks, *Our Beloved Kin: A New History of King Philip's War* (Yale University Press, 2018), esp. 3–4, 131. On *Virginia and Maryland* see Christian J. Koot, *A Biography of a Map in Motion: Augustine Herrman's Chesapeake* (New York University Press, 2018).

14. MacMillan, "Sovereignty 'More Plainly Described,'" 426–32.

15. The cartouche closely resembles Theodor de Bry, "Their manner of fishynge in Virginia," in Thomas Hariot, *A briefe and true report of the new found land of Virginia* [Frankfurt], 1590. A high-resolution image is available at: https://jcb.lunaimaging.com/luna/servlet/detail/JCB~1~1~778~640005:Their-manner-of-fishynge-in-Virgin; Michael Gaudio, *Engraving the Savage: The New World and Techniques of Civilization* (University of Minnesota Press, 2008), xii–xiii; Ute Kuhlemann, "Between Reproduction, Invention and Propaganda: Theodor de Bry's Engravings After John White's Watercolours," in *A New World: England's First View of America*, ed. Kim Sloan (British Museum Press, 2007), 82–92. See also Martin Brückner, "The 'New England' Cartouche: Tablets, Tableaux, and Theatricality in Eighteenth-Century Cartography," in *New Views of New England: Studies in Material and Visual Culture, 1680–1830*, ed. Georgia Barnhill and Martha McNamara (Colonial Society of Massachusetts, 2012), 227–32.

16. Journal of Lords of Trade and Plantations, April 16, 1681, *Calendar of State Papers: Colonial Series, America and West Indies, 1681–1684*, ed. J. W. Fortescue (H. M. Stationery Office, 1898), 31, quotations (hereafter *CSPC*). Richard R. Johnson, *Adjustment to Empire: The New England Colonies, 1675–1715* (Rutgers University Press, 1981), 11–19, 30–31, 45–46, 50–57; Stanwood, *The Empire Reformed*, 29–31, 49–53. The complaints against Massachusetts were most clearly articulated in Edward Randolph's reports to the Lords of Trade. See, for example,

those Randolph submitted in April 1681 to the Lords of Trade as extracted in *CSPC, America and the West Indies, 1681–1684*, ed. Fortescue, 31–33.

17. See most recently Owen Stanwood who situates the dispute at the intersection of Stuart imperial reforms and religious politics. Stanwood, *The Empire Reformed.*

18. "The Charter of Massachusetts bay," 1637 in Francis Newton Thorpe, ed., *The Federal and State constitutions: colonial charters, and other organic laws of the States, territories, and Colonies, now or heretofore forming the United States of America* (Government Printing Office, 1909), vol. 3: 1847.

19. Jenny Hale Pulsipher, *Subjects unto the Same King: Indians, English, and the Contest for Authority in Colonial New England* (University of Pennsylvania Press, 2005), 39–42, 56; Johnson, *Adjustment to Empire*, 19–24.

20. Richard S. Dunn, "John Winthrop, Jr., and the Narragansett Country," *William and Mary Quarterly* 13, no. 1 (1956), 68, 74–77.

21. Pulsipher, *Subjects unto the Same King*, 39–42, 46–59, 63–68, 83; Black, *The Blathwayt Atlas*, 2:63–71; Johnson, *Adjustment to Empire*, 19, 21–22; Dunn, "John Winthrop, Jr. and the Narragansett Country," 82–86. The fourth article of the commissioner's instructions encouraged them to try to negotiate border disputes. "Instructions to the King's Commissioners to visit Massachusetts," 23 April 1664, f. 106r, no. 51, CO 1/18, TNA.

22. Edward Randolph to Secretary Coventry, 17 June 1676, ff. 46v–47r, CO 1/37, TNA; Johnson, *Adjustment to Empire*, 19, 21–22, 28–33; Pulsipher, *Subjects unto the Same King*, 195–99, 244–46.

23. Report of Lords Chief Justices Sir Richard Raynsford and Sir Francis North to the Lords of Trade and Plantations, 17 July 1677, f. 53r, CO 1/41, TNA. The Lords of Trade discussed the Mason and Gorges petitions and the Massachusetts replies nine times between March and September 1677. *CSPC, America and the West Indies, 1677–80*, 17, 44, 56, 79, 99, 102, 110, 118–19, 135, 147–48. The next spring the issue was back before them. Journal of the Lords of Trade, 8 April 1678, *CSPC, America and the West Indies, 1677–80*, 233–34; Governor Andros answer to Enquiries of the Council of Trade, April 9, 1678, *Documents Relative to the Colonial History of the State of New York*, ed. E. B. O'Callaghan (Weed, Parsons, & Co., 1853), 3:262–63 (hereafter *DRCNY*). Mason was awarded a patent for New Hampshire in July 1679. Mr. Robert Mason's agreement with the King about New Hampshire, 1 July 1679, no. 80, CO 1/43, TNA. The final decision, however, was delayed as Massachusetts appealed. Petition of Robert Mason to the Lords of Trade and Plantations, 6 August 1680, f. 424, CO 1/45, TNA. The discussion about the two colonies' borders and limits of their authority resumed in 1681 and 1682: *CSPC, America and the West Indies, 1681–1684*, 129–30, 138–39, 198–99, 200, 240, 254, 288–91, 349. The Narragansett Country returned to the lords' agenda in 1678 and remained through 1681. *CSPC, America and the West Indies, 1677–1680*, 306–7, 309, 462–63, 492, 499, 500, 520–21, 575–76, 578–79, 589–90, 592–94, 609–11, 617–18; *CSPC, America and the West Indies, 1681–1684*, 126–27, 270. For a discussion on the Stoughton-Bulkeley map see Black, *The Blathwayt Atlas*, 2:63–65, 69–71.

24. Journal of the Lords of Trade, April 8, 1678, *CSPC, America and the West Indies, 1677–80*, 233–34; Governor Andros answer to Enquiries of the Council of Trade, April 9, 1678, *DRCNY*, 3:262–63; Philip J. Schwarz, *The Jarring Interests: New York's Boundary Makers, 1664–1776* (State University of New York Press, 1979), 5–34; John Miller, *James II*, 3rd ed. (Yale University Press, 2000), 87–109.

25. For the argument that Bostonians saw their city as in effect Massachusetts and that it should be seen as such see Mark Peterson, *The City-State of Boston: The Rise and Fall of an Atlantic Power* (Princeton University Press, 2019).

26. On the authority of print see Stephen Orgel, *The Reader in the Book: A Study of Space and Traces* (Oxford University Press, 2015), esp. 9; William B. MacGregor, "The Authority of Prints: An Early Modern Perspective," *Art History* 22, no. 3 (1999): 389–420.

27. There is a wide body of scholarship about this map that focuses on its authorship and timing but little work on how it pictures space. David Woodward, "The Foster Woodcut Map Controversy: A Further Examination of the Evidence," *Imago Mundi*, 21 (1967), 50–61; Matthew H. Edney and Susan Cimburek, "Tell the Traumatic Truth: William Hubbard's 'Narrative' of King Philip's War and His 'Map of New-England,'" *William and Mary Quarterly* 61, 2 (2004), 317–21. Most scholars refer to this map as the "Foster map" after John Foster who cut the wooden block it was printed from in Boston, but I follow Edney and Cimburek who have definitively demonstrated that it was Hubbard who designed it. Edney and Cimburek, "Tell the Traumatic Truth," 317–48.

28. *Records of the Governor and Company of the Massachusetts Bay in New England*, vol. 4, pt. 2 (Boston, 1854), 183; Black, *The Blathwayt Atlas*, vol. 2, *Commentary*, 65–66.

29. This map is in the collection of the John Carter Brown Library, Providence, R. I., William Reed. [New England, showing Massachusetts' boundaries], [1678]. A high-resolution image is available at: https://jcb.lunaimaging.com/luna/servlet/detail/JCBMAPS~1~1~1572~101870004:-New-England%2C-showing-Massachusetts. Black, *The Blathwayt Atlas*, vol. 2, *Commentary*, 63–67; Peter Benes, *New England Prospect: A Loan Exhibition of Maps at the Currier Gallery of Art* (Boston University for the Dublin Seminar for New England Folklife, 1981), 13–14, 28–29.

30. Mark Peterson, too, characterizes Massachusetts's leaders as seeing the wider region as part of an expansive Massachusetts rather than as New England, but he attributes this vision to Boston itself and he makes no cartographic argument. Peterson, *The City-State of Boston*, esp. 139–63, 169.

31. A high-resolution image of Richard Daniels, *A Map of ye English Empire in ye Continent of America* is available at: https://jcb.lunaimaging.com/luna/servlet/detail/JCBMAPS~1~1~1123~100770003:A-Map-of-ye-English-Empire-in-ye.

32. Coolie Verner, "John Seller and the Chart Trade in Seventeenth-Century England," in *The Compleat Plattmaker: Essays on Chart, Map, and Globe Making in England in the Seventeenth and Eighteenth Centuries*, ed. Norman J. W. Thrower (University of California Press, 1978), 127–57; Sarah Tyacke, *London Map-Sellers: 1660–1720* (Map Collector Publications, 1978), xii–xv.

33. A high-resolution image of John Seller, *A chart of the sea coasts of New-England, New-Jarsey, Virginia, Maryland and Carolina* (London, 1680) is available at: https://digitalcollections.nypl.org/items/510d47db-b042-a3d9-e040-e00a18064a99.

34. On this map see Brückner, "The 'New England' Cartouche," 224–230.

35. This assessment is based on study of the maps listed in Barbara Backus McCorkle's comprehensive carto-bibliography, *New England in Early Printed Maps 1513 to 1800: An Illustrated Carto-Bibliography* (John Carter Brown Library, 2001).

36. Coolie Verner, Bibliographical Note, in *The English Pilot. London 1689: The Fourth Book* (Theatrum Orbis Terrarum, 1967), v–xx; Verner, "John Seller and the Chart Trade," 139–40.

37. These maps are John Thornton, *Part of New England* (London, 1689–1713), John Thornton *Part of New England, New York, East Jersey and Long Island* (London, 1689–1713),

and John Thornton, *A chart of the seas coast of New Found Land, New Scotland, New England, New York, New Jersey, with Virginia and Maryland* (London, 1698–1749), and they are maps 689.6, 689.7, and 698.6 in McCorkle, *New England in Early Printed Maps.*

38. A high-resolution image is available at: https://www.loc.gov/resource/g3300.mf000041/.

39. S. Max Edelson, *The New Map of Empire: How Britain Imagined America Before Independence* (Harvard University Press, 2018), 3.

40. Quotation from text printed on map. Edelson, *The New Map of Empire*, 55–56.

41. See note 6 above.

CHAPTER 3

Lines on a Map

Crafting and Contesting Borders in Guillaume Delisle's and Herman Moll's Early Eighteenth-Century Maps of North America

ALEX ZUKAS

Two prominent eighteenth-century cartographers provide a rich archive for examining how early modern Europeans applied an understanding of cartographically expressed political space to the landscape of a continent where such an epistemology was unknown and where it did not fit the political realities on the ground.[1] Herman Moll (ca. 1654–1732), a German engraver who immigrated to London in the late seventeenth century and became Geographer to King George I of Great Britain after 1715, and Guillaume Delisle (1675–1726), who became First Geographer to King Louis XV of France in 1718, were the foremost European mapmakers of the early eighteenth century.[2] Both were strong proponents and propagandists, respectively, of British and French imperial prerogatives. Their maps foregrounded Britain's and France's foreign economic and political interests in an age of fierce imperial rivalries and sought to legitimize British and French territorial claims in North America in what became a "map war," which Delisle initiated in 1703 and which lost steam after Moll's final cartographic response in 1720. Both Delisle and Moll engaged in what American studies scholar Rich Heyman terms "re-territorialization (that is, a reconfiguring of territoriality)," an "attempt to territorialize what was *already* territorialized, namely native landscapes."[3] Such reterritorialization first involved deterritorialization, the elimination of boundaries that signified sovereign control cartographically, and focused on reterritorializing the land claims of their imperial adversaries.

Borders on most seventeenth- and eighteenth-century colonial maps produced in Europe appeared as deceptively precise boundary lines "meant to establish clear-cut territorial jurisdiction and legal and political sovereignty confirmed cartographically."[4] Visualizing definitive borders reflected the wishes of an emerging cartographic public of politicians, lawyers, merchants, and intellectuals—Moll's and Delisle's customers—rather than the realities and actual limits of European state power in an Indigenous-colonial setting and the further reality that borders have never been "unchanging, uncontested, and unproblematic."[5] In contrast to assertions of political sovereignty coded cartographically by borders, interracial, interethnic, and interpolitical contact zones, what Samuel Truett in this volume calls "spaces of entanglement," often occurred in what historians Colin Calloway, Claudio Saunt, and Juliana Barr call Native "homelands" and what historical anthropologist Robbie Ethridge calls "shatter zones" brought on by European and Indigenous rivalries.[6]

This chapter will argue that, to support the expansive claims of their respective governments, Moll and Delisle defined and contested territorial space, boundaries, and border-making in North America in a number of crucial maps that focused on *European* claims to territorial sovereignty and the deterritorialization of Native realms as borderless regions, thereby coding such realms as cartographically and sovereignly open.[7] However, for their colonies to be secure and prosperous, Britain, France, and even Spain needed to form and maintain alliances with the powerful Native nations that lived in or near territories that they claimed for their European sovereign. This political and economic reality made the boundaries engraved on maps of North America far more porous than they appeared, especially with outline coloring.[8] It is worth noting in the context of European imperial conflict that, in the early and mid-eighteenth century, independent Native Americans still had effective control of half of the Western Hemisphere and were a force to be considered in any imperial calculation. Consequently, they were not erased from these maps but how they appeared on them aligned with the colonial purposes and presence of the respective imperial power.[9] In many respects both cartographers, and likely their readers and promoters, succumbed to the logical fallacy that maps produce territories, that maps permit territorial control, reproducing longstanding problems in the history and historiography of cartography.[10] A related issue concerns the idea of an anticipatory geography that "served to frame colonial territories in the minds of statesmen and territorial speculators back in Europe. Maps were the first step in the appropriation of territory."[11]

Both Delisle and Moll consciously developed the maps that are the subject of this chapter as symbolic resources to support the empire-building of their respective nations and weaponized them as part of an ongoing ideological struggle over colonial borders and European sovereignty in North America. Compiling and revising their partisan maps involved a recursive practice that was in a dialectical relationship with the lived world of physicality, politics, and political economy. French and British travelers, agents, and explorers culled and condensed the vast amount of information that they acquired daily in explorations of the physical world of North America in the seventeenth and eighteenth centuries into *memoires* and reports that Delisle and Moll translated into cartographic representations. This translation was made easier by the fact that the men who compiled and wrote these reports were instructed to determine their geographic locations astronomically and to narrate their travels with geographic-territorial ends in mind.[12] Based on their respective French and British sources, Delisle's and Moll's cartographic representations circulated in European political and diplomatic arenas, and their "map wars" were bound up in complex ways with actual wars, highlighted the current political, economic, and military struggles in North America, and expressed the expansionist dreams of French and British leaders in the early eighteenth century.[13] All in all, contrary to poststructuralist claims, Delisle's and Moll's struggles show that, with all their border-making and terraqueous claims, maps on their own do not produce territory.

Delisle's *Carte du Mexique et de la Floride* (1703)

In 1703 Delisle compiled information collected by French explorers, especially René-Robert Cavelier, sieur de la Salle, Pierre Le Moyne, sieur d'Iberville, and Pierre-Charles Le Sueur to publish a *Map of Mexico and Florida and English Territories and the Islands of the Antilles*.[14] It was issued two years into the War of the Spanish Succession (1701–1714; Britain involved 1702–1713), known by the English in the North American theater as Queen Anne's War and by the French as the Second Intercolonial War. The map located the course and delta of the Mississippi River more accurately than previous maps, represented Delisle's interpretation of the colonial claims of Spain and England (an emphasis

Figure 3.1. Guillaume Delisle, *Carte du Mexique et de la Floride*. Paris, 1703. Courtesy, David Rumsey Map Collection, David Rumsey Map Center, Stanford Libraries.

of its title), placed the western border of Spanish Florida at the Pecos and Rio Grande Rivers, and triggered a "map war" soon after the War of the Spanish Succession ended. While the major pitched battles of the war occurred in western and central Europe and war in North America took the form of destructive, mostly hit-and-run raids by colonial militias and their more numerous Indigenous allies and privateers in the Caribbean and coastal Atlantic, the war's underlying cause was the possibility that the Spanish monarchy and territories could become merged with France's monarchy and overseas territories. That is, while its ostensible cause involved dynastic issues, what led to warfare were imperial issues that entailed the disposition and composition of the Spanish empire in North America, the supply of slaves to that empire, and the borders of the Spanish, French, and British colonial claims there. In 1702, as the French were allied with the Spanish and positioned themselves to intervene in the affairs of Spain's empire, the Spanish Crown granted the French Guinea Company the *Asiento*, an exclusive contract to import enslaved Africans into Spain's American colonies.[15] As historian Phillippe R. Girard notes,

"French-British rivalries were nothing new, but a new development was the growing importance of imperial rivalries as a motive for going to war and [colonies] as a theater of operations."[16]

Delisle's map was an early expression of the official French position that the entire watershed of a river or lake in Indigenously settled lands was encompassed by the right of Christian, European "discovery," which deterritorialized Native peoples.[17] He applied this doctrine to reterritorialize the claims of all three colonial powers. For the French-allied Spanish, it meant that the borders of Spanish Florida extended as far as west, east, and north as the length of the rivers that drained into the Mississippi south of the Ohio River, thus creating a region that extended from the Ohio River to the Gulf of Mexico and included most of Texas and the southern plains, while English claims extended as far as west as the headwaters of rivers that drained from the Appalachians to the Atlantic. The French based their claims to the region around the Great Lakes and the St. Lawrence River on the proclamation of Simon François Daumont, Sieur de St. Lusson (d. after 1677), in 1671 that, having traversed the region from Montreal to Lake Superior by water, he took possession of Lakes Huron and Superior and "all other Countries, rivers, lakes and tributaries, contiguous and adjacent thereunto, as well as discovered as to be discovered," which included the Ohio River and upper reaches of the Mississippi River. St. Lusson's extension of this doctrine to undiscovered lands of course made a mockery of its premise that discovery was necessary to legitimate a territorial claim.[18]

This theory of *de jure* terraqueous colonial possession suited the French situation, which required travel by North American rivers and lakes and eliminated the need to actually traverse, occupy, or even "discover" terrain far away from river banks in order to claim it. Estimates of the European population in Canada in 1700 was 6,200 persons, not enough to make much of an impression on the landscape or "discover," claim, or defend much of it.[19] Consequently, Delisle peppered the landscape of Florida, New Mexico, and New France's northern Mississippi Valley with the French names of Indigenous peoples, whose labor power and military acumen were crucial to the *de facto* success (or defeat) of the French colonial enterprise, concentrating their names along or near rivers while indicating the existence of only three Native bands in the British colonies. He deterritorialized them by not drawing borders around any Indigenous names that would have signaled to his readers that they held inviolate sovereign territories protected against European territorial claims, including the Natchez, Choctaw, and Chickasaw whose settled way of life he indicated with teepee-like symbols, or the numerous "wandering" people in what is now Texas. Delisle placed the Haudenosaunee (Iroquois) around the Finger Lakes, which he located within the borders of New France, and he gave no indication that they controlled territory further west in the Ohio River valley. The absence of borders around Indigenous names in the Delisle map conveyed de facto occupancy or presence, not the sense of Haudenosaunee dynamism or expansion that Charles Prior maintains concerning the Mitchell map he discusses in this volume. The attentive map reader would notice, however, that Delisle indicated Haudenosaunee residence with the symbol for town that he used for European settlements, rather than the teepee symbol he used for other Indigenous peoples, subtly distinguishing them from other Native Americans.

The string of forts and outposts on his map highlighted the fact that French, British, Spanish, and Portuguese imperial holdings were "islands of occupation surrounded by a [Indigenous] sea of land"[20] within "a networked nodal world of trade, diplomacy, dispossession, coercion, and cross-cultural mixing and violence."[21] France's nodal approach to spatial control was an imperial geography that relied both on control of strategic trading and military routes, rather than absolute control of large swaths of territory, and on negotiation with Native peoples to gain access to Indigenous trade networks and military resources.[22] Delisle did show the "Route that the French take to get to Carolina" from Illinois, which followed the Tennessee River from the Mississippi to the Appalachians. Much of the information on the map came from Native American sources transmitted by French missionaries' and traders' reports to the French state, which Delisle could access via his position at court and as a member of the French Royal Academy of Sciences who had direct contact with government ministers who requested specific works from him.[23]

To support French claims to Florida, Delisle's map showed La Salle's abandoned fort ("Fort François") on Baye St. Louis (today Lavaca Bay in Texas), recently founded French forts in Mobile and Biloxi Bays, and an unnamed fort (Fort de La Boulaye) in the Mississippi River delta. Encoding all these features, including explicitly labeling the "Mouth of the Mississippi" River, gave cartographic shape to La Salle's idea in the early 1680s and Iberville's from the early 1700s that French control of the mouth of the Mississippi River would provide a suitable location for attacking Spanish provinces in northern Mexico and gaining control of its productive silver mines through an alliance with the Native peoples in the area who were indignant over their enslavement by the Spanish.[24] Indeed, Delisle noted the location of numerous silver mines across the border in Mexico, perhaps to incentivize French adventurers to ally with

Native Americans who raided those provinces on a regular basis, implying that such a "paper border" was permeable, negotiable, and impermanent.[25] In the context of a major inter-imperialist war, Delisle's titular focus on Florida, Mexico, and the British colonies, and their prominent boundaries on the map, showed where future French colonial interest lay.

Moll's *A New and Exact Map of the Dominions of the King of Great Britain on ye Continent of North America* (1715)

British elites, however, did not accept "discovery" or French riverine theory as a foundation for territorial claims. They maintained that in addition to letters patent from the Crown, symbolic markers, formal declarations of British sovereignty, and occasional purchases of land from "those too powerful to displace by force of arms or who might have stronger claims to their lands in the eyes of the English," physical settlement by (non-Indigenous) people and the establishment of permanent (non-Indigenous) settlements legitimized a European claim to American territory, setting up a confrontation with France over their respective colonial borders.[26] This argument played to Britain's strength: English colonies had an estimated European and African population of 284,000 people in 1700 with 12,000 people alone in tiny South Carolina, where most of the place names had European derivation.[27]

In 1715 Herman Moll published *A New and Exact Map of the Dominions of the King of Great Britain on ye Continent of North America* as a direct response to Delisle's 1703 map. This map expressed the vocal concerns of colonial interests that France was confining Britain to the Atlantic littoral while setting up a vast empire in the interior of the continent for itself.[28] In 1713, the Treaty of Utrecht ended the War of the Spanish Succession for Britain, but failed to define the boundaries in the St. Lawrence region and between Carolina and Florida. Moll's map fanned the dispute between France and Great Britain over the unsettled boundaries. Moll asserted a particular postwar reterritorialization on the map by provocatively extending New Scotland's (present-day Maine, New Brunswick, and Nova Scotia), New England's, and New York's borders to include French-claimed territories on the south bank of the St. Lawrence River, which the English had long coveted but not settled. He did show how settled the region claimed by Britain was by naming numerous towns throughout New England, port cities along the coastline, and various inland counties.

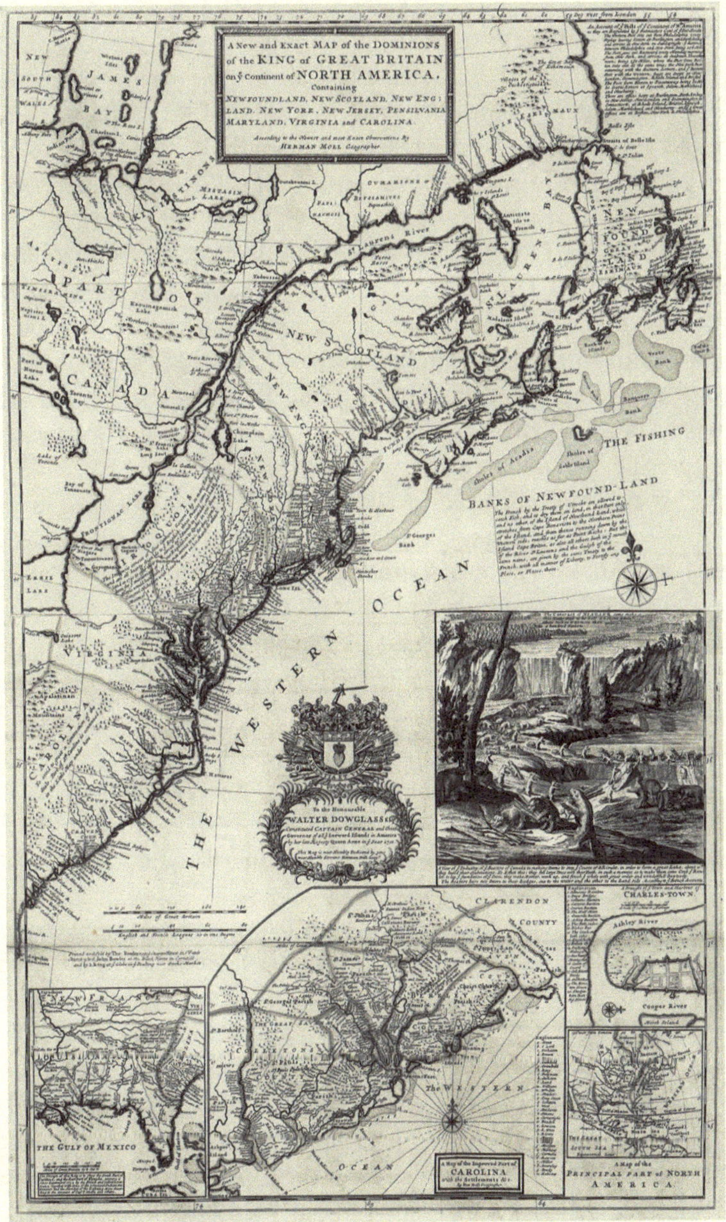

Figure 3.2. Herman Moll, *A new and exact map of the dominions of the King of Great Britain on ye continent of North America*. 5th state, London, 1731 (1715 orig.). Courtesy, David Rumsey Map Collection, David Rumsey Map Center, Stanford Libraries.

One border did not show up on Moll's map. While he indicated the location of "Susquehannock, An Indian Fort," which in French and Haudenosaunee understanding marked the western border of Pennsylvania, he drew no border there.[29] Beyond the fort was the homeland of the Haudenosaunee, which Moll deterritorialized by placing it within the clearly engraved borders of Pennsylvania and Maryland, indicating that it lay within the bordered "Dominions of the King of Great Britain." In the late seventeenth and early eighteenth centuries, European geographers were uncertain about the extent and location of the Haudenosaunee. A 1701 geography text, which Moll knew quite well and for which he engraved the maps, conveyed a generally negative and condescending attitude toward the Haudenosaunee that did not suggest any notion of sovereignty and stated that "*Irocoisia*, or *Iroquois*, [is] a Country, whose Situation is differently express'd in several Maps."[30]

On his map Moll engraved "Iroquois" in large letters in an arc southeast of Frontenac Lake (now Lake Ontario) and composed a long note under their name that he took great care not to extend beyond the northern border of Pennsylvania and the southern border of Maryland.

Moll's note filled their homeland with English words and purpose, highlighting their value as military allies in imperial conflict with the French: "The Iroquois consist of four Cantons, Govern'd by so many Kings and are all hearty friends to ye English: those Princes came into England in 1710 to offer their service agt [against] ye French in Canada and had it not been for ye miscarriage of our Expedition to Quebec in 1711, those People would have been of great service to us, for they joyn'd General Nicholson with 2000 men on his March to attack Monreal [Montreal]."

The episode with the "Iroquois kings" is instructive regarding the differences between official and general public understandings of colonial affairs. Attempts in the early eighteenth century to forge links between Britain and the Haudenosaunee had mixed success with colonial officials and the public at large. Moll mentioned the major instance in the long notation on this map cited above. The 1701 geography text with Moll's maps was republished and expanded in 1709 and the descriptions of the Haudenosaunee became more xenophobic and overwrought while mentioning their councils and highlighting their long-distance military expeditions and purported cruelty.[31] This view of the Haudenosaunee circulated widely in Britain on the eve of the visit of the "Four Iroquois Kings" whom Moll memorialized in that long note. A detailed study of the episode concluded that, in popular media, "the figures of the kings were ambiguous, contradictory, and half realized; the new view of Indians as legitimate agents of state power jostled uncomfortably with the older idea that Indians were simple and savage."[32] Colonial governors organized the visit in the hope of getting royal support for an attack on Canada and did all they could to create the impression of an official state visit, but what is significant is that the organizers "had little hope of finding allies among the leading sachems of the confederacy. Instead they found supporters where they could and then shamelessly falsified their credentials." All were without influence at home and four of the five nations of the Haudenosaunee did not even participate. As the historian of the event noted, "There was an element of masquerade attached to the visit" and "From the start, readings of the visit were contested."[33] Moll's map heightens the masquerade since he knew full well that there were five Haudenosaunee nations, not four, but since only four "kings" showed up in Britain, it may have been awkward to explain on the map why some opted out.[34] The popular essayist and writer Daniel Defoe, whom Moll knew well and counted among his circle of friends and associates, published a view of the visitors that aligned with the geography books whose maps Moll engraved. Defoe wrote that they were from "a small Nation of *Savages* in the Woods, on the back of our two Colonies of *New-England* and *New York*, the same from whence our four pretended *Indian* Kings came lately on their own Fools Errand; they were always esteem'd as the most Desperate, and the most Cruel of the Natives of *North-America*."[35] The visit's diplomatic fig leaf worked, nevertheless: The governors got Crown support for an invasion of Canada but it miscarried, as Moll lamented on his map.

As the editors of this volume contend, maps are material artifacts that express arguments and a range of desires, and so a great deal depends on who commissions or develops a map and who the intended audience is. Moll's 1715 map differs greatly from Mitchell's map forty years later that Prior analyzes in part because they were responding to different political and military environments, had different origins, and were making different arguments as a result. Richter has characterized the different periods of the long eighteenth century in North America as ones of violent regional wars (1670s to 1710s), regional equilibrium (1720s to 1750s), and the return of violent regional wars (1760s to 1815), each requiring different interpretations.[36] Moll created his map in the first period on the cusp of the second, Mitchell produced his map in the second period and on the cusp of the third, and Moll's map was a private commercial venture designed to sell as many imprints as possible while Mitchell's map was commissioned by the president of the Board of Trade "who sought to advance his own aggressive agenda with respect to the empire in North

America in the run-up to the Seven Years' War."[37] Moll's and Mitchell's maps expressed different arguments for the literate public that was their audiences and, in Mitchell's case, a clear government purpose. Unlike Mitchell's map, Moll's map was not centered on Iroquoia. Indices that were present in Mitchell's map were absent in Moll's and Delisle's maps: There were no blocks of text positioned around the map that narrated episodes in the diplomatic history of the interior and no explanations of migrations and territorial transactions. Moll's map presented the idea that the Haudenosaunee were valuable military allies subject to the Crown of Great Britain, while Mitchell contended they were valuable military allies and had a greater degree of sovereignty than Moll allowed. After all, Moll composed his map in light of Clause XV of the Peace and Friendship Treaty between France and Great Britain at Utrecht in 1713 that ended the War of the Spanish Succession, which considered "the five nations or cantons of Indians, subject to the dominion of Great Britain," and his map reflected that public understanding.

Moll acknowledged the military importance of the Haudenosaunee as allies but not much else. Unlike Mitchell, Moll contained the Haudenosaunee polity within Britain's colonial borders and, consistent with his general practice, he withheld from the Haudenosaunee the cartographic code, a border, which signaled sovereignty.[38] This downplaying of the sovereignty of the Haudenosaunee was as fanciful as the many borders Moll did draw, but the imperial and colonial context of his era was quite different from Mitchell's and there is no evidence that Moll worked with the Board of Trade to promulgate particular state-sanctioned diplomatic and imperial goals with his map. As a commercial cartographic entrepreneur seeking profit above all else, Moll appealed to the tastes of the literate general public, which, as we have seen, diverged greatly from the preferences of governmental elites regarding Native American allies. Moll was granted the title of Geographer to the King, which was as a royal warrant to supply goods and services to the royal household rather than an office of state, which was Delisle's position in France. The warrant served as an endorsement of quality that elevated the prestige of the tradesperson who produced the good or service. As a private citizen, craftsman, and small entrepreneur, however, Moll had no privileged access to confidential state documents, deliberations, or directives. His information came from the public domain and his circle of acquaintances, some of whom, like Defoe, had connections to various ministers.[39]

Moll produced a map centered on European imperial antagonisms that took place on the homelands of Indigenous peoples and highlighted their importance as martial resources in inter-imperial conflict. The map extended New York's and New Scotland's borders, which now contained French forts in the area around Lake Champlain and "French Settlements" along the St. Lawrence River. While acknowledging French occupancy, Moll appeared to be pushing the British colonial border to the southern bank of the St. Lawrence. He placed "The Great Fall of Niagara" at the northwestern corner of the state of Maryland and made it the aqueous boundary between French and British claims in that part of the Great Lakes region. He also made it the focus of a beautiful inset against the right margin of the map. The inset was a visual fable about the productivity and bounty of North American nature, especially beavers, a lucrative resource worth claiming and contesting, as attested by the intermittent conflicts between the Haudenosaunee, other Native peoples, and the French that occurred after 1640 and which are sometimes called the Beaver Wars (ca. 1641–1701).[40] Canada was a French colony so this inset was an invitation to challenge French borders in the heart of North America. Moll's dedication of the map to Walter Douglass (1670–1739), whom Queen Anne had appointed as Captain-General of the Leeward Islands in 1711 and whose coat of arms bristled with weaponry suitable to defend and extend territorial borders, reinforced a bellicose attitude toward French claims.

In an inset along the bottom margin of the map Moll indicated a British colonial border along the Appalachian Mountains with New France and Louisiana to the west and a fortified Charleston, which functioned as a border garrison against the Spanish and Yamasee people to the south in a second inset.[41] The largest inset, "The Improved Part of Carolina," presented a patchwork of peoples in the colony and the presence of so-called "settlement Indians" who forged a political and economic accommodation with the rulers of Carolina: a "Santee Indian Fort" near the northern border of the colony, "Wando Indian" and "Indian Fort" northeast of Charleston, and "Indian Settlements" (including "Cayawah" [Kiawah] occupying a large island) along the coast southwest of Charleston. Moll did not engrave any borders around these deterritorialized names but showed them living within the well-defined borders of parishes. Disease and warfare reduced their numbers and many disappeared from colonial records after 1743.[42] A fourth inset at the bottom left corner of the map showed a zone of increasing imperial and Indigenous conflict in southeastern North America, a region Claudio Saunt characterized as an imperial borderland *and* Indigenous homeland where imperial conflicts manifested themselves in proxy wars fought by strong regional powers allied with Britain or France but not controlled by either imperial power. Europeans had little presence until the English established a colony at Charleston in 1670 and the French established outposts in the Biloxi-Mobile area around 1700. In 1715, no Europeans resided in these Indigenous territories; however, about 3,000 English and 8,600 slaves lived in South Carolina and about 300 French colonists and 100 slaves resided on the Gulf Coast.[43] France obtained control of Florida in September 1712 and made it part of Louisiana, a fact Moll acknowledged in a notation at the bottom of the inset. However, British colonists and newly coalesced Indigenous polities like the Creek, Cherokee, Chickasaw, and Choctaw contested French possession on the ground.[44] While overall the inset emphasized the permeability, volatility, and bellicosity of this imperial borderland, which included Florida and Carolina, to stress the commercial vitality of the region along with its permeable borders, Moll engraved the Carolina Trading Route, which led from Charleston across the land of the Creeks (Muscogee) to the land of the Chickasaw on the Mississippi River.

Moll based this inset on Thomas Nairne's *A Map of South Carolina*, which decried French presence and promoted a more vigorous British role in the southeast.[45] Interested in expanding the colony of South Carolina farther to the west, Nairne made an overland reconnaissance trip to the Mississippi River from Carolina in 1708 to firm up alliances with the Chickasaw and to fight and enslave Native Americans in Florida and Louisiana for the colony of Carolina. His map claimed all the territory from the Atlantic coast to the Mississippi River between the 29th and 34th parallels for South Carolina. Moll did not follow Nairne in extending the border of Carolina to the Mississippi in this inset, but he did follow Nairne in highlighting the fact that French territory was populated by scores of Native nations, some of whom, like the Chickasaw, allied with the British against the French. Moll identified Indigenous settlements with small hut-like symbols and noted how many fighting men they had according to Nairne and other sources. Like his comment about the Haudenosaunee as "hearty friends," this calling out of Indigenous martial power signaled to his audience that the significance of Indigenous people lay in their military ability to aid or thwart British designs on a region where waging war appeared increasingly endemic. The French held a similar view of "our Indians" versus "their Indians" and armed their Indigenous allies to fight the Indigenous allies of the British, who reciprocated. Moll's map and insets supports Richter's contention that "Indian alliances were vital to Britain's conflicts with French and Spanish rivals" since Native American allies would do most of the fighting.[46]

Richter writes, "The importance of the Cherokees to British Imperial interests was driven home in 1730, when the eccentric Scottish baronet Sir

Alexander Cuming . . . recruited six alleged chiefs for a highly publicized London interview with the Board of Trade."[47] Moll memorialized this visit with a note next to the name "Cherecies" that "One of ye Kings of this Nation was in England in 1730." During that visit a delegation signed a treaty that referred to the Cherokee twice as the children and subjects of the British king who agreed to be his allies and fight against Britain's enemies.[48] Historian Kristofer Ray contends that "Carolinians . . . believed that Cherokees provided both a bulwark and a means of expanding via the Tennessee River" but "Cherokees were not simple pawns in a European imperial chess match": They "consistently manipulated both French and British officials" while "rejecting Anglo definitions of nationhood" since the Cherokee had a decentralized polity in which no one individual or individuals could bind all the clans in one treaty.[49] Indeed, "no Native polity in 1725 would have recognized the blanket assertions of singular nationhood" that British Carolinians imposed on a map drawn by Fanni Mingo: "By adding them British Carolinians overlaid an imagined political world on the one described by Fanni Mingo." The lines on Mingo's map, representing trading paths and rivers, focused on mobility rather than exclusive jurisdictional claims, the focus and import of the border lines on Delisle's and Moll's maps of an imagined, imperially centered political world.[50]

Delisle's *Carte de la Louisiane et du cours du Mississippi* (1718)

Guillaume Delisle's 1718 *Map of Louisiana and the Course of the Mississippi River* (see Figure 3.3) certainly read like a French response to Moll's 1715 map of British claims in North America. The title fully captured the official French position that Louisiana and the watershed of the Mississippi River were coextensive. This claim dates to La Salle in his 1682 proclamation claiming Louisiana for Louis XIV and rested on the theory that the entire watershed of a river or lake was encompassed by the right of "discovery," which meant that the furthest extent of the Mississippi and its tributaries became the boundaries of Louisiana. La Salle's claim fulfilled the imperial vision of Jean-Baptiste Colbert (1619–1683), who "perceived the great river as the axis of both offense and defense in North America," thwarting English moves west and Spanish moves east and securing "a French empire in North America."[51] The French applied the principle consistently: The spine of the Appalachians created two watersheds, one for the French to the west and the British to the east. With

Figure 3.3. Guillaume Delisle, *Carte de la Louisiane et du cours du Mississippi.* 1718. Courtesy, David Rumsey Map Collection, David Rumsey Map Center, Stanford Libraries.

the War of the Spanish Succession over, Delisle eliminated "Florida" from his map since he no longer needed to consider the territorial claims of a wartime ally. The French had long been interested in Florida and Delisle absorbed it into Louisiana following the French claim that their exploration of hundreds of miles of the Mississippi River meant that they could claim the land drained by it and all its tributaries, including those in what had been Florida. With this "amphibious" approach, to use a concept coined by Truett in this volume, Delisle's map showed how terrestrial and aqueous spaces were entangled in French territorial claims. As indicators of persistent conflict with Native peoples and imperial rivals, Delisle had French forts dotting the colonial interior of New France and Louisiana from Lake Champlain to Lake Ontario, down the Mississippi River, and on the Gulf Coast.

For Delisle, the border of British colonies began just south of Charleston and ran north and then northeast along the spine of the Appalachian Mountains, turning east to Fort Susquehannock in Pennsylvania and then north again but remaining south of Lake Champlain. An "old fort of the French" in the southeast, La Salle's short-lived fort in "Texas" and his 1685–1687 explorations in that region, the 1710 and 1713 journeys by Louis Juchereau de St. Denis (1676–1744) and that by Jean Baptiste le Moyne, Sieur de Bienville in 1717 shown on the map completed these French claims of "discovery." He also showed the new French settlements of Natchitoches and New Orleans. While contesting Spanish paper claims to a region it never occupied, Delisle acknowledged those claims by engraving the route of Hernando de Soto's (1497–1542) and Luis de Moscoso Alvarado's (1505–1551) expedition of 1540–1543, Alonso de León's (ca. 1639–1691) 1689 route from northern Mexico to confront La Salle's (abandoned) fort, and the location of Nuestro Padre San Francisco de los Tejas which, in 1716, provided Spain a sustained physical presence for the first time. Delisle placed Louisiana's western border at the Rio Grande and Pecos rivers and its northern border at the Great Lakes. At the bottom of the map Delisle provided a key to some of the map's symbols. A hut represented "Indian dwellings" or towns, a star "disturbed Indians," and a cross "destroyed Indians." The hut symbols were quite numerous while the others were quite sparse. As he did on his earlier maps, he noted the names of the major Indigenous nations in "Louisiana" including the Iroquois, Miami, Cherokee, Choctaw, Osage, Caddo, and Apache, but he also noted somewhat sensationally that there were "nomadic and cannibalistic Indians" along the Gulf Coast of Texas.[52] A notation that Delisle placed in Carolina asserted that the French discovered, took possession, and named the area in honor of Charles IX back in the sixteenth century. Delisle placed the Iroquois outside the boundaries of British colonies and within French-claimed domains between the Great Lakes and the headwaters of the Ohio River. He noted the Indigenous nations "destroyed by the Iroquois" including the Erie, Huron, and others, indicating the extent of Haudenosaunee military forays into the Ohio River Valley and the Ontario Peninsula, but he did not place these areas under Haudenosaunee territorial control or dominion. As in his previous map, he did not draw any borders around Native nations as indications of territorial sovereignty. His map deterritorialized Native peoples and reterritorialized Spanish and British claims to the area in favor of the French.

In 1717 only twenty-seven French families lived in Louisiana but the French were on the move.[53] To impede Spanish colonization of the region, St. Denis built a fort on the Red River in 1714, staking a claim to the region.[54] In 1718 Bienville founded New Orleans to regulate use of the Mississippi River and to create a more permanent presence and alliances with the powerful Native peoples of the region like the Choctaw and Natchez. Delisle's placement of the border of Louisiana at the Rio Grande and Pecos rivers in his map was an aggressive move encouraged by officials in Paris.[55] Although the French did not have enough settlers or military presence to enforce such a boundary, the Spanish and British took French imperial imaginings seriously.[56] Delisle made Louisiana look like a unified territory with specific political boundaries.

The publication of his map with its expansive view of Louisiana's borders coincided with a publicity campaign that accompanied the launch of John Law's joint-stock Company of the West and a French economic boom. Delisle produced the map for the company and noted mines containing lead and copper as well as routes indicating possible trade relations in northern Louisiana.[57] The company issued shares to investors and promised to develop the new territory. Investors expected fantastic returns based on rumors of Louisiana's natural resources (gold, other precious metals, pearls) and prospects for economic development.[58] British elites were disturbed by the Company of the West, and Delisle's significant reduction of the western and southern borders of British colonies along the eastern seaboard and appending that territory to French Louisiana only inflamed matters.[59] The fact that a few months after he published his map he was awarded the unique title of *First* Geographer to the King (with an annual salary) made it appear that the French state gave its official blessing to the disposition of French territorial holdings in North America that Delisle had proposed in his map. Delisle's reputation as a methodical and meticulous geographer added scientific heft to what was clearly a political act.

Kim Gruenwald's idea of a riverine West neatly encompasses the terraqueous region that the French claimed in the interior of North America, a region drained by the Mississippi River and its tributaries. In French theory and practice, rivers were not so much borders as arteries or fluid conduits of connection; being connective and expansive rather than delimiting and restrictive, they justified French territorial claims, and no map expressed that idea more than Delisle's map of Louisiana, confirming Gruenwald's insight that the Atlantic world made "landfall" along rivers leading to the interior of the continent.[60] Delisle's map and Moll's response to it show that rivalries of the Atlantic world spread to the continental interior.

Moll's *New Map of the North Parts of AMERICA claimed by FRANCE* (1720)

Delisle's territorial claims and border renderings incensed English colonial elites like Richard Beresford (d. 1724), who wrote to the Commissioners for

Figure 3.4. Herman Moll, *A New Map of the North Parts of AMERICA claimed by FRANCE under the names of Louisiana, Mississippi, Canada and New France with the Adjoyning Territories of England and Spain*. 1720. Courtesy, David Rumsey Map Collection, David Rumsey Map Center, Stanford Libraries.

Trade and Plantations in 1717 that "the French had a design to dispossess us of all our plantations" in North America.[61] The governor of New York, William Burnet (1688–1729), wrote to the Commissioners for Trade and Plantations in 1720 that Delisle's new map made even greater encroachments on British territories than his earlier maps and was drafted in the same spirit. Complaining that the French had set up a trading post near Niagara Falls and were working to reduce British influence in the region, he urged the British government "to break the neck of [French] trade."[62] For their part, in August 1719 the Commissioners for Trade and Plantations requested maps from the English colonies since there were few reliable maps of colonial territory that could support British claims to territory and specific borders.[63]

Attuned to the political and commercial opportunities of the situation, Moll issued a map with a very long title explicitly challenging Delisle's portrayal of colonial territories and the borders that distinguished them: *A New Map of the North Parts of AMERICA claimed by FRANCE under the names of Louisiana, Mississippi, Canada and New France with the Adjoyning Territories of England and Spain* (see Figure 3.4). As in his earlier map, Moll was at pains to show in this map that the conflict over borders was a conflict over claims to territory *and* to the Indigenous military and economic resources needed to defend and exploit them, in particular the trade in beaver and deerskins and enslaved human beings.[64] Moll enumerated two areas of conflict in the title cartouche. The first area concerned colonial boundary-drawing and Delisle's contraction of the western border of British claims in North America, especially Carolina. Moll's map was not an exact reproduction of Delisle's 1718 map. Instead, he placed Delisle's borders on a new base map: "the South West part of Louisiana is done after a French Map Published at Paris in 1718 and we give you here the Divisions or Bounds according to that Map, which Bounds begin 30 Miles S West from Charles Town in Carolina and run on to ye Indian Fort at Susquehannock 30 Miles West of Philadelphia &c." and, in case his map readers doubted him, noted just to the right of the title cartouche, "N.B. The French Map mention'd in the Title is done by Monsr. Delisle and Publish'd by him at Paris in June 1718, which I am ready to shew to any Gentlemen that desires it." The second area of conflict concerned Delisle's placing the Iroquois and Cherokee within the French colonial sphere. Moll insisted to his map readers that the powerful Iroquois and Cherokee people were staunch and reliable military allies of the British, "the Bulwark and Security of all their Plantations in North America." His sentiment echoed the understanding of the role of Native Americans in maintaining European colonies that English colonists like William Byrd II expressed: "The Indians are very numerous on the branches of this river [the Mississippi], and if the French find means to gain them, it will render the English plantations very unsafe."[65] As graphic reinforcement of the importance of Indigenous allies in shifting imperial borders, Moll placed a view of Fort Susquehannock, which the Haudenosaunee (with English support) sacked in the late seventeenth century, in the upper left-hand corner of his map. While the French thought that the Indigenous fort marked the western colonial border of Pennsylvania and the eastern border of French-claimed territory, Moll's view emphasized its link with Indigenous peoples rather than its boundary-making quality, perhaps because by 1719 the fort was a ruin and represented the power to shift borders in Britain's favor.[66]

Even though Moll placed information from Delisle's map on his own map, including the names of Indigenous peoples without territorial boundaries, he did not copy Delisle's notations of the destruction of other Indigenous peoples by the Haudenosaunee, and he could not bring himself to trace Delisle's boundary line very far because it showed French claims to western Maryland, Pennsylvania, New York, and Maine (New Scotland), so he angled it to more closely match British territorial claims. He expanded the boundaries of Maryland and Pennsylvania westward and included the Haudenosaunee within them. He did show Delisle's diminution of Carolina to stoke British outrage, but he could not resist putting the proposed-but-never-founded British colony of Azilia within French-claimed territory and placing many Cherokee villages within it.

Moll's map memorialized imperial conflicts that occurred in Native homelands and involved transgressing imagined borders during the War of the Spanish Succession. For instance, Nairne mounted a successful raiding party deep into Florida that he undertook with his Yamasee allies in 1706 to capture Spanish-allied Native Americans and bring them back to Charleston to sell into slavery. Moll showed Nairne's band transgressing the border with Florida and provided an "Explanation of an Expedition in Florida" in a note just east of the Yucatan Peninsula, which included a numeric key identifying different places on Nairne's route into and out of the Everglades and a comment that his raiding party captured thirty-five Indians and "took and killed 33 Men." A note at the top of Palaxy Bay, which Moll copied from Delisle, that the English destroyed Santa Maria de Palaxy, a Spanish-Apalachee mission in 1705, represented a clear act of aggression across colonial borders that was part of a sustained British campaign with Yamasee and Creeks allies to remove Spanish impediments to English expansion southward.[67] For his British audience who likely would never see Delisle's map, Moll replicated the forts on that map and added more to heighten the imperial threat posed by the French. Two insets on the right margin of the map showed sites of great strategic importance during and after the war. Both involved maritime borders and "islands of occupation." The first displayed the fortified harbor of Annapolis, formerly Port Royal, taken from the French in 1710, which helped secure British territorial claims in Nova Scotia. The second depicted Mobile Bay and the mouth of the Mississippi River, which offered the French access to native trade routes in the interior and blocked British and Spanish advancement in the region.

Below the compass rose Moll engraved an Advertisement that referenced Cabot's "discovery" of Florida in 1498. The reference served to justify extending the border of Carolina southward to include northern Florida. The dotted line above it was another possible border based on King Charles II's 1663 authorization for the English to settle and farm Carolina, whose western border he extended to the Pacific Ocean, thereby encompassing large chunks of Louisiana, Texas, and New Mexico. In Moll's view, it was easy to see the extent of French and Spanish "encroachments" on good British claims to Native

American homelands. His map served to reterritorialize French and British claims even as it deterritorialized Native Americans.

Conclusion

Borders on Moll's and Delisle's maps were deceptively precise boundary lines meant to establish clear-cut territorial jurisdiction and claims to legal and political sovereignty. However, their maps also showed borders to be unstable because they treated the territorial claims of Native peoples and other European powers as contestable and provisional and as a source of future conflict. In contrast to assertions of political sovereignty represented by borders on these maps, much of North America consisted of interracial, interethnic, interpolitical contact zones where borders were contested and where no single group ruled supreme.

Sometimes map historians claim too much for maps when they contend, like John Pickles (following Denis Wood and Jean Baudrillard) and Matthew Edney, that maps produce the territory: "Maps and mapping precede the territory they 'represent.' . . . Territories are produced by the overlaying of inscriptions we call mappings" and regions "are human concepts that are created by being mapped."[68] There is a way, of course, that this philosophically idealist claim is true: Maps have a discursive function and create a graphic image of a territory, an abstraction that is not the territory itself but a visualization of a physical space that allows an idea of the territory to be constructed as part of a larger political or social discourse. As mentioned earlier, political borders on maps encode the idea of territory for map viewers but cartographic codes do not create territory on the ground. That is because maps, as discursive representations and claims to territory, were, are, and can be contested and voided. We have seen claims and counterclaims to territory in the maps by Delisle and Moll, who often did not deploy the same toponyms for the same mapped features. What territory was "produced" simply by the acts of being mapped and claimed? Claiming and mapping land for France did not by itself make that land "French territory," especially when Native peoples and the British and Spanish contested the claims and toponyms. While maps mediate our understandings of territory and borders, it is a fallacy to conflate them with the territory mapped, much less with producing territory or dominating it. The last fallacy is one Edney identifies.[69] After all, the map is not the territory and paper claims on maps did not produce territory.

"Louisiana," as Delisle outlined it in his 1718 map, was mostly imaginary and while imagining is a first step to acquiring, it is insufficient for possessing or "producing" territory. If borders on maps become authoritative, it is less on the "authority" of a map than a consequence of political and often military action in the physical/material world. Maps are "paper claims" and "paper encroachments" that can stir up passions and indignation, but they mean little without political or military or economic action to enforce those claims and encroachments in the physical world to which many maps, like those that are the focus of this chapter, refer. In the end, it is human labor-power, human activity "on the ground," that produces territory, not merely human imaginings, even if these imaginings and the map war it produced in the early eighteenth century helped frame different public understandings in France and Britain and called these publics' attention to specific points of contention regarding French and British colonial claims in North America during an era of intense imperial conflict and outright war.[70] While Brian Harley endorsed the idea that the map precedes the territory, he also mentioned an important caveat: Such maps were an "anticipatory geography." Anticipation may not lead to realization, so his further claim that "maps were the first step in the appropriation of territory" is only true if the appropriation of territory actually happened. It is not even clear that maps were a necessary "first step" for colonial territorial claims since neither Colbert nor La Salle had any idea of the extent of the Mississippi River and certainly no map of its geographic reach before laying claim to all the land it touched. La Salle named it Louisiana before the name ever appeared on a map, so the idea that maps precede the territory must be relativized to specific cases and not declared a general maxim regarding the power of maps to create territory.[71]

In his 1720 map, Moll confronted the visual and textual rhetoric of Delisle's 1718 map that asserted the French claim to vast territory based on the discovery of a watershed with a visual and textual rhetoric of his own that asserted the British view that only European settlement justified territorial claims as exemplified by showing major colonial cities like Charleston, Jamestown, Philadelphia, New York, Boston, Quebec, and Montreal. Moll's 1720 map repeated much of the information available on his Beaver Map. He reproduced the borders on Delisle's map in order to mock them and rile up his compatriots. Where Delisle emphasized the importance of the Mississippi River and its tributaries on his map and deemphasized municipal settlements in the French colonial sphere, Moll deemphasized riverine features and emphasized municipal settlements in the British colonial territories. They both engraved the names of Native American nations across the landscape, but they gave little indication of Native political sovereignty despite the clear indication that they occupied the land near their name. Their maps also made clear that European-style political borders in North America in the early eighteenth century were convenient fictions, asserted and rearranged according to colonial agendas. Despite being fictions, borders asserted claims to territorial jurisdiction and legal and political sovereignty that were often followed up with action, with some polities, like European colonies, "deserving" the cartographic symbol for political sovereignty, a border, while others, like Indigenous polities, did not.

Notes

1. Michael Biggs, "Putting the State on the Map: Cartography, Territory, and European State Formation," *Comparative Studies in History and Society* 41, no. 2 (1999): 374–405; Nancy Shoemaker, *A Strange Likeness: Becoming Red and White in Eighteenth-Century North America* (Oxford University Press, 2006); Juliana Barr, "Geographies of Power: Mapping Indian Borders in the 'Borderlands' of the Early Southwest," *William and Mary Quarterly* 68, no. 1 (January 2011): 5–46.

2. Dennis Reinhartz, *The Cartographer and the Literati: Herman Moll and His Intellectual Circle* (Edwin Mellen Press, 1997), 1; Seymour I. Schwartz and Ralph Ehrenberg, *The Mapping of America* (Wellfleet Press, 2001), 133; Nelson-Martin Dawson, *L'Atelier Delisle: L'Amérique du Nord sur la table à dessin* (Éditions Septentrion, 2000), 43–47; Mary Sponberg Pedley, *The Commerce of Cartography: Making and Marketing Maps in Eighteenth-Century France and England* (University of Chicago Press, 2005), 33; Christine Marie Petto, *When France Was King of Cartography: The Patronage and Production of Maps in Early Modern France* (Lexington Books, 2007), 41, 59; Christine Marie Petto, *Mapping and Charting in Early Modern England and France: Power, Patronage, and Production* (Lexington Books, 2015), 45, 54, 71n3.

3. Rich Heyman, "Locating the Mississippi: Landscape, Nature, and National Territoriality at the Mississippi Headwaters," *American Quarterly* 62, no. 2 (June 2010): 303–33 (quote 331).

4. Michiel Baud and Willem van Schendel, "Toward a Comparative History of Borderlands," *Journal of World History* 8, no. 2 (Fall 1997): 211–42, esp. 215. See also Charles Maier, *Once Within Borders: Territories of Power, Wealth, and Belonging Since 1500* (Harvard University Press, 2016), 76; Jordan Branch, *The Cartographic State: Maps, Territory, and the Origins of Sovereignty* (Cambridge University Press, 2014), chapters 4–5.

5. Baud and van Schendel, "Comparative History," 211, 216.

6. See Samuel Truett's chapter in this volume, "Amphibious Tales: Villagers and Strangers in a Border-Crossing World," 292; Colin G. Calloway, *One Vast Winter Count: The Native American West Before Lewis and Clark* (University of Nebraska Press, 2007), who argues on page 2 that Native Americans may have witnessed "zones of interaction" but these "formed, overlapped, and re-formed around" Indigenous homelands where they had the upper hand in the seventeenth and eighteenth centuries; Claudio Saunt, "'Our Indians': European Empires and the History of

the Native American South," in *The Atlantic in Global History, 1500–2000*, ed. Jorge Cañizares-Esquerra and Erik R. Seeman (Pearson Prentice Hall, 2007), 61–75; Barr, "Geographies of Power," 8–9, 14, 19, 21, 29, 38, 43; Robbie Ethridge, "Creating the Shatter Zone: The Indian Slave Traders and the Collapse of the Southeastern Chiefdoms," in *Light on the Path: The Anthropology and History of the Southeastern Indians*, ed. Thomas J. Pluckhahn and Robbie Ethridge (University of Alabama Press, 2006), 207–18; Robbie Ethridge, "Introduction," in *Mapping the Mississippian Shatter Zone: The Colonial Indian Slave Trade and Regional Instability in the American South*, ed. Robbie Ethridge and Sheri M. Shuck-Hall (University of Nebraska Press, 2009), 1–62.

7. For border-making by Native peoples and the erasure of those borders on European maps see Barr, "Geographies of Power," 6–7, 8–11, 16–17, 21–22, 26–27, 29–30, 37–38, 43–46. On the power of Native peoples, see Chad Anderson, "Rediscovering Native North America: Settlements, Maps, and Empires in the Eastern Woodlands," *Early American Studies* 14, no. 3 (2016): 478–505. On the reliance of Europeans on Indigenous geographic knowledge and mapping skills as well as Indigenous expression of territorial sovereignty see Jeffers Lennox, *Homelands and Empires: Indigenous Spaces, Imperial Fictions, and Competition for Territory in Northeastern North America, 1690–1763* (University of Toronto Press, 2017), chapters 1 and 2.

8. See Rachel St. John, "Imperial Spaces in Pekka Hämäläinen's 'The Comanche Empire,'" *History and Theory* 52, no. 1 (February 2013): 75–80, where she discusses the porous nature of imperial boundaries and the nodal, rather than broad territorial, nature of imperial control: "The presumption of territorial authority that [was] represented in official documents and maps often far exceeded the reality of imperial power on the ground" (78). Most control was nodal and over specific places, enclaves, and corridors, not over broad swaths of contiguous land.

9. David J. Weber, "New Spain and Its Borderlands," in *Major Problems in the History of North American Borderlands*, ed. Pekka Hämäläinen and Benjamin H. Johnson (Wadsworth, 2011), 179; Barr, "Geographies of Power," 44; Pekka Hämäläinen, "The Shapes of Power: Indians, Europeans, and North American Worlds from the Seventeenth to the Nineteenth Century," in *Contested Spaces of Early America*, ed. Juliana Barr and Edward Countryman (University of Pennsylvania Press, 2014); Gilles Havard, *The Great Peace of Montreal of 1701: French-Native Diplomacy in the Seventeenth Century*, trans. Phyllis Aronoff and Howard Scott (McGill-Queens's University Press, 2001). Reversing a Eurocentric field of vision, Calloway contends that Native Americans were not "people in between" European empires but that Europeans were "people on the edge" of Native territories. Calloway, *Winter Count*, 313. With their tidy border markings, Moll's and Delisle's maps openly maintained that European empires claimed all of the North American continent that was known to Europeans but, upon closer viewing, those maps also demonstrated the marginal place of those "empires" whose "core" areas lay on the edges of the continent.

10. The logical fallacy involves confusing a conceptual model of territory with the territory itself, a confusion of epistemology with ontology. On maps and the illusion of territorial control, see Matthew Edney, *Cartography: The Ideal and Its History* (University of Chicago Press, 2019), 76–77.

11. J. Brian Harley, "Rereading the Maps of the Columbian Encounter," *Annals of the Association of American Geographers* 82, no. 3 (September 1992): 522–36 (quote 531).

12. In Delisle's case it would be reports from the journeys of René-Robert Cavelier, sieur de la Salle (1643–1687), Pierre Le Moyne, sieur d'Iberville (1661–1706), and Pierre-Charles Le Sueur (ca. 1657–1704); and in Moll's case it would be reports from the journeys of William Dampier (ca. 1651–1715), William Funnell (fl. 1700–1707), Woodes Rogers (ca. 1679–1732),

and Thomas Nairne (fl. 1695–1715). See Petto, *King*, 102–3; Petto, *Mapping*, 127–29, Reinhartz, *Cartographer*, 77–83, 118–22, 130.

13. While the term does not appear in the body of her article, I borrow the concept "map war" from the title of Mary Pedley, "Map Wars: The Role of Maps in the Nova Scotia/Acadia Boundary Disputes of 1750," *Imago Mundi*, 50 (1998): 96–104, to characterize the acrimony of the cartographic conflict between Moll and Delisle and to link it to the territorial and military disputes of the period. For a brief discussion of mutual "paper encroachments" and the tension between Moll and Delisle with a focus on the period after 1718, see Petto, *King*, 105–6; Petto, *Mapping*, 125–33.

14. Guillaume Delisle, *Carte du Mexique et de la Floride, des Terres Angloises et des Isles Antilles, du Cours et des Environs de la Riviere de Mississipi. Dressée sur un grand nombre de memoires principalemt. sur ceux de Mre. D'Iberville et le Sueur. Par Guillaume Del'Isle, Geographe de l'Academie Royale des Sciences. A Paris, chez l'Auteur sur le Quai de l'Horloge, (avec) Privilege du Roy po. 20. ans, 1703.* Call # G1015 .L5 [1731] FF, https://purl.stanford.edu/fr132rs4988 (accessed July 4, 2021).

15. While there is no thorough, English-language book-length study of the war in its entirety, or of its North American theater, there have been some specialized studies including John Hely Owen, *War at Sea Under Queen Anne, 1702–1708* (Cambridge University Press, 2010); Renger E. de Bruin et al., eds., *Performances of Peace: Utrecht 1713* (Brill, 2015); Wooter Troost, "Leopold I, Louis XIV, William III and the Origins of the War of the Spanish Succession," *History* 103, no. 4 (357) (October 2018): 545–70; as well as those that place the war in a larger context like Bruce P. Lenman, "Colonial Wars and Imperial Instability, 1688–1793," in *The Oxford History of the British Empire, Volume II: The Eighteenth Century*, ed. P. J. Marshall (Oxford University Press, 1998), 151–68.

16. Phillippe R. Girard, "Europe and the World," in *A Companion to Eighteenth-Century Europe*, ed. Peter H. Wilson (Blackwell, 2008), 402–17 (quote 410).

17. Sara Stidstone Gronim, "Geography and Persuasion: Maps in British Colonial New York," *William and Mary Quarterly* 58, no. 2 (2001): 373–402, esp. 382. See also V. Y. Mudimbe, *The Idea of Africa* (Indiana University Press, 1994), 33–36; Patricia Seed, *Ceremonies of Possession in Europe's Conquest of the New World, 1492–1640* (Cambridge University Press, 1995), 51–68. For a discussion and critique of the European discourse and application of *terra nullius*, in which non-European peoples had no legal title to the land they occupied, see Alex Zukas, "Terra Incognita/Terra Nullius: Modern Imperialism, Maps, and Deception," in *Lived Topographies and Their Mediational Forces*, ed. Gary Backhaus and John Murungi (Lexington Books, 2005), 49–76, where I conclude (67), "Until its eventual collapse . . . the norm of *terra nullius* justified the immoral and illegitimate exercises of imperial power which the maps of the period helped validate, mediate, and memorialize."

18. A translation of the official minutes of the ceremony is in E. B. O'Callaghan, ed., *Documents Relative to the Colonial History of the State of New York* (Weed, Parsons and Co, 1855), 9:803–4.

19. Schwartz and Ehrenberg, *Mapping*, 133; Samuel Truett, "Settler Colonialism and the Borderlands of Early America," *William and Mary Quarterly* 76, no. 3 (July 2019): 435–42.

20. Tamar Herzog, *Frontiers of Possession: Spain and Portugal in Europe and the Americas* (Harvard University Press, 2015), 1.

21. Truett, "Settler Colonialism," 437–38.

22. On French dependence on Native peoples, see W. J. Eccles, *The Canadian Frontier, 1534–1760*, rev. ed. (University of New Mexico Press, 1983), 1–156; Richard White, *The Middle Ground: Indians, Empires, and Republics in the Great Lakes Region, 1650–1815* (Cambridge University Press, 1991), 1–140. On the issue that "partial control—irregular borders, layered loyalties, fragmented spaces—may well be a prevalent imperial condition" in the early modern world, see Pekka Hämäläinen, "What's in a Concept? The Kinetic Empire of the Comanches," *History and Theory* 52, no. 1 (February 2013): 81–90 (quote 86).

23. Patricia Galloway, "Debriefing Explorers: Amerindian Information in the Delisles' Mapping of the Southeast," in *Cartographic Encounters: Perspectives on Native American Mapmaking and Map Use*, ed. G. Malcolm Lewis (University of Chicago Press, 1998), 223–40. Delisle was a tutor to the dauphin and became a member of the state-sponsored Académie Royale des Sciences in 1702. Pedley, *Commerce*, 38–39; Petto, *King*, 41, 153, 158, 181.

24. David J. Weber, *The Spanish Frontier in North America* (Yale University Press, 1992), 148. On the productivity of the silver mines of northern Mexico and their importance to the global economy in the seventeenth and eighteenth centuries, see John Tutino, *Making a New World: Founding Capitalism in the Bajío and Spanish North America* (Duke University Press, 2011), 1–208, and John Tutino, *The Mexican Heartland: How Communities Shaped Capitalism, a Nation, and World History, 1500–2000* (Princeton University Press, 2018), 1–101. Tutino's account helps explain the desire of the French and British to possess these mines.

25. Weber, "New Spain," 180–81; Pekka Hämäläinen, *The Comanche Empire* (Yale University Press, 2009).

26. Patricia Seed, "Taking Possession and Reading Texts: Establishing the Authority of Overseas Empires," *William and Mary Quarterly* 49, no. 2 (1992): 194. See also Patricia Seed, *American Pentimento: The Invention of Indians and the Pursuit of Riches* (University of Minnesota Press, 2001), 32–34, and Zukas, "Terra Incognita," 54.

27. Schwartz and Ehrenberg, *Mapping*, 133; Samuel Truett, "Settler Colonialism," 435–42.

28. Cecil Headlam, ed., *Calendar of State Papers, Colonial Series, America and West Indies, Volume 28, August 1714–December 1715* (H. M. Stationery Office, 1928), 116, 11 March 1715, #271.

29. David H. Landis, "The Location of Susquehannock Fort," *Journal of the Lancaster County Historical Society* 14, no. 3 (1910): 2–3, 19–20.

30. *A System of Geography, or a New & Accurate Description of the Earth In all its Empires, Kingdoms and States. Part the Second, Containing the Description of Asia, Africa, and America. Written in Latin by Joan. Luyts Professor in Acad. Ultraj [Utrecht]. English'd with large additional Accounts of the East-Indies, and the English Plantations in America. Illustrated with Maps, Fairly Engraven on Copper, according to the latest Discoveries and Corrections, by Herman Moll.* (Printed for Tim. Childe, at the White-Hart in St. Paul's Churchyard, 1701), 159.

31. *Thesaurus Geographicus, or The Compleat Geographer. Part the Second. Being the Chorography, Topography, and History of Asia, Africa, and America. Faithfully extracted from the Best Modern Travellers, and most esteem'd Historians: and Illustrated with Maps, Fairly Engraven on Copper, according to the modern Discoveries and Corrections by Herman Moll*, Third Edition very much Enlarg'd, v. 2 (A and J. Churchill, and T. Childe, 1709), 235.

32. Eric Hinderaker, "The 'Four Indian Kings' and the Imaginative Construction of the First British Empire," *William and Mary Quarterly*, 53, no. 3 (July 1996): 487–526 (quotes, 488).

33. Hinderaker, "Four Indian Kings," 490–491, 496.

34. On the map of *The English Empire in America* that he engraved for *System*, 161, Moll wrote "The 5 Nations of the Iroquois" along the southern edge of the eastern Great Lakes so he, and likely his readers, knew how many "cantons" there were, but the "mistake" does raise the question of the legitimacy and public perception of the visit.

35. Defoe's diatribe appeared on March 15, 1712, at the beginning of *A Review of the State of the British Nation* (volume 8, number 153), a periodical he published. The quote appears in Hinderaker, "Four Indian Kings," 524. On Moll's relationship with Defoe, see Reinhartz, *The Cartographer and the Literati*, vii–ix, xii–xiii, 3, 6–8, 25–26, 81–87, 113–126, 140–146, 151.

36. Daniel K. Richter, "Native Peoples of North America and the Eighteenth-Century British Empire," in *The Oxford History of the British Empire, Volume II: The Eighteenth Century*, ed. P. J. Marshall (Oxford University Press. 1998), 347.

37. Matthew H. Edney, "John Mitchell's Map of North America (1755): A Study in the Use and Publication of Official Maps in Eighteenth-Century Britain," *Imago Mundi* 60, pt. 1 (2008), 63–85, esp. 70–78 (quote, 63).

38. On their maps of Asia and Africa Moll and Delisle engraved borders for empires and kingdoms like China, the Mughal Empire, Congo, and Angola, which Europeans recognized as sovereign states; however, for people from whom Europeans withheld such recognition, he engraved their names across a borderless landscape. It might be objected that Moll and Delisle did not draw boundaries around Haudenosaunee or other Indigenous lands because they did not know their actual extent, but lack of such knowledge did not prevent them from proposing borders for China, Congo, and Angola.

39. Pedley, *Commerce*, 33, 39; Petto, *King*, 41, 59, 158; Petto, *Mapping*, 132–133.

40. William K. Boyd, ed., *William Byrd's Histories of the Dividing Line Betwixt Virginia and North Carolina* (Dover, 1967), 292, 294. See also Daniel K. Richter, *The Ordeal of the Longhouse: The Peoples of the Iroquois League in the Era of European Colonization* (University of North Carolina Press, 1992), 50–66, 74, 98–99, 144–149, 161–164, 189, 256; Matthew D. Zarzeczny, "Beaver Wars," in *Shaping North America*, edited by James E. Seelye Jr., Shawn Selby, and Christine Eisel, vol. 1 (ABC-CLIO, 2018), 97–100.

41. Schwartz and Ehrenberg, *Mapping*, 135; William P. Cumming, *British Maps of Colonial America* (University of Chicago Press, 1974), 6. Moll based this inset on a map by Edward Crisp that appeared in his 1711 *A Compleat Description of the Province of Carolina in 3 Parts.*

42. Ethridge, "Introduction," 34, 37 and Eric E. Bowne, "'Caryinge awaye their Corne and Children': The Effects of the Westo Slave Raids on the Indians of the Lower South," in Ethridge, *Shatter Zone*, 107–8. On Indigenous population decline, see Peter H. Wood, "The Changing Population of the Colonial South: An Overview by Race and Region, 1685–1790," in *Powhatan's Mantle: Indians in the Colonial Southeast*, revised and expanded edition, ed. Gregory Waselkov, Peter H. Wood, and Tom Hatley (University of Nebraska Press, 2006), 72–74.

43. Saunt, "'Our Indians,'" 61–62.

44. See Ethridge, "Introduction," especially 2, 13–19. For population estimates, see Wood, "Changing Population," 60–61, 72, 84–85, 95–97, 105–106.

45. Alexander Moore, "Thomas Nairne's 1708 Western Expedition: An Episode in the Anglo-French Competition for Empire," *Proceedings of the Meeting of the French Colonial History Society* 10 (1985): 47–58.

46. Richter, "Native Peoples," 349. On the French use of their Native allies to attack the Native allies of the British and the resulting regional destabilization and devastation of Indigenous societies, see Saunt, "'Our Indians.'"

47. Richter, "Native Peoples," 359.

48. Headlam, *Calendar of State Papers, Volume 28*, 116, 11 March 1715, #271.

49. On diplomatic theater involved in the negotiations and final treaty and Cherokee expectations, see Gregory D. Smithers, "'Our Hands and Hearts Are Joined Together': Friendship, Colonialism, and the Cherokee People in Early America," *Journal of Social History* 50, no. 4 (Summer 2017): 609–29. That members of the Board of Trade believed, perhaps erroneously, that the Cherokee had acknowledged subjection to the British king, see Verner W. Crane, *The Southern Frontier, 1670–1732* (Duke University Press, 1928), 299–302. Kristofer Ray, "Cherokees and Franco-British Confrontation in the Tennessee Corridor, 1730–1760," *Native South* 7 (2014), 33–67, considers Crane's work "an older but still useful interpretation" (59). Quotes are on page 36.

50. Kristofer Ray, "Interpreting *Native* Trans-Appalachia, 1670–1770: Or, How I Stopped Worrying and Learned to Read Fanni Mingo's Map," *Revue de la Société d'études anglo-américaines des XVIIe et XVIIIe siècles* [Online], 78 (2021), paragraphs 3 and 4, accessed December 1, 2024, http://journals.openedition.org/1718/8090. Ray remarks in endnote 3 that the Cherokee "maintained similar religious and cultural beliefs across their five settlement regions. Spiritual, material, and cultural connections, however, are not the same as political and diplomatic, even if contemporary Europeans showed a maddening willingness to conflate them."

51. Richard Gross and Craig P. Howard, "Colbert, La Salle, and the Search for Empire," *Journal of the Illinois State Historical Society* 113, 2 (Summer 2020): 68–101 (quotes, 73, 83, 88). See also, Heyman, "Locating the Mississippi."

52. Cannibalism is not one of the attributes Juliana Barr ascribes to the nomadic peoples of the Texas Gulf Coast whom she discusses in "Geographies of Power," 14–16, 19.

53. Schwartz and Ehrenberg, *Mapping*, 139.

54. J. R. Edmondson, *Alamo Story: From Early History to Current Conflicts* (Taylor Trade Publishing, 2000), 10. Delisle incorrectly attributed the founding to Bienville in 1717.

55. Petto, *Mapping*, 128–130.

56. Petto, *Mapping*, 130.

57. Petto, *King*, 103. William Cumming, *The Southeast in Early Maps*, 3rd ed., rev. and enlarged by Louis de Vorsey Jr. (University of North Carolina Press, 1998), 212.

58. Antoin E. Murphy, *John Law: Economic Theorist and Policy-Maker* (Oxford University Press, 1997); Malcolm Balen, *The Secret History of the South Sea Bubble* (Fourth Estate, 2003), 45–67; Richard Dale, *The First Crash: Lessons from the South Sea Bubble* (Princeton University Press, 2004), 56–72; Kenneth J. Banks, *Chasing Empire Across the Sea—Communications and the State in the French Atlantic, 1713–1763* (McGill-Queen's University Press, 2008).

59. Balen, *Secret History*, 50–52, 58–61, 64–65.

60. Kim M. Gruenwald's chapter in this volume, "When a River Is a Border: Rivalries and Commercial Networks in the Riverine West," 144.

61. Headlam, ed., *Calendar of State Papers, Volume 30, August 1717–December 1718* (H. M. Stationery Office, 1930), 119–20, 7 December 1717, #238.

62. Headlam, ed., *Calendar of State Papers, Volume 32, March 1720–December 1721* (H. M. Stationery Office, 1933), 202–7, 26 November 1720, #303.

63. Headlam, ed., *Calendar of State Papers, Volume 31, January 1719–February 1720* (London: H. M. Stationery Office, 1933), 187–88, 26 August 1719, #354.

64. See for example, Alan Gallay, *The Indian Slave Trade: The Rise of the English Empire in the American South, 1670–1717* (Yale University Press, 2002); Kathryn E. Holland Braund,

Deerskins and Duffels: The Creek Indian Trade with Anglo-America, 1685–1815, 2nd ed. (University of Nebraska Press, 2008), 1–70.

65. As quoted in Margaret Beck Pritchard and Henry G. Taliaferro, *Degrees of Latitude: Mapping Colonial America* (The Colonial Williamsburg Foundation and Harry N. Abrams, 2002), 123.

66. Moll copied the view from Arnoldus Montanus's 1671 *Description of New Netherland*. Landis, "Susquehannock Fort," 2–3, 19–21.

67. John E. Worth, "Razing Florida: The Indian Slave Trade and the Devastation of Spanish Florida, 1659–1715," in Ethridge, "Shatter Zone," 299–306; John E. Worth, "English Conquest of Georgia," *The New World*, September 18, 2011, https://thenewworld.us/english-conquest-of-georgia/ (accessed March 21, 2021).

68. John Pickles, *A History of Spaces: Cartographic Reason, Mapping and the Geo-Coded World* (Routledge, 2004), 5; Edney, *Cartography*, 55, and J. Brian Harley, "Rereading the Maps of the Columbian Encounter," *Annals of the Association of American Geographers* 82, no. 3 (September 1992): 522–36, esp. 531. The postmodern/poststructural theorist of simulacra and the hyperreal, Jean Baudrillard, writes, "The territory no longer precedes the map, nor survives it. Henceforth, it is the map that precedes the territory—*precession of simulacra*—it is the map that engenders the territory." Jean Baudrillard, *Simulacra and Simulation*, trans. Sheila Glaser (University of Michigan Press, 1995), 1. (Original French version published in 1981.)

69. See David Turnbull, *Masons, Tricksters and Cartographers: Comparative Studies in the Sociology of Scientific and Indigenous Knowledge* (Routledge, 2000), 98, on this point. Edney makes a related point that confuses ontology with epistemology when he asserts that it is unacceptable (for mapmakers, map users, or anyone) to assume that a region exists before it is mapped. Edney, *Cartography*, 50–51, 55, 57, 76–77.

70. See, for example, the discussion in Hämäläinen, "What's in a Concept," of the idea of a "kinetic" empire that the Comanches created by movement and activity (raiding, trade fairs, political assemblies, control of shifting economic nodes) over a large geographic area that had as its aim territorial integration rather than uncompromised territorial control. See also St. John, "Imperial Spaces," 78–80. On the dynamic, dialectical, and mutually constituting relationship between the material (tangible) and immaterial (intangible) in the creation of borders and territories see Olga Demetriou and Rozita Dimova, eds., *The Political Materialities of Borders: New Theoretical Directions* (Manchester University Press, 2019) and Stuart Elden, "Terrain, Politics, History," *Dialogues in Human Geography* (August 2020): 1–20, https://doi.org/10.1177/2043820620951353, where he argues that territory is continually "made and remade by human agency and physical processes" (7). On the "performative" rather than the purely "representational" understanding of territory, Turnbull contends that written descriptions of territory lasted well into the eighteenth century. See Turnbull, *Masons*, 12–13, 126, 133–64.

71. Harley, "Rereading," 531–32. In fact, mapmakers like French royal geographer Nicolas Sanson (1600–1667) in the 1650s and 1660s gave the toponym "Florida" to the area between the Ohio River and the Gulf of Mexico, clearly a case where an "anticipatory geography" of Spanish claims fell flat.

CHAPTER 4

Data Maps of Downeast Maine

Missionary Records from the Early Republic Borderlands

JOHN MORTON

In 1804, the Reverend Alfred Johnson took a missionary trip to the far eastern corner of the District of Maine.[1] Dozens of other Congregational ministers had taken similar trips in the previous decade and a half. The population of the district was expanding rapidly, and missionaries were doing their best to reach the far-flung settlements. They traveled from town to town, bringing sermons, books, and baptisms to communities that had yet to build permanent churches or hire settled preachers. When these ministers were feeling optimistic, they attempted to make inroads with Maine's Catholic Indians—the Penobscot and Passamaquoddy—who consistently rejected their advances. So most missionary energy was spent on the white settlers who had recently moved north from southern New England.

Almost all of these itinerant ministers were employed by a missionary society. This meant they needed to keep a journal of their travels and the services rendered in order to be paid. Alfred Johnson was working for the Society for Propagating the Gospel Among the Indians and Others in North America (SPGNA).[2] This Boston-based Congregational group was the first American missionary society to begin work after the Revolution, and it was the best funded, thanks to an annual grant from the Massachusetts Legislature. For years, the SPGNA had been sending ministers up to Maine with relatively simple instructions. Early on, they were simply asked to visit "eastern settlements." Later, instructions got a bit more specific, referencing, for example, "the vacant towns and Plantations in the Counties of Hancock and

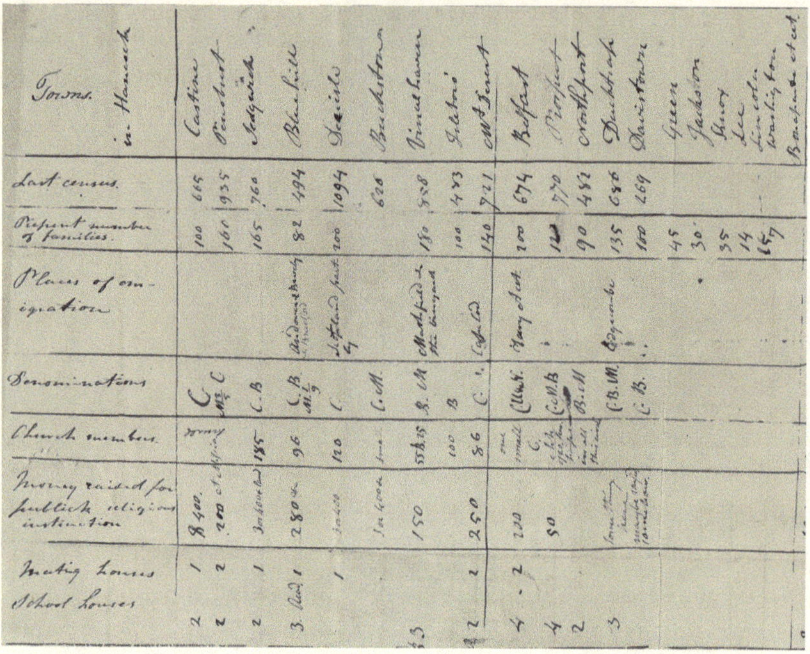

Figure 4.1. One section of the Reverend Alfred Johnson's chart of towns in Hancock and Washington Counties, drawn up during his missionary trip to the District of Maine in 1804. The full chart contains data on more than fifty towns. Detail of Rev. A Johnson's Ecclesiastical Chart for Hancock & Washington, 1804, Society for Propagating the Gospel Among the Indians and Others in North America Records, MSS 48, box 3, folder 1. Phillips Library, Peabody Essex Museum.

Washington."[3] Itinerants were also asked to expand their missions by distributing books and pamphlets along their way. Having to carry increasing quantities of printed material around the frontier was a logistical challenge, and more demanding for the ministers than simply providing sermons and baptisms. But the fundamental mission—to bring religious services to as many needy families as possible—remained the same.

The year 1804 brought a significant reorientation. The SPGNA asked the Massachusetts legislature for more money in order to begin building a network of schools. The state agreed. And so the society asked Johnson and its other itinerants to go beyond their traditional role and to "map" the missionary field. It had previously been enough to preach at every available opportunity and distribute books. Specific data about the people one encountered was less important than covering a lot of ground. Now a more precise accounting was necessary. To create schools for the state the society needed to know who belonged to the state and what resources they already had. Ministers were instructed to "inform yourself as far as you can of the general state of the people . . . [and] the particular state of the several towns and plantations through which you pass" and assemble the resulting data into a chart.[4]

Reverend Johnson's chart—filed away in the SPGNA papers at the Phillips Library in Rowley, Massachusetts—is not a traditional map. It does not provide the topography of the region, nor does it trace the borderline and the New Brunswick towns on its far side. And yet it provides an incredibly detailed picture of the eastern part of the District of Maine. Johnson laid out more than fifty towns and townships—providing the approximate population for each, the number of families in residence, where they had emigrated from, the local religious denominations, how many people were church members, how much money was raised for "public religious instruction," the number of school and meeting houses, and the name of the local minister, if one existed. He was particularly meticulous about settlements in and around Passamaquoddy Bay, directly alongside the border. Johnson's chart listed not only large communities—like Eastport on Moose Island, with 562 people—but also the tiniest of unnamed townships, like township one (twenty families and a little over 100 people) and township eleven (twenty-seven families). This chart, together with others produced by Johnson's contemporaries, provided the SPGNA with a precise and comprehensive data map of the Maine missionary field.[5]

One striking thing about these missionary documents is their clarity regarding the boundary dividing Maine from New Brunswick. It appears that in 1804, to the people living in and visiting borderland communities, the boundary was a fixed reality. And yet on a diplomatic level this was not the case at all. Throughout the first two decades of the nineteenth century the precise location of the line was hotly contested. In a 1798 address to Congress, President Adams specifically highlighted Passamaquoddy Bay as a place where the border had yet to be negotiated. During the Jefferson administration—in the same years that SPGNA missionaries were creating their data maps—American and British diplomats argued over rights to the various Passamaquoddy Bay islands. The Americans emphasized the Royal Commission of 1763, which had attempted to clarify boundaries after the French and Indian War, to buttress their claims. The British pointed to the wording of early seventeenth-century land grants, which seemed to assign all the disputed territory to Nova Scotia.[6] The British then tried to render the whole

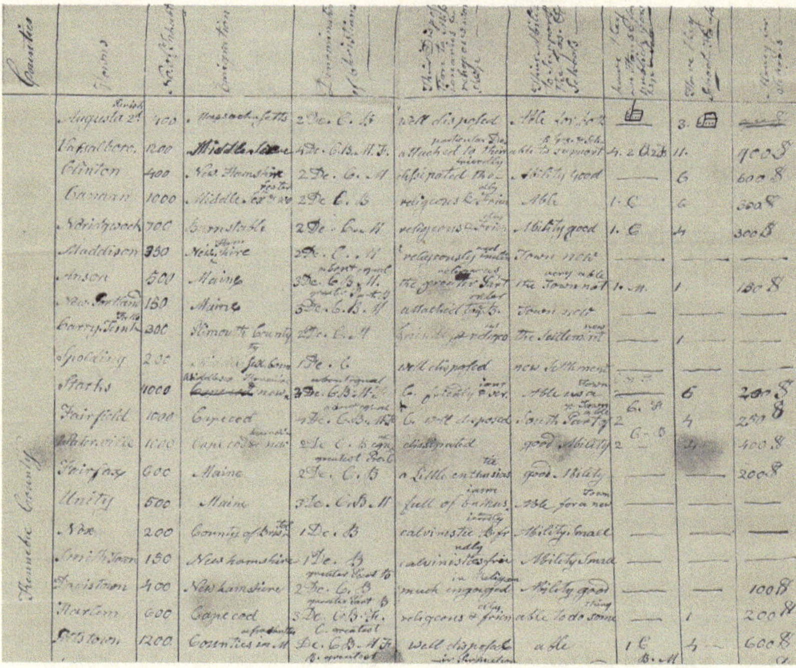

Figure 4.2. One section of the Reverend Benjamin Chadwick's chart of towns visited on his mission trip to the District of Maine in 1804. The full chart contains data on forty-four towns. Detail of Mr. Chadwick's Schedules for Kennebeck and Lincoln, 1804, Society for Propagating the Gospel Among the Indians and Others in North America Records, MSS 48, box 1, folder 17. Phillips Library, Peabody Essex Museum.

controversy irrelevant by invading and annexing eastern Maine during the War of 1812, only to be forced by the war's conclusion to return to *status quo ante bellum*.

Even war and the resulting peace treaty did not settle matters. London and Washington continued to argue over the islands, especially the largest, Grand Manan, and the most populated, Moose Island. Traditional topographical maps were often enlisted as tools. An 1818 pamphlet promoting settlement in the British provinces featured a detailed map of Grand Manan, highlighting its strategic importance at the mouth of the Bay of Fundy. To protect New Brunswick and Nova Scotia, its author wrote, the island had to be kept out of American hands. He was concerned, though, as he believed the island's residents were potentially "tinctured with American manners and principles."[7]

American mapmakers, for their part, simply started including all the disputed islands in maps of Maine. An American atlas published in 1824 made Grand Manan clearly a part of Washington County, Maine (and gave the state a much more generous northern border than what it would eventually receive).[8] It was a battle of the maps: British subjects created maps that made the islands British, and American citizens created their own maps to do the opposite.

One prominent study of borders and frontiers in American history defines borderlands as "contested boundaries between colonial domains," in which competition between rival states allows the inhabitants "room to maneuver and preserve some element of autonomy." Another, focusing on the United States and Canada, characterizes a singular "borderland" as a place where a political boundary divides "people with common social characteristics."[9] In the postrevolutionary period, the area shared by the northeastern United States and British North America fit these definitions well. Stretching from the Bay of Fundy to the Great Lakes, it was a large zone of borderlands in which people frequently contested national boundaries. In northern Maine, for example, no permanent line would be drawn until 1842. In Vermont, the need to ship goods into the Atlantic via the St. Lawrence River led to a great deal of transnational negotiation and political intrigue.[10] This chapter focuses on the easternmost part of the zone, a borderland shared by the District of Maine and the Province of New Brunswick.[11] The American citizens and British subjects living in this region certainly had common social characteristics. Many on both sides had moved north from southern New England before the war, and almost all were Protestant English speakers.

There are two distinct kinds of documents in the archives that deal with this borderland, and they tell conflicting stories. State Department papers and traditional maps of the region from the first decades of the nineteenth century present a flexible border. The data maps created by Maine missionaries during the same years present a fixed border. This dissonance arises from the way the two different groups saw the region and the questions they asked about it. On the one hand diplomats, and others looking from afar, saw this borderland in an abstract way, as a place of competing claims and precedents. Their borders were aspirational. There is no evidence that diplomats used missionary documents to learn about conditions on the ground, because the experiences of the people who lived in the borderland mattered less than the legal case that could be made for drawing the line a certain way. This was about using maps to make bold claims. Missionaries, on the other hand, were more concerned with the people than with the land. As they focused more on books and schools than on sermons, they dealt with finite resources. What was the most efficient way to distribute those resources? Who actually belonged to the District of Maine, and what did they need? Legal justifications for potential adjustments of the border were not their concern.

Missionary papers also offer a window into the attitudes of borderland residents toward the flexibility and permeability of the line. The Jefferson administration made its claims, and patriotic mapmakers drew the maps they wished, but these people lived with the border. Where did they understand it to be? Who could cross it? Was shifting it actually a possibility? Missionary letters and journals consistently show a border that was less flexible than diplomats and traditional mapmakers made it out to be. Though residents on both sides often encouraged itinerant ministers to cross the line, everyone agreed on its location. The people of Moose Island knew they were in Maine, regardless of what the British claimed. The people of Grand Manan Island knew they were in New Brunswick, and thus American ministers did not bother trying to serve them.

As the missionaries traveled through Maine and created their data maps, they began to serve as border-makers themselves. The charts they submitted were used to determine resource allocation, and so by collecting and presenting data they were helping to solidify the boundary as they experienced it. Communities they mapped onto Maine through their charts saw more visits from itinerants, more books, and eventually schools. Communities they failed to include were ignored. This phenomenon—the border becoming fixed through the actions of the people visiting it and living around it—reverses a common understanding of border formation. Borders are often seen as structures imposed from above on a resistant populace. According to this thinking, diplomats draw lines on maps, and borderland people adapt to those lines. Multiple studies of Maine adopt this approach, arguing that post-Revolutionary British and American authorities drew and enforced a meaningful border, while nonstate actors like privateers and smugglers resisted and kept the border flexible.[12] Missionary society sources from eastern Maine complicate this narrative. They present a situation in which distant authorities sparred over a boundary they (incorrectly) assumed was still negotiable, while itinerant ministers and their parishioners fixed the line in place. The diplomats should have paid more attention to what was happening in the communities they were haggling over. While they argued about the wording of old treaties, missionaries and their data maps were building a permanent border along the coast of Maine.

* * *

Debate over the precise location of the northeastern boundary began immediately after the Revolution and continued for decades. According to the 1783 Treaty of Paris, the eastern boundary of the United States was the "river Saint Croix, from its Mouth in the Bay of Fundy to its Source." This phrasing may have seemed straightforward to negotiators in far-off France, but it quickly led to controversy in Maine. First, locals did not call any of the rivers "Saint Croix," so that provoked immediate disagreement. Second, all of the rivers in question emptied into Passamaquoddy Bay, not the Bay of Fundy. And third, there were dozens of islands in the area to consider. The treaty awarded the United States "all islands within twenty leagues of [its] . . . shores," while noting that this did not include "such islands as . . . heretofore have been within the limits of . . . Nova Scotia." This provision again sounded simple but was in fact extremely complicated. The British and the Americans had conflicting ideas about which islands had "heretofore" been part of Nova Scotia.[13]

The question of the "true" river Saint Croix was settled first, though it took years of testimony from locals as well as archaeological excavation. Then the debate moved on to the islands. There are many islands in and around the bay, but only three really mattered to both sides: Moose, Campobello, and Grand Manan. For the Americans, the debate came down to proximity and common sense. All three islands are much closer to Maine than New Brunswick. Moose Island—the only one of the three the Americans would ultimately retain—is so close that it is now linked to the mainland by a causeway. Though Campobello is today part of New Brunswick, visitors arrive over a bridge from the town of Lubec, Maine. And Grand Manan is about seven miles off the coast of Maine, but eighteen miles from New Brunswick. When American diplomats presented their case for the islands, they almost always focused on this proximity. The US minister to Great Britain in 1803 argued that Campobello was obviously westward of any "suitable boundary line." Moose Island was even farther west. Both would obviously be better administered from the United States. As for Grand Manan, the Americans pointed to the Royal Commission of 1763, which had divided Massachusetts from Nova Scotia by drawing a line from Cape Sable, across the Bay of Fundy, to the western tip of Passamaquoddy Bay. Grand Manan is west of that line. It was "clearly within the general limits of the United States," according to Secretary of State James Madison, who asserted that "no just title can . . . be alleged on the British side."[14]

The British argument was even more straightforward than the American. Every island belonged to New Brunswick, period. This position was most clearly articulated in a courtroom battle of 1805, in which an American citizen attempted to reclaim his seized ship from New Brunswick authorities. Ebenezer Lock's sloop, *Falmouth*, had been picking up a load of gypsum in what Lock called the "principal channel" of the Saint Croix, between Campobello and Moose Islands, when it was captured by New Brunswick customs officials. *Falmouth*, Lock maintained, had been anchored in American waters, or shared waters at the very least. The British response attempted to settle the matter once and for all. Though Americans were trying to argue that there were still islands in dispute, they were wrong. There was no shared channel that needed negotiating. The whole of Passamaquoddy Bay was one arm of the Bay of Fundy. Any "principal" channel was limited to the river proper. The borderline thus ran along the mainland on the western side of the bay, and every island, no matter how close to Maine, was part of New Brunswick. If American ships had previously managed to trade unmolested in parts of the bay, it was "not from any doubt of His Majesty's right to the islands, but because the Americans were in the actual possession of some of them, and it was not thought prudent . . . to dispossess them." The New Brunswick court, unsurprisingly, ordered the *Falmouth* and its cargo to be forfeited.[15]

Not long afterward, American and British diplomats began again to negotiate the issue in London. Americans James Monroe and William Pinkney "argued in vain" for the American claim to Grand Manan Island. The British announced that they were in possession of that island and had been for years, and "possession . . . was a reasonable ground upon which to presume everything which constituted title." When the Americans pointed out that this position ignored the actual stipulations of the treaty of peace, the British retorted that the American claim to Moose Island was also "very questionable." For the time being, questionable claims could cancel each other out. At this point, the secretary of state asked his men to ease off, as he was concerned that trade between the United States and British North America was "more beneficial to the United States than to those colonies." Around the time of this dispute, the Americans were importing somewhere between thirty and fifty tons of gypsum per year from the Maritime colonies to use as fertilizer. The *Falmouth* had been participating in this trade when it was seized for straying into British waters. Madison worried that a provoked New Brunswick might shut down the gypsum trade entirely. Monroe and Pinkney reluctantly agreed to "leave the case of Grand Manan for future adjustment."

There the negotiations ended for the time being, with the line unsettled, but a profitable trading relationship preserved.[16]

Part of the reason the British were willing to delay settling the border issue for so long was that they had designs on more than just the islands. Many in British North America felt that the Penobscot River was a more reasonable boundary for Maine. That would attach about a third of today's Maine to New Brunswick and provide for a much more direct connection between Nova Scotia and Quebec. During the War of 1812, Lord Bathurst, the secretary for war and the colonies, sent orders to Nova Scotia to "rectify" the boundary by occupying "that part of the District of Maine which at present intercepts the communication between Halifax and Quebec."[17] In August 1814, British forces invaded and occupied everything east of the Penobscot River.

The British obviously hoped the new boundary would be permanent, but even if it was not, they were determined to keep the Passamaquoddy Bay islands. Soon after the invasion, all men on Moose Island were forced to swear an oath of allegiance to the king. In the mainland part of the occupied zone men had a choice: They could swear the oath to the king, or a different oath that pledged them only to behave "peaceably and quietly," and refrain from carrying arms.[18] Even after the war, when Americans took back control of Maine, British forces refused to evacuate Moose Island. They occupied it for three more years before finally giving up and departing in 1818. This evacuation created some uncertainty on the British side. If Americans had managed to wrest possession of Moose Island from New Brunswick, what else might they do? Was any island safe? Meanwhile, American mapmakers became more assertive, boldly including all the Passamaquoddy Islands in maps of the newly independent state of Maine.[19]

As late as the 1820s, those looking at the region from afar saw the boundary as unfixed. It was a line on a map, which had the potential to be drawn and redrawn, altered by force, or left for future adjustment. To people who lived there, however, that line became fixed much earlier. Though the border was permeable, and crossing or trading over it was routine, no one seems to have expected it to shift. And as missionary societies supplied ever more resources to Maine communities and mapped out the limits of their mission fields more precisely, changes to the boundary became more and more unlikely.

As diplomats maneuvered, the region they were attempting to delineate saw flourishing missionary activity. The population of northern New England was increasing exponentially, and the District of Maine was the epicenter. It had about one hundred thousand residents in 1790. By 1810 that more than doubled, and by 1820 it tripled.[20] Similar, if slightly slower, growth took place in New Hampshire, Vermont, upstate New York, and the Ohio country. In many cases, the fastest growth happened in the most remote settlements, directly alongside the border: Passamaquoddy Bay in Maine, the Champlain Valley in Vermont, the Genesee country in New York, and the Western Reserve in Ohio.

Members of Congregational Church organizations in southern New England were greatly concerned by this growth. The settlers were "our fellow Citizens," and without preaching, Bibles, or schools for their children, they might lose all "principles of Religion and morality."[21] In the late 1780s, Boston's SPGNA began to hire ministers to itinerate in Maine, and the Missionary Society of Connecticut arose to serve that state's emigrants in Vermont and New York. Both organizations also promoted the need for outreach to the Indians, but in private decided that outreach was impractical. Native American communities in Maine were Catholic, having been converted by the French in the seventeenth century, and they flatly rejected all overtures from Protestants. The Connecticut society sometimes tried sending ministers to Native towns in New York, but they failed to make much progress either. The societies responded by devoting almost all their resources to white settlers in "the back settlements." Their successes led to a flood of similar societies, all collaborating on the same goal. The Berkshire and Columbia Missionary Society, the Massachusetts Missionary Society, the Evangelical, Hampshire, New Hampshire, and Maine Missionary Societies—each added more funding and ministers to the network. By 1810 these societies together had achieved remarkably thorough coverage, reaching even the newest and most remote communities in the northeastern states.

The papers of these missionary societies provide a wealth of ground-level detail on life in the northeastern borderlands. Itinerant ministers were required to submit detailed descriptions of their travels before receiving payment. In some cases—the SPGNA, for one—these letters and journals themselves survive. In others, summaries of missionary activity, complete with excerpts from letters, were published in annual reports. As time went on the societies began to focus more on sending printed material to the settlements, and so they required ever more precise accounting. Ministers were asked to detail exact numbers of each title, how many copies went to each town, and what contributions they collected from locals. This change brought a certain level of defensiveness about border crossing, as well. A minister providing only sermons might not need to justify a quick trip over the line in search of an audience.

Once he was assigned a particular county, however, and was entrusted with books, time spent in New Brunswick rather than Maine required an explanation. Ministers wrote of locals asking them to cross over, or of the locals crossing themselves in search of preaching. The result is an extensive collection of reports from the ground in borderland communities, in which the permeability of the border can be tracked over time.

It is clear that borderland residents believed there was a line, knew where it was, and understood that it demarcated their rights as American citizens or British subjects. But they often dismissed the idea that the border should serve as a hard barrier for preachers. One reason was the wide variety of preachers operating in the northeast in the late eighteenth century. When emigrants from southern New England first began to move to what became Nova Scotia, New Brunswick, and Maine in the 1760s, they brought their Congregational Churches with them. As settlement accelerated after the Revolution, however, religious enthusiasm and itinerant preaching grew in tandem. Even though the Congregational was the established church of Massachusetts, and the Anglican the established church of British North America, all sorts of ministers were circulating. Methodists, Presbyterians, Quakers, and Baptists all vied to reach the burgeoning settlements. In Nova Scotia, a particularly charismatic preacher named Henry Alline inspired the "New Light" movement. Even Catholic priests made their rounds to serve Native communities and the small groups of Irish Catholics that were beginning to arrive. By the 1780s and 1790s the region had become a thriving religious marketplace. As an example, one study counted at least twenty-six distinct itinerant preachers visiting the community of Liverpool, Nova Scotia in those years. There were fifteen Methodists, eight New Lights, two Quakers, and a Presbyterian.[22]

Most of these men paid little attention to the international boundary after it was created. Many in British territory were originally from New England. They had emigrated north as children and saw all English-speaking settlers as souls to be saved, regardless of whether they were subjects or citizens. And though settlements on both sides of the border were growing, northern New England was growing the fastest. Henry Alline (born in Rhode Island) set an example by crossing to Maine just after the peace in 1783. He aimed to itinerate all the way to Boston, but took ill and died while preaching in New Hampshire. By the 1790s, at least five Nova Scotia New Lights were regular visitors to eastern Maine: James and Edward Manning, James Murphy, Daniel Shaw, and Elijah Estabrooks. Murphy and the Manning brothers even managed to establish a New Light church in Machias, Maine—by harassing the local

Congregational minister and "publicly . . . mention[ing] him, in prayers, as an unconverted person," until he left. The people of Machias apparently did not discriminate against these preachers from British territory. Good preaching was welcome, regardless of what direction it came from.[23]

As a result, when Congregational itinerants began to visit the region from southern New England, they crossed the border as well. Ministers preached wherever they could find an audience, even if it was in British territory. The instructions they received in those years also tended to encourage a laissez-faire attitude toward the mission fields. The first public statements from the Connecticut Missionary Society were quite vague, specifying only "destitute settlements" populated by "brethren emigrants."[24] Some men demanded more guidance. When Rev. Daniel Staniford was asked to attend to "the settlements on Connecticut River, adjacent to . . . Lower Canada," he complained about receiving "no particular instructions from the Society."[25] When the SPGNA assigned Levi Frisbie to "the eastern settlements," Frisbie requested "further Instructions respecting the extent and bounds of my circuit."[26]

Crossing the border was especially simple in eastern Maine, because of the constant communication between British and American towns by water. It must have been refreshing, after struggling along between tiny settlements served by poor roads, to be able to simply row across the bay from one large town to the next. Jonathan Fisher of Blue Hill, Maine, was one preacher who took advantage. On his first mission trip, in 1801, Fisher preached his way up the coast to Passamaquoddy Bay, and then carried on over the line into British territory. In his journal, Fisher mentioned big crowds in New Brunswick. He preached in at least three different towns and claimed to have drawn "probably more than 200 persons, of four if not five denominations" for one sermon.[27] Another Maine minister, Jotham Sewall, spent years serving the eastern towns of the district for the Massachusetts Missionary Society, before eventually deciding that he too should go ahead and cross the border. He visited Reverend Fisher on at least one of his trips eastward, which might have affected his thinking. Perhaps Fisher told him about his sizable New Brunswick audiences. On Sewall's second or third trip to the borderland, he tested the waters with a quick visit over the line. The following year he did much more, preaching multiple times in just about every sizable community on the British side of the bay.[28]

Another factor playing into this border crossing was pressure from the parishioners themselves. As mentioned above, there was constant communication between British and American settlements over the water. Locals were sometimes frustrated by missionaries' reluctance to cross such an easily crossable border. As missionary societies provided more circumscribed mission fields, ministers sometimes wrote that they were being pressured to ignore the border. Alexander Maclean found himself in this sort of situation in the summer of 1803, as he was performing a mission for the SPGNA. After weeks of successful itinerating in western Maine and along the coast, Maclean headed for the "plantations bordering on New Brunswick" in mid-August. When he arrived, however, he had trouble finding a crowd. He began visiting one home at a time—directed from family to family, most often by the woman of the house. A Mrs. Goddard referred him to her sister Mrs. Brewers, who then suggested that Maclean cross the border and visit the Methodist preacher Duncan McColl. This provided quite a reversal of fortune. After lots of "edifying discourse," McColl recommended that Maclean preach on the British side instead. McColl claimed that he regularly crossed to speak on the American side, and that people from both sides of the river would probably attend. Maclean "cheerfully consented . . . as there appeared to me no prospect of having an assembly [in Maine]." The next day "a very solemn and attentive" audience showed up for his sermon, "tho very few from the American side of the river."[29]

The day after Maclean's sermon, a man named Cales from the New Brunswick town of St. Andrews introduced himself. Cales was originally from Essex County, Massachusetts—and so not all that different from the many Massachusetts settlers on the American side of the river. He asked Maclean to come preach in St. Andrews too. Maclean said no, at first. He was uncomfortable and noted that it was "entirely out of my power & inconsistent with my engagement." Cales was disappointed, but asked when the minister might reach Robbinston, Maine. Robbinston was easily accessible from St. Andrews, and if the minister would not cross over, the listeners themselves would. Maclean was a bit frustrated by the whole experience. His next sermon on the American side of the line "had but few hearers." Then he departed, as he wrote to the SPGNA, "having received the kindest treatment from the people on the British side of the river," and nothing but disinterest from the Americans.[30]

This was not the last that the SPGNA heard from the people of St. Andrews. The town had one minister in residence—an Anglican, coincidentally named Reverend Andrews—but he was getting older and could not preach every week. The parishioners were obviously quite aware of ministerial services on the other side of the river. A few years after Maclean's visit, a Reverend Oliver spent much of 1805 and 1806 traveling around eastern Maine. It appears from his letters that while visiting Eastport he made several visits to St. Andrews. Then, in 1817, Reverend Daniel Lovejoy provided the SPGNA with an explanation of what was happening. Lovejoy wrote that he knew the society wanted his services "principally performed in Robbinston," and also that he was not supposed to preach in places where there was a settled minister. The trouble was that people from St. Andrews kept attending his services and pressing him to preach on the British side. When he finally agreed, "the people attended remarkably well . . . I had many more hearers in this place than any other, where I preached during my mission." So Lovejoy began visiting St. Andrews every Sabbath; it appears he preached in the morning in Maine, and in the afternoon or evening in New Brunswick.[31]

Rev. Lovejoy, like Alexander Maclean, felt defensive about the border crossing. He took pains to mention that he knew St. Andrews was "within the British dominions," but that the aged Anglican clergymen had given his blessing to the visits. He also noted that the cash donation from the people of St. Andrews—twenty dollars—was the second largest total that he managed to collect in his travels. The St. Andrews parishioners apparently also sent a letter to explain the situation and confirm Lovejoy's story. Alexander Maclean had adopted a similarly defensive tone in complaining about his American audiences a few years earlier. Many people on the New Brunswick side were originally from America anyway: "The inhabitants on the [eastern] side . . . are partly Americans partly British, [and] . . . a kind hospitable, religious people." And yet neither the cash donations nor the ambiguous nationality of these borderland people seems to have swayed the SPGNA. After 1804, they were committed to collecting information on Maine settlements so as to better distribute their resources in Maine. Though they sent Lovejoy back to Robbinston repeatedly over the following years, he never again mentioned crossing the border. The SPGNA may have asked him to stop, or he might have decided that the St. Andrews visits were frowned upon and kept them unofficial.[32]

Documents exploring this dynamic from the British provinces into the United States are a little harder to come by. This dearth arises in part because the New England missionary societies sent many more preachers to the borderland than did British missionary societies. The few sources that do exist, however, seem to confirm a similar experience. The London Missionary Society (LMS) provides the most direct comparison with the various New England missionary societies. The LMS and the New England societies were all Congregational organizations ("dissenters" in England), they communicated with each other, and they shared essentially the same goals. The LMS was outnumbered by the New England societies, but it was large and wealthy and aimed to serve places as far afield as the Cape of Good Hope and the South Pacific. Although the LMS saw British North America as a minor part of its field, it did occasionally send ministers there. It appears from their letters that crossing the border was just as easy from the British side, and borderland Americans just as welcoming to visiting preachers.[33]

In 1800, the LMS sent two men, Clark Benton and John Mitchell, to Lower Canada (today Quebec). The LMS was responding to a letter from Protestant English speakers in Quebec City, who were dissatisfied with their preaching options. Quebec was overwhelmingly Catholic and French speaking, but the city had one Anglican and one Presbyterian minister to serve Protestants. Apparently, there was a market for a dissenting minister. Benton and Mitchell were instructed to go where they were needed within the province, so upon arrival Benton chose to stay in Quebec City, and Mitchell ended up in New Carlisle, a small fishing community on the Gulf of St. Lawrence. In Quebec, Benton hired a room for Sunday services and attempted to assemble a congregation. By about a year later he claimed fifty or sixty weekly listeners, though he also complained about a lack of books and communication from London.[34]

Benton had a difficult time in Quebec City. The established Anglican and Presbyterian communities were unfriendly and attacked him for attempting to perform marriages and baptisms. In 1803 he decided to cross the border temporarily to get a pamphlet published that responded to his critics. Benton was thrilled to be in the States, where apparently he finally felt welcomed. The Connecticut Missionary Society had been serving small towns in the Champlain Valley of Vermont and New York for years, so the arrival of a random preacher was a common occurrence. Benton made his way south, preaching in Middlebury and Cornwall, Vermont, and Albany and Troy, New York. Around Albany he had his pamphlet printed, attended a couple of ordinations, and was even invited to preach to an assembly of ministers. His letters made pointed mention—in an ironic reversal of Maclean's experience in Passamaquoddy Bay—of how nice and welcoming the Americans were compared to the decidedly unfriendly atmosphere in Lower Canada. Benton took his time heading back north, itinerating around Vermont for a while and preaching in a few more towns. Finally he made his way to Quebec City, presented his (apparently quite pointed) pamphlet, and was promptly arrested and imprisoned for libel.[35]

Very few British ministers ventured as far into the United States as Benton did. Those who briefly popped over the border, though, in the same manner as SPGNA ministers around Passamaquoddy Bay, seem to have been received well. Methodist Church records from New Brunswick indicate that Duncan McColl did indeed preach in Maine regularly, just as he told Alexander Maclean. In 1804, McColl reported to superiors in Halifax that he "frequently spent some months . . . in the U. States." A few years later he claimed to have sparked a minor revival in Calais, Maine, just across the river from his home in St. Stephen. Some American women had asked him to come preach, which led to "powerful conversions."[36] Even Rev. Andrews in St. Andrews crossed over to Robbinston once in a while, though mostly to baptize babies rather than to preach.[37]

This freedom to cross the border and maximize one's audience aligned well with the goals of the New England missionary societies when they were first organized. In their earliest mission statements most societies were quite expansive. They announced plans to serve all in need of the gospel, wherever they might be. When Connecticut Congregationalists first began to organize their missionary efforts, they wrote of the "destitute state of the settlements now forming in the wilderness to the Westward and N. Westward of us . . . many of whom are our Brethren Emigrants."[38] There was no reason for this missionizing to exclude settlements in British territory, which were also populated by many Connecticut emigrants. The SPGNA put its broad vision directly into its name, specifying "Indians and others in North America," rather than the United States. It did not intend at first to focus on the District of Maine; that situation developed over time, as the society collected more money from the Massachusetts legislature. In its early years, the SPGNA occasionally sent ministers west instead of east and provided services to at least one Indian community in New York. Another group, the Maine Missionary Society, was even more ambitious. Due to the "great objects which the society have in view," it investigated the possibility of sending missionaries west to the Mississippi country, or even east to Africa. The stated goal of the Maine society was to "send the glorious gospel to those who are destitute of . . . religious instruction." This description applied to people just about anywhere.[39]

The SPGNA was the first of the missionary societies to align itself explicitly with state government and treat state boundaries as the limits of its mission field. It began that process in 1791, when it secured an annual grant from the Massachusetts legislature with the promise to "disseminate the principles of religion and morality" in the District of Maine.[40] The SPGNA used that grant to pay the salaries of the ministers itinerating in the district, which freed its remaining funds for books and pamphlets. The society purchased massive numbers of books—not just Bibles and testaments, but also spelling books, primers, and other works for children. By 1800 SPGNA ministers had distributed close to nine thousand books in communities throughout the northeast. The focus on books had a side effect of reducing border crossing as well; the word of God was free to all, but books cost money and were intended for Maine citizens. Jonathan Fisher, now working for the SPGNA, wrote of delivering a "horse load" of books along a 118-mile circuit of eastern Maine between Penobscot Bay and the border. Though he may have remembered his big New Brunswick audiences fondly, he did not report bringing them any books.[41]

In 1804 the SPGNA raised the stakes in its partnership with the Commonwealth of Massachusetts. Instead of promising to spread morality and civilization in a general way, the SPGNA offered itself more explicitly as a partner in state-building. The society and the state, it turned out, had a great deal in common. Maine was experiencing an increasing amount of political and religious turbulence. As one SPGNA missionary put it in 1797, "the religious state of Maine is not as one could wish." The district was crawling with "zealous" Baptists, Methodists, New Lights, and many others, preachers who lacked training and "scriptural knowledge" and who led the unwary into "a religious [*sic*] of the imagination & into strange & very hurtful opinions."[42] Furthermore, many Maine settlers distrusted educated preachers, labeling them "grand folks." The Congregationalists felt that they were losing ground in the spiritual marketplace, but they started to see schools as a way to reverse this trend. In towns where missionaries served as both preacher and schoolmaster, they were more appreciated in the latter role. One missionary in the town of Ellsworth made it clear that the townspeople only tolerated his presence for the sake of their children's education. An Ellsworth committee wrote to the missionary society with the same message, noting that the whole town wanted Rev. Nurse for a teacher, but only half wanted him for a preacher.[43]

This focus on schools coincided with a gradual narrowing of the expansive missionary work the societies had anticipated. Though the kingdom of God was vast, and the need for missionaries global, there was a lot of demand to be met in the American backcountry. There was also a growing recognition that British organizations like the London Missionary Society were partners rather than rivals. The Connecticut Missionary Society exchanged letters with the LMS and published excerpts in its periodical. According to one such letter, there was "a glorious space to traverse" between British and American territory, and itinerants from both sides would eventually meet in the middle: "The Atlantic flowing between prevents not our united hands and hearts from cordially meeting in this work." Readers knew that the LMS sent ministers to Canada and other parts of British North America, and so those places were covered. New Englanders could feel free to concentrate on their own backyard, where those troublesome Maine settlers were being seduced by unlettered Baptists.[44]

While the Congregationalists had religious anxiety about Maine, the Commonwealth of Massachusetts had political anxiety. The families moving to the northeastern frontier needed to be reminded that they fell under Boston's authority. The 1790s had seen a resurgence of backcountry resistance to large landowners, especially in the newly settled towns of central Maine. Massachusetts legislators did not know what was more troubling, the chaos caused by these "liberty men" in their attacks on land surveyors, or the possibility that they would give up on settling in Maine. If the state failed to provide, these people could move west—or worse, to British territory. The SPGNA directly addressed this fear in its petition for more funding. Expanding the mission to include schools would give Massachusetts an edge by inducing "people of sober life and conversation" to choose the District of Maine "in preference to going into New York, the Canadas and other places without the Commonwealth." Schools were particularly necessary in the borderland around Passamaquoddy Bay, since those towns were "situated near the people of a foreign province, the sovereign of which, regards the conditions of his subjects in the wilderness with a tender eye." Massachusetts certainly did not wish to be outmaneuvered by the tender eye of George III.[45]

Some people in Massachusetts were concerned about the possibility of Maine becoming an independent state. The missionary society was asking for a thousand dollars, which was double its previous grant. Why spend all this money to set up schools in a place that might break away from the Commonwealth entirely? The SPGNA petition addressed the issue by arguing, essentially, that state boundaries mattered less than national boundaries. Whether Maine became independent or not, the people there would always be "our brethren." Should "events of a future day sever them from this state . . . they [would] remain a part of the nation." And Massachusetts would always have an interest in the development of Maine's economy and commerce.[46] These arguments seem to have quieted any doubts about the wisdom of funding operations in Maine. A legislative committee responded warmly, thanking the SPGNA for its "essential service . . . with incalculable benefits to the inhabitants and to the Commonwealth."[47] The state approved the increased grant.

And so the SPGNA asked its ministers to expand their work. They had been preachers; now they became census takers of a sort. What the SPGNA asked for, however, went beyond the simple survey that was the US census. Under the Constitution, the decennial census simply counted the population, sorting people based on age, sex, and status as free or enslaved. The SPGNA asked for not only numbers, but also "the general state of the people . . . [and] the particular state of the several towns and plantations through which you pass." Ministers were supposed to paint a detailed picture with this data about who exactly these people were and how willing and able they were to support churches. The SPGNA may also have provided more specific guidance in letters that have not survived, because several ministers submitted charts with extremely precise detail. Benjamin Chadwick, after spending the summer of 1804 serving Kennebec and Lincoln counties, produced a neat grid detailing the forty-two towns he visited, their populations, where their residents had emigrated from, what denominations were present, the "religious disposition" and ability to support schools, and the number of existing schoolhouses and houses of worship.[48] The same year in Hancock and Washington counties, Alfred Johnson contributed his aforementioned chart of more than fifty towns and townships, with a similar level of detail.[49]

SPGNA ministers were particularly careful with their accounting of communities close to New Brunswick. Alfred Johnson's chart divided Washington County into two parts: a "North Circle" around Machias and an "East Circle" directly on the border. The largest settlement in the area was Eastport, on Moose Island, which had 100 families and over 500 people. Johnson seems to have taken for granted that Moose Island was part of Maine, and he spent a significant amount of time there. But Johnson also accounted for the tiniest of unnamed townships, many with only a handful of families. Another unsigned and undated chart from the SPGNA records (likely from the same year) divided the borderland towns into "About Passamaquoddy" and "On Schoodic River." The Schoodic was the local name for the Saint Croix, so towns "on Schoodic" were directly across the river from New Brunswick. This person traveled up the river to unnamed townships number six and seven, reporting that only ten families lived there, and that they had no money to support schools or the gospel. The anonymous itinerant, like Johnson, was much more bullish on Moose Island and Eastport, writing that it had considerable prospects and would be a place of great consequence in trade.[50]

These charts of the Maine borderland were intended to facilitate the establishment of schools, but they did more than that. Compiling the data maps changed the way Congregational missionaries understood their field. In their earliest missions, preachers had been asked to look for destitute settlements and people in need. They were supposed to find an audience. It was not that difficult to convince Alexander Maclean to cross into New Brunswick because he felt that his job, at least in part, was to reach the maximum number of people. Now, in 1804, SPGNA employees were being asked to take a step back. Instead of assembling an audience, they were to provide a statistical picture of the district. The picture they provided altered the work of itinerants even once the focus returned to preaching.

Ephraim Abbott accepted a SPGNA mission to Passamaquoddy Bay in 1811. Like his predecessors, he did most of his preaching in the larger towns of Robbinston, Schoodic (Calais), and Eastport. Yet Abbott also made several trips upriver to preach in those tiny numbered townships recorded in the 1804 charts. He mentioned specific visits to townships one, three, six, seven, twelve, and seventeen.[51] Abbott could easily have devoted some time to the New Brunswick towns on the other side of the river. He could also have visited Campobello or Grand Manan Island, which were close by and had lots of settlers, many of whom were recent emigrants from the United States. Islands in the bay were far more accessible than tiny, remote, upriver townships—and the crowds would have been better. But neither island had been included in the data maps compiled by SPGNA missionaries. These maps helped the border evolve into more of a barrier for Abbott than it had been in the past. The society and its missionaries knew exactly which communities they would be serving. These changing ideas about the border also help explain why Daniel Lovejoy was so apologetic in his letters of 1817, explaining that he knew the town of St. Andrews was British, and that he was not really supposed to be preaching there. SPGNA ministers were supposed to be improving the lives of Maine settlers. Preaching to "brethren" emigrants wherever they might be was no longer the mission.

* * *

Borders are sometimes fixed in place through intense, high-level diplomatic processes. Empires and nations challenge each other with conflicting maps, fighting to lock down their peripheries. But this is not always how border-making works. Sometimes lines can be fixed through collaboration, rather than conflict, and this is what happened in the early nineteenth-century District of Maine. Itinerant American ministers came to an understanding with their British counterparts. They asked for and received state funding to allow them to expand their book distribution and open schools. Then they built the data maps they needed to serve their borderland parishioners and began distributing resources accordingly. Though diplomats continued for years to draw and debate lines on maps, the actual work was done on the ground. Missionaries mapped the data, and the data maps created a meaningful border.

Notes

1. It was called the "District" of Maine because until 1820 Maine was part of Massachusetts.

2. The SPGNA should not be confused with the Society for the Propagation of the Gospel in Foreign Parts (SPG), an Anglican organization. The British SPG could no longer operate in the United States after the Revolution. Boston Congregationalists explicitly modeled their society on the SPG's example, but they were not Anglican and the two groups had little to do with one another.

3. Levi Frisbie Letter, Folder 11, Box 2; Samuel Eaton Letters, Folder 5, Box 2, Society for Propagating the Gospel Among the Indians and Others in North America papers, Phillips Library, Rowley, MA (hereafter SPGNA, PL).

4. Instructions to Missionaries, 1804, Folder 1, Box 8, SPGNA, PL.

5. Alfred Johnson Letter, Folder 1, Box 3, SPGNA, PL.

6. Francis M. Carroll, *A Good and Wise Measure: The Search for the Canadian-American Boundary, 1783–1842* (University of Toronto Press, 2001), 38–40.

7. Anthony Lockwood, *Brief Description of Nova Scotia with Plates of the Principal Harbors; Including a Particular Account of the Island of Grand Manan* (Printed for the Author by G. Hayden Brydges St. Covent Garden, 1818).

8. *Map of Maine, Finley's New General Atlas* (Jos. Perkins, 1824).

9. Jeremy Adelman and Stephen Aron, "From Borderlands to Borders: Empires, Nation-States, and the Peoples in Between in North American History," *American Historical Review* 104, no. 3 (1999): 814–41; Lauren McKinsey and Victor Konrad, *Borderlands Reflections: The United States and Canada* (Borderlands Project, Canadian-American Center, 1989).

10. For example, the Ira Allen canal affair, in which Vermonter Allen attempted to organize a French invasion of Canada because London would not finance a canal project. For the whole saga, see J. Kevin Graffagnino, "'Twenty Thousand Muskets!!!': Ira Allen and the Olive Branch Affair, 1796–1800," *William and Mary Quarterly* 48, no. 3 (1991): 409–31.

11. Maine, at the time, was still part of the state of Massachusetts. It would become an independent state in 1820. New Brunswick was one of the newest provinces of British North America, having split off from Nova Scotia in 1784.

12. Edward J. Martin, "The Prize Game in the Borderlands: Privateering in New England and the Maritime Provinces, 1775–1815" (PhD diss., University of Maine, 2014); Joshua Smith,

Borderland Smuggling: Patriots, Loyalists, and Illicit Trade in the Northeast, 1783–1820 (University Press of Florida, 2006).

13. The Paris Peace Treaty of September 30, 1783, *The Avalon Project*, Yale University, accessed May 22, 2019, http://avalon.law.yale.edu/18th_century/paris.asp. Samuel Champlain dubbed a river in the area "Saint Croix" during his explorations of the early seventeenth century, so European maps included that name.

14. James Madison to James Monroe, 15 May 1806; Monroe and Pinkney to Madison, 25 April 1807, in *Diplomatic Correspondence of the United States: Canadian Relations, 1784–1860*, ed. William Manning (Carnegie Endowment for International Peace, 1940), 172, 589.

15. "The Question of the Rights of the U.S.A. to the Islands in Passamaquoddy Bay, by virtue of the Treaty of 1783, considered in the case of the Sloop Falmouth, in the court of Vice Admiralty, for the Province of New Brunswick, in the year 1805." Printed by J. Ryan, Printer to His Majesty, Folder 24, Box 1, Barclay Collection, Maine Historical Society (hereafter MEHS), Portland, ME.

16. What seems to have worried Madison was the potential for a slew of cases like that of the *Falmouth*. Gypsum fertilizer was important to the economies of the northeastern states. The gypsum trade involved a lot of small ships on both sides and depended on weak enforcement of trading regulations. British decisions to enforce the letter of the law would have seriously affected the US economy. Madison to Monroe and Pinkney, 17 May 1806, in Manning, *Diplomatic Correspondence*, 173; Monroe and Pinkney to Madison, 25 April 1807, in Manning, *Diplomatic Correspondence*, 595–96; Gerald Graham, *Sea Power and British North America, 1783–1820: A Study in British Colonial Policy* (Harvard University Press, 1941), 171.

17. Robert Dallison, *A Neighbourly War: New Brunswick and the War of 1812* (Goose Lane Editions, 2012), 90–96; George F. W. Young, *The Capture and Occupation of Downeast Maine, 1814–1815/1818* (Penobscot Books, 2014), 12–13.

18. Proclamation, Governor John Sherbrooke, *Nova Scotia Gazette*, 5 October 1814, in Young, *Capture and Occupation*, 106–11.

19. See for example Anthony Finley's 1827 map of Maine: https://commons.wikimedia.org/wiki/File:1827_Finley_Map_of_Maine_-_Geographicus_-_Maine-finley-1827.jpg.

20. Publius Research, "Population Since 1741," Maine: An Encyclopedia, accessed June 18, 2019, https://maineanencyclopedia.com/population-since-1741/.

21. *Acts and Laws of the Commonwealth of Massachusetts 1791–1792* (Boston: Wright and Potter, 1869–1886), 563.

22. Elizabeth Mancke, *The Fault Lines of Empire: Political Differentiation in Massachusetts and Nova Scotia, 1760–1830* (Routledge, 2005), 122; for the religious marketplace in the United States, see Nathan Hatch, *The Democratization of American Christianity* (Yale University Press, 1989); Stephen A. Marini, *Radical Sects of Revolutionary New England* (Harvard University Press, 1982). For the marketplace in Maritime Canada, see George Rawlyk, *Wrapped Up in God: A Study of Several Canadian Revivals and Revivalists* (Welch Publishing, 1988); Rawlyk, *The Canada Fire: Radical Evangelicalism in British North America* (McGill-Queen's University Press, 1994).

23. Stephen Jones to Edward Manning, *The Newlight Baptist Journals of James Manning and James Innis*, ed. D. G. Bell (Lancelot Press, 1984), 92; Mancke, *Fault Lines*, 117.

24. Meeting minutes, 22 June 1775, Folder 15, Box 1, Congregational Churches in Connecticut General Association. Records, Connecticut Historical Society, Hartford, Connecticut.

25. Daniel Staniford Letter, Folder 43, Box 1, New Hampshire Missionary Society Papers, New Hampshire Historical Society, Concord, NH.

26. Levi Frisbie Letter, Folder 11, Box 2, SPGNA, PL.

27. Jonathan Fisher Journal, Mss Boxes F, Jonathan Fisher Papers 1791–1826, American Antiquarian Society, Worcester, MA.

28. Rev. Jotham Sewall (Jr.), *A Memoir of Rev. Jotham Sewall of Chesterville, Maine* (Tappan and Whittemore, 1853), 120–50.

29. Alexander Maclean Journal, Folder 7, Box 3, SPGNA, PL.

30. Ibid.

31. Daniel Lovejoy Letters, Folder 6, Box 3, SPGNA, PL.

32. Ibid., and Alexander Maclean Journal, Folder 7, Box 3, SPGNA, PL.

33. London Missionary Society, Continent of America, Incoming Correspondence, Box 1B: Canada, SOAS Archives and Special Collections, London, UK.

34. Ibid.

35. Ibid.

36. McColl to William Black, 6 August 1804, and Black to Robert Smith, 11 July 1814, MC990, Microfilm Reel #9720, Wesleyan Methodist Church Foreign Missions Fonds, Provincial Archives of New Brunswick (hereafter PANB), Fredericton, NB.

37. St. Andrews, NB, Anglican Church records, Microfilm Roll #F1082, PANB.

38. Meeting minutes, 22 June 1775, Folder 15, Box 1, Congregational Churches in Connecticut General Association. Records, Connecticut Historical Society, Hartford, CT.

39. Trustee Records 1807–1829, Vol. 1, Maine Missionary Society Records, MEHS; "Address of the Maine Missionary Society to the Public," 1808, ECL 5.26, MEHS.

40. *Acts and Laws of the Commonwealth of Massachusetts 1791–1792*, 563.

41. Jonathan Fisher Journal, Mss Boxes F, Jonathan Fisher Papers 1791–1826, AAS.

42. Paul Coffin letter, 1797, Folder 22, Box 1, SPGNA, PL.

43. Nurse to Thayer, 21 May 1811; Committee to Thayer, 26 January 1811, Folder 2, Box 1, Evangelical Missionary Society in Massachusetts records, 1808–1914, Andover-Harvard Theological Library, Harvard Divinity School, Cambridge, MA.

44. *Connecticut Evangelical Magazine*, Volume 1, Number 1, July 1800, Connecticut Historical Society, Hartford, CT. For the idea of a vast kingdom of God, see Emily Conroy-Krutz, "The Vast Kingdom of God," *William and Mary Quarterly* 78, no. 2 (April 2021): 222–29.

45. Petition, 30 January 1804, Folder 1, Box 6, SPGNA, PL; for liberty men and backcountry resistance, see Alan Taylor, *Liberty Men and Great Proprietors: The Revolutionary Settlement on the Maine Frontier, 1760–1820* (University of North Carolina Press, 1990).

46. Petition, 30 January 1804, Folder 1, Box 6, SPGNA, PL.

47. Response from the Legislature, Folder 1, Organizational Records and Miscellaneous Bills and Receipts, 1791–1807, Box 8, SPGNA, PL.

48. Benjamin Chadwick Letter, Folder 17, Box 1, SPNGA, PL.

49. Alfred Johnson Letter, Folder 1, Box 3, SPGNA, PL.

50. Oversized Folder 1, Financial Papers 1789–1830, SPGNA, PL.

51. Ephriam Abbott Journal, Folder 1, Box 1, SPGNA, PL.

PART II

Creating Place

CHAPTER 5

Depicting and Defining the Plantation in the Early English Caribbean, 1625–1675

PAUL MUSSELWHITE

Despite being an intensely contested space, maps of the early English Caribbean depict few borders. They are, though, pockmarked with settler claims in the form of "plantations." Richard Ligon's famous 1657 map of Barbados was an early example of the form; Ligon depicted the island as an unbordered space interspersed with hills, mounted horsemen, escaping enslaved people, and draft animals, but along the leeward coast he marked a string of small rectangular symbols and names (see Figure 5.1).[1] Modern viewers instinctively recognize these symbols as plantations and understand them immediately through the lens of a rich scholarship that has identified the plantation as a private space of commercial agriculture and brutal racialized labor exploitation. Nobody stops to notice the absence of lines delimiting these places, even though surveying and enclosure are recognized as a critical precursor of this Atlantic form of capitalist agriculture. The omission, though, is significant. The absence of lines on a map, as much as their presence, can be revealing about the *way* a space was claimed. In this case it provides an important insight into what was actually a messy process of defining the plantation as a form of settler possession in the first half-century of English settlement in the region.

England's earliest Caribbean cartographers, such as Ligon, marked plantations as nodes, more akin to the towns and villages on contemporary European maps, rather than as the two-dimensional bounded spaces that were beginning to be charted by contemporary property surveyors in England and even in other colonial zones such as Ireland and Bermuda. English surveyors

Figure 5.1. Richard Ligon, *A topographicall description and admeasurement of the yland of Barbados in the West Indyaes: with the mrs. names of the seureall plantacons*. 1657. Reproduction courtesy of the Norman B. Leventhal Map & Education Center at the Boston Public Library.

certainly had the necessary skills and experience to quickly master the surface area of comparatively small Caribbean islands; in fact, they began drawing plats for individual deeds by the late 1640s. Yet, mapmakers chose not to even gesture at such lines when they drafted maps of the English islands. They sought to portray plantations as unbounded nodes of settler power rather than bounded acreages granted by the state. Even as "plantations" became spaces of naked capitalist exploitation and surveyed boundaries began to appear in land transactions, planters cultivated and adapted the plantation's associations with the public good and public authority in the service of retaining far more commercial and legal independence than they would have enjoyed as mere freeholders. Making plantations appear like towns, with power over elusive hinterlands, in the cartographic evidence that reached England was part of an effort to define exactly what kind of place a plantation was.

The absence of boundaries on early maps, and this larger effort to define the plantation, has been overlooked because historians have long considered the plantation as a precociously private space. One of the foundations for the standard account of the rise of the plantation complex is the drive for private property. Scholars have long argued that, among European invaders in the Americas, English colonists were uniquely fixated on private landownership, property boundaries, and profit, and that this mindset made the emergence of slavery and plantation agriculture seem rational and inevitable in Anglo-America.[2] Recent historiography has challenged this picture of inevitability by recovering other English Caribbean models built around local production, inter-imperial trade, and privateering.[3] However, even much of this analysis assumes a clear dichotomy between plantation and nonplantation models. Accepting the definition of the private, capitalist plantation, it assumes that every place described as a "plantation" in the Caribbean, and every node marked on Ligon's map, reflected at least the aspiration to establish nakedly acquisitive commerce, overlooking the many "plantations" elsewhere in the English Atlantic that were organized around purposes as diverse as gathering congregations of the elect or drying cod.[4] The assumption that the plantation represented one simple path of development flattens the picture of the emergence of capital-intensive, slave-powered agriculture. In fact, the plantation was a completely new form of colonial place that was still being defined and redefined in ways that mediated the struggles that historians have identified over the growth of capitalism and state power.

This essay argues that cartography itself served as a medium for that debate about the nature and purpose of the plantation. Ligon's map of Barbados reflected the aspirations of an emerging commercial and slave-owning class on the island in the 1650s, but the contours of the debate became clear in the wealth of maps depicting the newly acquired island of Jamaica in the 1660s and 1670s. In the aftermath of Oliver Cromwell's ill-fated Western Design to conquer the Spanish Indies in 1655, when the English sought to salvage their pride by "planting" Jamaica, they confronted head-on the ambivalent meanings inherent in that objective.[5] Under the governorship of Thomas Modyford there was an unprecedented outpouring of maps of Jamaica, representing an effort to define the nature of its plantations, in the space between nodes of public settler authority and bounded spaces of private commerce within the Restoration Empire. Parishes became bounded spaces while plantations remained nodes, albeit with new output-oriented details. While Jamaica would take decades to eclipse the sugar profits flowing from Barbados, its prodigious crop of cartography helped to refine the definition of plantation much earlier. They attest to the intellectual effort that went into constructing the plantation as a potent new conceptual and spatial technology of settler colonial appropriation and racial capitalism—a technology so powerful that generations of scholars have employed the English-derived term "plantation" to describe a whole system of slave-powered commercial agriculture that came to shape the modern world.

Conceiving the Plantation

Since the sixteenth century British thinkers had described colonizing efforts, particularly in Ireland, as a process of "planting civility" in barbarous frontier zones. Drawing from deep classical and Christian traditions they defined civility as manifest in towns and cities, and they were convinced that only arable agriculture, or "planting," would allow a sufficiently large and sedentary population to support these institutions. Early English ideas about "planting" and colonialism were, therefore, not about appropriating land for private individuals to plant particular crops, but rather about completely dismantling Indigenous pastoral societies that they viewed as *too* atomized and privatized; "planting" involved reconstituting colonial space with new social and political units structured around English settlers who would practice the right form of agriculture to underpin "civility." These new communities were initially described as colonies, seigneuries, or towns, but around the turn of the seventeenth century the novel term "plantation" began to be applied to places being constructed and claimed by the logic of "planting."[6]

Plantations, then, involved the appropriation of land, but the plantation was not a private landholding, but rather an exercise in commonwealth-building that Francis Bacon described as among the most "ancient, primitive, and heroical works."[7] The Crown claimed and then leased land to companies and individuals to engage in plantation—to allocate and manage a collection of land rights and jurisdictions with the aim of establishing a stable new settler community. Plantations therefore offered an opportunity for practical experimentation in different forms of public and private landholding that colonial projectors believed best suited to cultivating the civic virtue that supposedly underpinned the colonial endeavor.[8] Planners were certainly not naïve; they recognized that private gain was essential to incentivize individuals; as one English planter in Ireland noted, "this word Myne is a strong warrior, every man for his owne will adventure farre."[9] The key, though, for those schooled in Renaissance humanist thought, was how to balance the allocation of private land with other types of public property that would give colonists a shared stake in the venture and support public infrastructure.

Early English plantations differed considerably from one another in the structure of land distribution. Plans for plantations in Ireland ranged from neo-Roman colonies in which a community of small-scale landholders were centered on a city that would collectively manage public works, to hierarchical seigneuries in which manorial lords received land from the Crown and collected rents but were expected to provide public services to their tenants.[10] When the Virginia Company began granting private land in the 1610s they also established plantations as corporate communities charged with dividing the land in service of the public good and establishing commons. The rival faction of colonial investors who took control of the Bermuda Company in these same years favored a manorial model in which wealthy individuals established "families" of tenants on their "shares."[11] In all of these cases, though, the "plantation" was the spatial container for working out these questions of political theory.

Despite its reputation as a place for grasping opportunists, the geography of the early English Caribbean was also defined by plantations envisioned within the terms of this debate over public and private interests. When Sir Thomas Warner established the first successful English colony in the Caribbean on St. Christopher in 1623, he erected a fortified compound not unlike early English colonies elsewhere and termed this settlement "a Plantacon." Commissioning him as governor of the island a few years later, the Earl of Carlisle noted that, by establishing the plantation, Warner had been "directing his Studies and endevours wholy unto Piety and ye publique."[12] Notably, though, other groups of investors and colonists settling on St. Christopher settled their own beachheads on other parts of the island as separate plantations with distinct jurisdictions and landholding patterns.[13] Elsewhere in the Caribbean, English colonial planners experimented with other forms of landholding and land use. On Barbados the initial settlement by Sir William Courteen eschewed private landholding and operated the plantation as a shared commercial enterprise. The Providence Island Company, established in 1630, settled the whole of Providence Island as a single plantation divided into tenancies with colonists laboring in families. In all of these cases the plantation remained a community organizing land distribution.[14]

Gradually, over the next two decades, the idea of plantations under the primary direction of one resident planter emerged. In the mid-1630s the Earl of Carlisle, who secured the proprietorship of most of the English Caribbean islands, began granting land as tenancies to those who would acknowledge his ultimate authority over the archipelago. Many colonists gratefully took the grants in an effort to secure their title to land amid a welter of conflicting authorities. It quickly became evident, though, that they rejected Carlisle's claim to collect rents and regulate commerce in the islands. In that context many colonists chose to begin describing their private lands, which were now ostensibly tenancies from the proprietor, as "plantations"; by the late 1640s new arrivals to Barbados, including Richard Ligon and a group of French refugees, were being advised to purchase a "plantation." English Caribbean settlers retained the term because it still carried public connotations that implied that their effort had vested their property with a legitimacy and authority that rivaled Carlisle's proprietary claim rather than being derived from it. The result was a confusing array of people and authorities across the English Caribbean claiming vastly different sized territories as plantations built upon a hybrid public-private identity, all participating in a broader debate about how land should be used to balance private interests with the public good.[15]

The increasingly precise art of private estate surveying, which was taking root in England during the seventeenth century, was quickly introduced to try to settle questions of property and authority, but surviving evidence suggests that it only served to inscribe some of the contradictions. A uniquely surviving Barbadian land survey from the 1640s, undertaken by John Hapcott for property purchased by Thomas Middleton, depicts a nuanced mixture of public and private spaces on the landscape. Dating from the very start of the consolidation of large-scale sugar production, Hapcott's survey delineates an abstract acreage of land for private conveyance, and it originated at a moment of unprecedented land transactions on the island, as smaller properties were traded and consolidated into larger acreages ideal for sugar cultivation. However, the description states that it represented "the fforme of three hundred acres of Land" *within* "the Ffort Plantation." The name "Ffort Plantation," which dated to the original settlement of the island in 1627, referred to the prior existence of the "Plantacion forte" nearby. St. James Church also stood within the bounds of the land being patented, as did three smaller parcels of land that are designated as belonging to tenant farmers. The spatial designation of "plantation" was therefore clearly tied to these public buildings and institutions, and it implied a community with a complex set of relationships that were underpinned by various forms of private landholding. The plantation remained something more than a mere enclosed private farm.[16]

Despite the huge profits being made from the nascent sugar industry on Barbados and the consequent incentives for wealthy investors to amass land for cane cultivation, the political and ideological forces at play in the English Atlantic during the 1640s and 1650s kept the idea of the plantation as a civic community central to imperial discourse. Radical republican ideas during the era of the Wars of the Three Kingdoms encouraged some leading Parliamentarians to pursue plans to strengthen local civic institutions across England and its possessions. In the colonial sphere this meant doubling down on the idea of the plantation as a self-governing community. In Ireland, following the brutal suppression of Catholic resistance, some English leaders were particularly anxious that new plantations of soldiers and settlers on seized land should be self-governing communities. They encouraged those receiving land to settle in towns, and they sought to attract new colonists to take over leadership of ruined Irish towns. One vocal army veteran, the Baptist Richard Lawrence, envisioned each "English Plantation" as a community of former soldiers who would receive grants of land as a means of incentivizing them to establish a militia to defend the area against Irish counterattacks.[17]

Similar connotations of plantation as a public and communal venture informed aspirations for the English Commonwealth's expansion in the Western Hemisphere. When Barbadian colonist and future Jamaica governor Thomas Modyford proposed establishing new English plantations in Guiana in 1652 he described it as a civic community structured around shared resources and common defense. He suggested transporting former Barbadian servants and settling them "in Townships of at least 50 men in a towne [with] small parcels of land at first assigned them." Modyford described these places as "towneships and plantacons" that would be defensible communities with "liberties" for the white male property-holding citizens who would emerge as patriarchal household heads with responsibility for militia service.[18] As late as the 1650s, then, plantations were still understood across the English Atlantic in ways that reflected the civic roots of the term. Although land was being carved up at pace in the Caribbean, and private investors were discussing buying and selling plantations, they were also envisioning new plantations as colonial beachhead communities and miniature commonwealths that defined American space through the communal construction of social and political authority rather than merely through abstract grants of private property doled out by an imperial bureaucracy.

This definition explains why plantations that appeared on early English Caribbean maps did so in ways that portrayed them as nodes of communal authority rather than as areas of private property. Depictions of Caribbean plantations were most reminiscent of towns, villages, or churches in other contemporary cartography. Richard Ligon's 1657 map of Barbados is an early example of this trend. This cartographic framework was reminiscent of well-known projects to map contemporary England, such as Christopher Saxton and John Norden's county maps in William Camden's *Britannia*. As Richard Helgerson demonstrated, this style of cartography depicting Britain served to bind discrete local communities across the kingdom into a larger political project.[19] When viewed side by side, therefore, the affinity of these maps clearly serves to equate Barbadian plantations not with agricultural acreage but with English communities. The similarity is even more striking in a manuscript map of English plantations in Surinam, dated from 1667; in this map the short-lived English colony is depicted as a series of "plantations" along the Suriname River, but each plantation is denoted by images ranging from one building to a small cluster around a church spire. The implication was clearly that these discrete plantations were substantial nodes of ordered English society rather than merely pieces of property.[20] These depictions represent a stark and revealing contrast to the contemporary trend in the mapping of the plantation landscape in Ulster that began with Josias Bodley's 1609 barony maps. Ulster plantation cartography featured increasingly careful delineation of property boundaries, geographic features, and architectural details, even on the large-scale official maps being sent to Dublin and London.[21] This Irish cartographic trend was in keeping with the Jacobean shift, identified by Kenneth Olwig, toward a new state-centric vision of landscape as a patchwork of resources and property lines intended to inform and also symbolize the state. The maps of Barbados and Surinam reflected a much more bottom-up sense of possession in which the plantation remained an inscrutable locus of power that allowed planters to shape their own landscapes, according to landscape's older definition as communal relations rooted in a particular place.[22]

Private ownership was certainly not entirely absent from early mapping of plantations. Both Ligon's Barbados map and those of Surinam did predominantly label the plantations using personal names such as "Butler" or "Knott." Yet a closer reading of Ligon's map complicates even this element of a privatized reading of the plantation. Ligon's title for his map noted that it contained "the M[aste]rs Names of the Severall plantacons," rather than the names of the plantations themselves. The implication was that the plantation itself had a distinct identity, such as the Ffort Plantation that had appeared on Hapcott's survey, and that Ligon was merely offering a shorthand by naming the "master" of the community. Furthermore, the named individuals were "masters" with broader authority over the community on the plantation rather than ownership of the land. Early maps, then, depict an ambiguous world of land tenure and authority in the English Caribbean, in which plantations could be controlled by private individuals but retained communal connotations.[23]

At the same time, though, competing conceptions about the plantation and its relation to the state were emerging in the English Atlantic in the mid-seventeenth century. Developing in London among the circle of thinkers around German émigré publisher Samuel Hartlib, ideas about "improvement" emphasized that application of natural philosophy could manipulate the natural world to generate new commodities, stimulate trade, and bring humanity closer to divine perfection. Improvement embraced projects to develop new manufacturing processes and even efforts at alchemical transmutation, but many focused on agricultural innovations on "plantations." One of Hartlib's correspondents, the horticulturalist Ralph Austen, was particularly invested in perfecting "plantations" of fruit trees in England, which he believed would combine spiritual perfection with pecuniary reward. The innovations in sugar production on Barbados were of deep interest to the Hartlib Circle because they seemed to perfectly fit their model of an innovative plantation.[24] In many ways the goal of establishing plantations within this worldview was no less public than the vision of plantations as civic communities; the Hartlibians argued that their plans were designed to make England (and more broadly the Protestant world) into a new Canaan, overflowing with abundance.[25] The critical difference, however, was that this form of plantation promised public benefit through individual "improvers" perfecting the relationship between humanity and the natural world rather than focusing on interpersonal relationships. It lent itself to the dehumanization of laboring people and the commodification of land. According to this definition the plantation fulfilled its imperial, and even eschatological, purpose by organizing crops and people with an eye to efficient extraction of resources and perpetual economic growth, rather than an orderly and virtuous civic community.

This alternative definition of plantation had two implications. First, planting employed a particular set of technical skills that people could acquire through study and experimentation, rather than civic virtue. In reality much of the knowledge that underpinned the intensification of sugar and tobacco production came from Indigenous and African sources or from rival empires, but by the 1650s colonists from Barbados, where sugar cultivation had boomed, were being distinguished as uniquely skilled "planters" particularly suited to the establishment of plantations elsewhere.[26] Second, planting strove for material gain that could accrue to the individuals and the state; Austen suggested that "By diligent planting . . . men would be stored with money, & all necessary & vsefull Commodities, & be much more able to beare all burthens, & Taxes."[27] English policymakers increasingly saw colonies as primarily sources of tax revenue, so they recognized the appeal of limiting plantations to innovative commercial units that promised infinite growth.[28] These ideas introduced a new overlapping definition of plantation that still connoted public benefit, but which did so through discrete private plantations that could be bought, together with material objects, draft animals, and the bondspeople who possessed knowledge of crop production. Ligon described a late 1640s plantation for sale in Barbados as "ready furnished, and stocked with Servants, Slaves, Horses, Cattle, Assinigoes, Camels, &c. with Sugar work, and an Ingenio."[29]

Cartography in the English Atlantic was beginning to be reshaped by these ideas. Most significantly, the Down Survey, which was overseen by Samuel Hartlib associate William Petty, represented a new comprehensive effort to map the conquered and sequestered land of Ireland in the 1650s. Petty's survey built upon the already more advanced surveying of Josias Bodley, but it was undertaken on an unprecedented scale and with minute attention to land quantity and quality. The maps that the Down Survey produced differed markedly from those of the mid-seventeenth-century English Caribbean. Carried out on a parish-by-parish basis and at a scale of 1:50,000, they carefully delineated the boundaries of every subdivision of the parish (known as a "townland"), noting its precise acreage and accounting for any land deemed unproductive due to bogs or rocks. They also sometimes included roads and depictions of the key buildings. The objective was to render the land legible as a quantifiable patchwork of resources and to allow for projections about its potential output and optimal level of human occupation—or the ratio of "hands to lands" as Petty called it—that would enable the English state to distribute it to conquering soldiers and investors in ways that would maximize productivity and revenue.[30]

Although not reflected in the maps of the English Caribbean, similar instincts were taking hold there; the increasingly detailed Barbadian land records from the 1650s attest to the fact that colonists were seeking precisely to delineate their property. Surveying skills were in demand in the region and archaeological evidence also points to linear boundaries on the landscape of some islands that may date to the mid-seventeenth century. Furthermore, although Richard Ligon's map of Barbados did not reflect this increasing interest in surveying and property boundaries, his description of the island was crowded with calculations and observations about the natural environment and the sugar production process that reflected new scientific methods and the pursuit of profit through investment in bound and enslaved labor. It was clear that plantation spaces were increasingly being charted and conceptualized in ways that emphasized property, commodity, and profit.[31]

Clearly there was a considerable distance between this invocation of private investment in land and enslaved laborers and the vision of compact settler communities depicted in planters' petitions and cartographic depictions of Barbados and Surinam. But in the mid-seventeenth-century English Atlantic, the language of plantation still bridged that gap, offering all colonial settlements the appearance of potential profit but also giving private agriculture a veneer of public service. Expansion and centralization of the English empire, though, would bring these contradictory definitions into conflict.

The Plantation in Jamaica

The ambiguity behind plantation had critical implications in the late 1650s when, as a result of Cromwell's ill-fated Western Design, English forces landed on Jamaica. Despite the disastrous outcome of the original mission to Hispaniola, the plantation remained at the heart of English plans. Cromwell had originally envisioned establishing plantations in the conquered Spanish Indies and had ordered his men not to pillage in hopes of preserving crucial public infrastructure. The English troops who straggled into Jamaica died at staggering rates from disease and malnutrition, and they faced persistent resistance and counterattacks from the displaced Spanish colonists and their Afro-Spanish allies, all the while hemorrhaging money and resources from the empire and engaging in bitter internal feuds. Because of their malleability, plantations promised to address all of these problems. "Nothing is of more important Considercon," Cromwell's advisors concluded after news of the arrival on Jamaica, "than that the business of planting bee ymediately and effectually applied unto." The voluminous papers about the island's future attest that almost everyone concurred that "planting" was critical to Jamaica's future. Col. William Brayne, one of the army commanders on the island, apparently "talked of nothing but planting."[32] Planting Jamaica seemed obvious because it simultaneously implied the establishment of orderly communities where former soldiers would develop virtue and willingly pursue self-defense, and also the development of profitable industry and financial returns.

The actual process of establishing plantations on Jamaica, though, forced the nascent English empire to confront the contradictions inherent in the idea of the plantation. Cromwell's initial proclamation in the autumn of 1655 ordered the "settling of a civil government" for Jamaica and encouraged individuals to "proceed upon plantations and settlements," suggesting that the plantations he had in mind were communities that would look like the nodes that dotted Ligon's map.[33] A little over a month after arriving in Jamaica, a group of army officers penned an appeal to Cromwell requesting the "allotment and distribution of land to the respective regiments" of the army who were to settle in communities. A few months later, seeking to "animate the soldiers," commanders ordered that every man receive "his particular land," but they issued strict instructions to plant "every regiment in the manner of a township," which was clearly reminiscent of Modyford's earlier plans for English settlement in Guiana.[34] The newly established places were frequently described as "Quarters," implying their continued connection to military encampments, but also either contiguous with a single "plantation" or consisting of multiple "plantations"; these were interspersed with savannas, which had previously been cleared by Spanish colonists for cattle grazing, and which were maintained as commons.[35] Col. Brayne was convinced that this distribution of land had had an effect by 1657, suggesting that most had "industriously imployed themselves to plant" and that the "Spirits of our people are much raised."[36]

The Protectorate state also focused upon attracting coherent communities of planters who would "transplant themselves" in groups. Daniel Gookin was dispatched to Boston to appeal to the "Townes and Plantations" of New England to migrate to Jamaica as "whole Churches," with promises of land, protection, public stocks, and other "Privileges and Immunities" for their discrete communities.[37] Although he did not succeed, another community of planters from Nevis was encouraged to migrate as a group and set up a self-governing plantation on the eastern tip of the island, far removed from the occupying army. After four years of brutal guerrilla conflict, army commanders also negotiated with some of the communities of formerly enslaved Africans on the island, noting that they were "endeavouring to make some plantations" and using that fact to support the idea of granting them a "charter" for self-government. It was clear that plantations were intended to be the building blocks of civil society on Jamaica.[38]

However, given the desperate state of the army that had limped into Jamaica, the practical business of growing food also took on a public valence.

Dispatches from Jamaica noted which officers were making the most progress in "planting," as a barometer of their success in raising food crops. Officers also framed their efforts to apply agricultural skills as part of a public commitment to address the depleted food stocks. One explained that he was using his "best interest to make those under [his] charge to plant, to the end they may be taken off from his highness['s] allowance." Admiral Goodson explicitly combined the expertise and civic models of planting when he appealed to the Protector to particularly send "ingenious publique speritted men of eminent virtue."[39]

Equally, though, some of the leading voices among the English occupiers and in London already had in mind the establishment of plantations predicated upon the commercial cultivation of nonsubsistence agricultural commodities. Some officers—particularly those who had experience of the sugar boom on Barbados over the previous decade—sent reports detailing "plantations of sugar, Cotton, Tobaccoe, Cocao." Col. Francis Barrington, who wrote detailed letters about the commodity output of the plantation under his management, was described as having a "genius much inclined to the way of plantation."[40] The form of plantation that Barrington exemplified required careful attention to the capacity of the land, knowledge of manufacturing techniques, an understanding of marketing exotic staples, and significant investment capital.

On Jamaica, the plantation as a space for perfecting nature also came to take over some of the legitimizing functions that the older discourse of plantation had served for early English expansion. When appropriating land that was owned by Irish or Indigenous inhabitants, or unsettled islands such as Bermuda or Barbados, colonists readily applied the established logic that plantations were justified because they introduced "civility." On Jamaica, though, the English were seeking to displace an established Spanish colony that boasted an urban center and a recognizable European civil society. In justifying this effort anti-Catholicism was obviously critical, but the conceit of introducing a new and more efficient agricultural regime to exploit the true potential of the island also played a role.[41] It would take many years for Jamaica's system of commercial agriculture and African slavery to reach maturity and become the centerpiece of the English Caribbean, but that did not stop colonists and officials from comparing their plantations favorably with the agricultural practices of previous Spanish colonists. John Ogilby's account of the island summarized this position by derisively describing the Spaniards' former *estancias* as "less considerable" plantations.[42]

Critically, though, this definition of the agriculturally efficient plantation also allowed officers to argue that they needed complete control over their men's labor to realize these objectives. Some officers fiercely resisted efforts to allocate ordinary soldiers private plots of land, arguing that "if they must plant, they shal plant only as . . . servants." The officers' efforts to organize the island into private commercial plantations under their control, with the ordinary soldiers laboring for them as opposed to being granted their land within communal plantations, attracted resentment, resistance, and even outright rebellion. Admiral Goodson recounted the story of Col. Richard Holdipp, who was ousted by his men when he attempted to manage them like servants; he noted that Holdipp "cannot be suffered to enjoy his plantation, which with some little encouragement . . . he might have brought to good perfection." Goodson's comments convey the resentment of Holdipp's men, but they also elucidate the attitude that provoked this anger—the assumption that the plantation was a result of Holdipp's single-handed effort to pursue "perfection" and that it was this effort that made the land into his property to "enjoy."[43]

Enslaved laborers quickly became the alternative to reluctant conscript labor. Cromwell's merchant advisors in England advocated incentivizing the shipment of servants and enslaved people that would encourage "planters" from elsewhere in the region to "enter upon Plantations" on the island. On the ground in Jamaica, acting Governor Edward D'Oyley heartily concurred with these conclusions, arguing that their greatest need was "Negroes, which are the life of Plantations." This choice of words spoke powerfully to the way that the distinct definitions of plantation were being melded together. It implicitly acknowledged that plantations were more than just bounded acreages of the newly conquered island. They were bodies that needed to be brought to "life" in order to fulfill their true purpose—a purpose that even former maroons were being credited with engaging with elsewhere on the island. Yet D'Oyley also clearly reflected the view that forced labor was essential to do the punishing and exploitative work of clearing land and mass-producing commercial commodities at low cost. The contradictions and tensions were palpable. The "life" of the plantation as a civic institution in the service of empire would now be generated by consuming Black lives.[44]

None of these plans made much progress in the chaotic final years of the interregnum, and so when Charles II returned to England in 1660, the question of Jamaica remained unresolved, leaving room for the king to articulate his own vision of planting. Desperately in need of sources of revenue, the king eagerly embraced the idea of encouraging commerce by cultivating new commodities. At the same time, though, Charles and his advisors were inherently skeptical of the state of the island. It was, after all, largely populated by the remains of an army that had been recruited by Cromwell. The Crown's response was to reframe planting once again by emphasizing patronage and hierarchical manorialism. Royal advisors, including army officer Thomas Lynch, encouraged Charles to appoint a member of the nobility, Lord Windsor, as governor, granting him 50,000 acres to begin a "plantation" and also giving him permission to endow manors of at least 500 acres on any worthy individual who invested in the island. Lynch also persuaded Charles to reserve for himself a 400,000-acre royal "pryvate plantacon" on Jamaica in order to help make planting "a Mode" at court. The royal "plantation" was to be rented out to communities of tenant farmers, allowing the king to paternalistically fulfill the civic goal of transforming poor English servants into productive tenants. Equally, though, building upon Charles's interest in new scientific innovations and his patronage of the Royal Society, the plantation would allow him to take on the persona of an "improver," sponsoring experiments with new commodities. Practical considerations led the Crown to abandon this plan, but this additional vision of the plantation as a royal estate illuminates yet another approach to reconciling the competing ideas of plantation in the mid-seventeenth century. Jamaica appeared to English colonists and planners to provide a blank slate on which these various visions of plantation could be imposed.[45]

Sorting and Mapping Jamaica

All of the conflict over the nature of plantation is critical to understanding the flurry of Jamaican map-making that followed over the next two decades; colonial leaders and their English allies sought to secure Jamaica's future in the empire by mapping it as a patchwork of plantations that were now defined in ways tailored to the ambitions of the Restoration empire. It is striking that, save for Ligon's map of Barbados, the English Caribbean had scarcely been mapped in any detail before 1660, yet the newly claimed island of Jamaica was depicted in at least four main maps (and numerous reprints) within less than twenty years of conquest. This cartographic explosion represented the efforts of some English colonists, and particularly Governor Thomas Modyford, to fit the plantation within a coherent vision of imperial jurisdictional space. It served to clarify the definition of the plantation as an independent but predominantly agricultural enterprise within a new, larger public geography of empire.

Interest in understanding Jamaica's spatial order developed quickly. Immediately upon his Restoration in 1660, Charles II prioritized Jamaica and specially requested that Gov. Edward D'Oyley provide a detailed survey of "what Quarters and Divisions the sevrall Souldiers, Planters and Inhabitants are now dispersed and settled [in]" as well as "a Register of the several Plantations which are or shall be settled."[46] The reports he received, though, were confused. In many places former garrisons were disbanded and the commanders were simply converted into justices and militia captains. When the colony's first assembly was called in late 1663, the delegates were elected from "precincts" that matched the previous garrison "quarters."[47] The remnants of the distinct plantation established by transplanted colonists from Nevis and the rapidly growing urban hub of Port Royal were each also given separate courts. One of the African communities was recognized by an order of the island's council that was to serve "as a Charter to them." Many of the savannas were still being treated as commons and there were also (probably notionally) four tracts, of 100,000 acres each, that were to form the royal plantation.[48] The initial result of the royal survey, therefore, was to reveal a patchwork of legal and institutional spaces in Restoration Jamaica, including some plantations, with no clear boundaries.

This complex political geography, though, did not endure. Thomas Modyford, who succeeded the short-lived governorship of Lord Windsor, was a Barbadian who had previously described himself as a "Planter Governor." He was charged with establishing "the more orderly and convenient settlement" of the island and with "persuading and encouraging the inhabitants . . . to turne their Labour and Industry to the . . . benefit of ye Plantacon." Before he left Barbados to take up his post he recruited landless men on that island with the promise of fifty-acre tracts on Jamaica, envisioning the establishment of land grants within counties, hundreds, and tithings. But when he arrived in Jamaica he quickly became reconciled to the virtues of large individual plantations. Modyford carefully calculated Jamaica's acreage and how much it might produce were it "well filled with People," demonstrating that he was thinking about land and productivity in the precise quantitative terms that were reminiscent of William Petty's Down Survey and its framing of "hands to lands." In addition, as factor for the newly established Company of Royal Adventurers into Africa, Modyford advocated for a steady supply of enslaved people to be shipped to the island.[49]

Modyford and his allies also looked for a way for the plantation's claims to civic authority to be wedded more firmly to the state. The solution they settled upon was the institution of the parish. The parish was already becoming an important unit of local government in Barbados and a few parishes had been laid out on Jamaica just before Modyford's arrival, but the new structure established under his watch nearly doubled their number so that they carved up almost the entire surface area of the island, rather than emerging organically as the zones around churches.[50] At first glance this decision seems curious, especially given that Modyford also advocated religious toleration.[51] However, the full range of the legislation Modyford and his allies developed clarifies the picture. The act establishing parishes was thin on details about spiritual matters, but it was accompanied by other acts restricting the movement of servants and slaves to within their "master's plantation" and developing infrastructure linking "plantations" to seaports; all of those acts leaned heavily upon the leadership of the parish vestry, which would be composed of leading Anglican conformist planters in the community. Parishes, therefore, provided units of local governance that could secure authority across the island in the hands of select individuals. In contrast to counties, towns, or quarters, though, parishes carried limited connotations of secular power and also deep symbolic ties to the reestablished English ecclesiastical structure that was becoming the marrow of loyalty to the restored Stuart monarch in the 1660s. The result was an effort to project a radically simplified geography of white power on Jamaica, consisting of two sets of spatial units—the "master's plantations," outside which enslaved individuals were not permitted to stray, and the parish whose vestry would enforce those rules.[52]

Modyford was anxious to portray this neat and comprehensive spatial relationship between parishes and plantations in a new map of Jamaica that he had surveyor John Man draw in the mid-1660s (see Figure 5.2). In his letter to the Duke of Albermarle, in which he reported the completion of the map, Modyford emphasized that it laid out "the division of the parishes & the several plantations." This delineation was achieved by the novel imposition of lines demarcating the parish boundaries, most of which lay across seemingly empty space, suggesting that they were state-structured containers just waiting to be filled with preordained plantation units. Modyford did retain hints of the previous definition of plantation, noting that the map conveyed "how all the knowen plantations relate to & are able to succour each other" as if they were interdependent communities capable of altruistic civic action. At the same time, though, the map itself began the work of narrowing the

Figure 5.2. John Ogilby, *Novissima et Accuratissima Jamaicae Descriptio*. 1671. Blathwayt Atlas, Map #35. Reproduction courtesy of the John Carter Brown Library, Providence, R.I.

definition of plantation and bounding it within lines drawn by the imperial state. It included a key of what it described as the "settlements" in each parish with notes on what commodities each location produced. In an important marker of toponymic distinction the list included some places with the descriptor "farme," and these locations were the only sites (save for towns and churches) that were not listed as producing one of the four commercial commodities (cocoa, indigo, sugar, and cotton). The implication was that "plantations," now defined by their market-oriented output, were the critical private subdivisions of each parish.[53]

Man's map of Jamaica did not represent a definitive watershed in the conception of plantation. John Ogilby's description of Jamaica that accompanied the first publication of the map notably did not follow its spatial structure, describing the island as consisting of three towns and seven "plantations" that corresponded to larger communities. Other maps produced in these years also hinted at alternative forms of space. The Vassal and Rogers map, completed in the early 1670s with the explicit intention of depicting the "settlements" of the island, reinscribed a wealth of Spanish toponyms, the locations of "villas," references to communities such as "Seaven Plantations," and to the African maroon communities.[54] Other Caribbean maps also continued to portray plantations in alternative ways.[55] Even Richard Forde's well-known 1675 map of Barbados offered a more complicated picture. It lacked the parish boundaries of the Man map, and it also contained a wealth of other neighborhood names such as "Four Square," "Scotland," and "Thickets" that were suggestive of alternative geographies beyond the plantation.[56]

Over the next half century, though, the neat spatial symmetry initially articulated in Man's map would take hold across large areas of the English empire, as the plantation was boiled down into a profit-maximizing machine reliant upon private property in land and enslaved labor. Numerous scholars have charted the surveying, agricultural, and labor exploitation techniques that facilitated the consolidation of smaller tracts into precisely surveyed 200–300-acre private estates on which dozens of enslaved Africans were confined and forced to produce and process sugar.[57] This consolidation was more than merely a set of technical innovations, though. It involved rationalizing diverse colonial spaces and systems of property ownership, such as proprietary tenancies and commons, into discrete units in which the authority of the planter and of the state were carefully demarcated.[58] The surviving letters of Christopher Jeaffreson, who inherited land on St. Christopher, reveal the rationalization of space that this involved. In one epistle soon after his arrival on the island he noted that he had "never had nor seen any deed" for most of his estate, and he had been forced to use hearsay and local knowledge to reconstruct the complex network of overlapping land claims, exchanges, and tenancies between friends and kin that had constituted his plantation. Jeaffreson, though, was intent on "settling" the plantation. For him, though, unlike his predecessors, this did not mean establishing civic communities of European agriculturalists on his lands, but rather securing official and unencumbered title to the land from the island's imperial officials, purchasing more enslaved laborers, and erecting a sugar works.[59] When English soldiers surveyed the spoils of their recent conquests from the Dutch in Surinam in 1667 they tallied up the "plantations & Sugar works," reflecting the fact that, in the Caribbean, the two had become synonymous. Colonists and imperial officials were adopting the logic of Man's map that reduced plantations to their output.[60]

Conclusion

Plantations retained a public identity that led to their continued appearance on new island-level maps that circulated in the metropolis, but their position was now clearly as a subordinate unit of governance within the empire whose public role was the supply of increasingly important raw materials. The description alongside Richard Forde's map of Barbados encapsulated this fact when it noted that the island was "divided into 11 Parishes . . . [and] subdivided into divers Plantations."[61] Forde captured the fact that plantations had retained their position as jurisdictions, but where they had previously been understood as miniature commonwealths they were now private jurisdictions that functioned as subdivisions of the state. Because they were now populated primarily by enslaved men and women they were no longer seen as potentially self-governing, but rather their "masters" were empowered by colonial law to police their inhabitants. Their bounds were now defined by property lines delineated and protected by the imperial state rather than by the community of the plantation itself acting out of civic virtue. This was a formula calculated to facilitate the exploitation of enslaved laborers for private profit and defer questions about the common good to the larger imperial sphere.

Lord Willoughby, proprietor of the English colony of Surinam in the 1660s, reflected this newly circumscribed assessment of the plantation. When he was confronted by imperial efforts to restrict his authority in Surinam to only the land he had physically settled (rather than the extensive boundaries of his patent), he argued that the move was unjust because his grant was for a "settlement" and it was "of another nature to a Plantation." The implication was that "plantation" now meant only a discrete private acreage.[62] For early English colonists, in the Caribbean and elsewhere, a plantation had been precisely as expansive and public as Willoughby's "settlement." But the debate over plantations in Jamaica had helped to clarify the potential of a simplified new geography of empire encapsulated in Man's map. It was a geography in which the plantation, once the fundamental civic unit of English colonialism, had become a subdivision of a parish—a privatized commercial space that retained jurisdictional authority (particularly over the enslaved women and men who labored there), but for the purpose of profit and within the superstructure of the state.

This bounded and privatized image of the plantation has come to dominate our modern imagination, but the descriptions, maps, and surveys of the early English Caribbean depict plantations as nodes rather than abstract spaces. In doing so, they preserve vestiges and markers of a fundamental debate about the nature of the plantation as a form of colonial place and a technology of colonization. They reveal a story not about how a "plantation model" came to become the dominant economic system of the English Caribbean, but how a particular economic model came to monopolize the definition and the accompanying civic and settler legitimacy of the "plantation." In a scramble for legitimacy, various English settlers staked claims to define and control "plantations" as a way to legitimize their appropriation of resources, position themselves in relation to the English state, and claim particular forms of authority over the bound and enslaved people in their midst.

Notes

1. Richard Ligon, *A topographicall [Description and] Admeasurement [of the yland of] Barbados in t[he West Indyaes] with the Mrs. [Names of the Seuerall plantacons]*, D657 L726t / 1-SIZE, John Carter Brown Library, Providence, RI (hereafter JCB), https://jcb.lunaimaging.com/luna/servlet/detail/JCBMAPS~1~1~1486~101320001:A-topographicall--Description-and--. See also Richard Ligon, *A True and Exact History of the Island of Barbados*, ed. Karen Kupperman (Hackett, 2011), 38–39.

2. Most scholarship on land and settler colonialism in English America still relies upon the picture of enclosure laid out by Joan Thirsk; see Joan Thirsk, *Economic Policy and Projects: The Development of a Consumer Society in Early Modern England* (Oxford University Press, 1978). For the English predisposition toward private property, see Patricia Seed, *Ceremonies of Possession in Europe's Conquest of the New World, 1492–1640* (Cambridge University Press, 1995), chapter 1. For the standard picture of plantation development referenced here, see Richard S. Dunn, *Sugar & Slaves: The Rise of the Planter Class in the English West Indies, 1624–1713* (University of North Carolina Press, 1972); Richard B. Sheridan, *Sugar and Slavery: An Economic History of the British West Indies, 1623–1775* (University of the West Indies Press, 1974); Gary A. Puckrein, *Little England: Plantation Society and Anglo-Barbadian Politics, 1627–1700* (New York University Press, 1984), chapters 2–3; Karen Ordahl Kupperman, *Providence Island 1630–1641: The Other Puritan Colony* (Cambridge University Press, 1993), chapter 5. A more complex picture of English agriculture in this era has emerged; see Andrew McRae, *God Speed the Plough: The Representation of Agrarian England, 1500–1660* (Cambridge University Press, 1996); Paul Slack, *The Invention of Improvement: Information & Material Progress in Seventeenth-Century England* (Oxford University Press, 2015). Allan Greer has recently and compellingly reversed the polarity of this argument by suggesting that regions of early America marked by slavery were precocious in developing ideas about private property in land, and this essay seeks to build upon and complicate the origins of that dynamic. See Allan Greer, "Owning Bodies, Owning Lands: Property Formation in the Early Plantation Colonies," *Historical Reflections/Réflexions Historiques* 50, no. 1 (2024): 22–42. See also Henry Jones, "Property, Territory, and Colonialism: An International Legal History of Enclosure," *Legal Studies* 39, no. 2 (2019): 187–203.

3. Christian J. Koot, *Empire at the Periphery: British Colonists, Anglo-Dutch Trade, and the Development of the British Atlantic, 1621–1713* (New York University Press, 2011); Kristen Block, *Ordinary Lives in the Early Caribbean: Religion, Colonial Competition, and the Politics of Profit* (University of Georgia Press, 2012); Jenny Shaw, *Everyday Life in the Early English Caribbean: Irish, Africans, and the Construction of Difference* (University of Georgia Press, 2013); Christian J. Koot, "Imagining the West Indies: A Seventeenth-Century English Map of Montserrat," *Winterthur Portfolio* 53, no. 1 (2019): 3–39.

4. For the relationship of public and private landholding elsewhere in the Americas, see Allan Greer, *Property and Dispossession: Natives, Empires and Land in Early Modern North America* (Cambridge University Press, 2018).

5. Carla Gardina Pestana, *The English Conquest of Jamaica: Oliver Cromwell's Bid for Empire* (Harvard University Press, 2017). Some recent scholarship has recognized the role of Jamaica in framing political economic debates, but it has largely portrayed planting as one undifferentiated option within this debate; see Abigail Swingen, *Competing Visions of Empire: Labor, Slavery, and the Origins of the British Atlantic Empire* (Yale University Press, 2015); Leslie Theibert, "Making an English Caribbean, 1650–1688" (PhD diss., Yale University, 2013).

6. James VI, *Basilikon Doron Devided into three books* (Edinburgh, 1599), 43. For civic humanism and civility in England, see Markku Peltonen, *Classical Humanism and Republicanism in English Political Thought, 1570–1640* (Cambridge University Press, 2004); Phil Withington, *Society in Early Modern England: The Vernacular Origins of Some Powerful Ideas* (Polity Press, 2010). For the impact on Ireland, see Nicholas Canny, *Making Ireland British, 1580–1650* (Oxford University Press, 2001), 121–34, 187–205; John Patrick Montaño, *The Roots of English Colonialism in Ireland* (Cambridge University Press, 2011); Hiram Morgan, "The Colonial Venture of Sir Thomas Smith in Ulster, 1571–1575," *Historical Journal* 28 (1985): 261–78.

7. John M. Robertson, ed., *The Philosophical Works of Francis Bacon* (New York: Routledge, 2013 [1905]), 776.

8. Robert C. Allen, *Enclosure and the Yeoman* (Oxford University Press, 1992); McRae, *God Speed the Plough.*

9. Thomas Blenerhasset, *A Direction for the Plantation in Ulster* (London: 1610), C1v.

10. Canny, *Making Ireland British*, 121–34, 187–205; Morgan, "The Colonial Venture of Sir Thomas Smith," 261–78 (esp. 274–78); Michael MacCarthy-Morrogh, *The Munster Plantation: English Migration to Southern Ireland, 1583–1641* (Oxford University Press, 1986), chapter 4; Withington, *Society in Early Modern England*, chapter 8.

11. Paul Musselwhite, "Private Plantation: The Political Economy of Land in Early Virginia," in *Virginia 1619: Slavery and Freedom in the Making of English America*, ed. Paul Musselwhite, Peter C. Mancall, and James Horn (University of North Carolina Press, 2019); J. H. Lefroy, *Memorials of the Discovery and Early Settlement of the Bermudas or Somers Islands, 1515–1685*, 2 vols. (Longmans, 1877–79), 1:104, 140.

12. "Warners Commission as Governor of St. Christopher, 29 Sept. 1629," f. 15, MSS Egerton 2395, the British Library (hereafter BL); see also discussions of discrete plantations in Anglo-French negotiations over St. Christopher, in ff. 3–17, MSS Egerton 2395, BL. For published sources on the settlement of St. Christopher, see V. T. Harlow, *Colonising Expeditions to the West Indies and Guiana, 1623–1667* (Hakluyt Society, 1925), 1–24.

13. Harlow, *Colonising Expeditions*, 4–10, 14–17. For Virginia's corporation structure, see Paul Musselwhite, *Urban Dreams, Rural Commonwealth: The Rise of Plantation Society in the Chesapeake* (University of Chicago Press, 2018), chapter 1.

14. Paul Musselwhite, "'Plantation,' the Public Good, and the Rise of Capitalist Agriculture in the Early Seventeenth-Century Caribbean," *Early American Studies* 20, no. 4 (Fall 2022): 597–618; Kupperman, *Providence Island,* chapter 5.

15. Musselwhite, "'Plantation,' the Public Good, and the Rise of Capitalist Agriculture." For the early settlement of Barbados, see James A. Williamson, *The Caribbee Islands Under the Proprietary Patents* (Oxford University Press, 1926). On the Courteen period, see Harlow, *Colonising Expeditions,* 36–42. For details of Carlisle's land grants, see [William Duke], *Some Memoirs of the First Settlement of Barbados* (Barbados, 1741), 51–62.

16. John Hapcott, "Estate Plan of 300 acres of land near Holetown, Barbados" (1646), Shelf Et647 1 Ms, JCB, https://jcb.lunaimaging.com/luna/servlet/detail/JCBMAPS~1~1~1159~106490001:This-plott-representeth-the-forme-o; Douglas V. Armstrong, Karl Watson, and Matthew Reilly, "The 1646 Hapcott Map, Fort (Trent) Plantation, St. James, Barbados: A Significant Resource for Research on Early Colonial Settlement in Barbados," *Journal of the Barbados Museum and Historical Society* 58 (2012): 137–54. For reference to the establishment of "The forte Plantacon" in 1627, see Harlow, *Colonising Expeditions,* 30; N. Darnell Davis, ed., "Papers Relating to the Early History of Barbados," *Timehri: The Journal of the Royal Agricultural and Commercial Society of British Guiana,* New Ser., 5 (1891): 55–58.

17. Richard Lawrence, *The Interest of England in the Irish Transplantation Stated* (London, 1655), 16. For Lawrence's civic vision, see Toby Barnard, *Improving Ireland? Projectors, Prophets and Profiteers, 1641–1786* (Four Courts Press, 2008), chapter 3. For the effort to encourage compact settlement, see "Commissioners to the Council of State, 8 Jan. 1652," in *The Manuscripts of His Grace the Duke of Portland, Preserved at Welbeck Abbey* (H. M. Stationery Office, 1891), 1:624. For focus on reviving town governance, see T. C. Barnard, *Cromwellian Ireland: English Government and Reform in Ireland, 1649–1660,* 2nd ed. (Oxford University Press, 2000), 62–71. For attracting New England migrants to Ireland, see Alison Games, *The Web of Empire: English Cosmopolitans in an Age of Expansion, 1560–1660* (Oxford University Press, 2008), chapter 8. For corporate structures in the interregnum empire, see Paul Musselwhite, *Urban Dreams,* chapter 3.

18. "Considerations Concerning the Settlement of the maine called Guiana to be presented to the committee of trade, 16 Feb. 1652," ff. 118–19, CO 1/11, the National Archives, London (hereafter TNA).

19. For England, see maps in William Camden, *Britannia* (London, 1607); Richard Blome, *Britannia, Or, a Geographical Description of the Kingdoms of England, Scotland, and Ireland* (London, 1673). For Spanish America, see maps in John Ogilby, *America: Being the Latest and Most Accurate Description of the New World* (London, 1671); Richard Helgerson, *Forms of Nationhood: The Elizabethan Writing of England* (University of Chicago Press, 1992).

20. For Surinam, see Unknown, "Surinam and Commewijne Rivers," Cabinet Blathwayt 38, JCB, https://jcb.lunaimaging.com/luna/servlet/detail/JCBMAPS~1~1~1614~102110002:-Surinam-and-Commewijne-rivers.

21. Annaleigh Margey, "Representing Plantation Landscapes: The Mapping of Ulster c. 1560–1640," in *Plantation Ireland: Settlement and Material Culture, c. 1550–c. 1700,* ed. James Lyttleton and Colin Rynne (Four Courts Press, 2009), 140–64.

22. Kenneth Robert Olwig, *Landscape, Nature, and the Body Politic: From Britain's Renaissance to America's New World* (University of Wisconsin Press, 2002), chapter 2.

23. Ligon, *A topographicall [Description.* The mixture of new private property with a communal project was also present in the contemporary development of "plantations," in England.

See Elly Robson, "Fen Plantation: Commons, Calvinism, and the Boundaries of Belonging in Early Modern England," *Journal of British Studies* 63, no. 1 (2024): 30–62.

24. R. A. Austen, *A Treatise of Fruit Trees* (London, 1653). On Austen, see James Grantham Turner, "Ralph Austen, 1612–1676," *Oxford Dictionary of National Biography Online.* On improvement, see Slack, *Invention of Improvement*, chapter 4; Paul Warde, "The Idea of Improvement, c. 1520–1700," in *Custom, Improvement and the Landscape of Early Modern Britain*, ed. Richard W. Hoyle (Ashgate, 2011). On the Hartlib Circle, see Mark Greengrass, Michael Leslie, and Timothy Raylor, eds., *Samuel Hartlib and Universal Reformation: Studies in Intellectual Communication* (Cambridge University Press, 1995). For connections to empire, see Richard Drayton, *Nature's Government: Science, Imperial Britain, and the "Improvement" of the World* (Yale University Press, 2000); Sarah Irving, *Natural Science and the Origins of the British Empire* (Pickering & Chatto, 2008). For Hartlibian interest in the Barbadian sugar industry, see Eric Otremba, "Enlightened Institutions: Science, Plantations, and Slavery in the English Atlantic, 1626–1700" (PhD diss., University of Minnesota, 2012), 39–51; John E. Crowley, "Sugar Machines: Picturing Industrialized Slavery," *American Historical Review* 121, no. 2 (2016): 403–36; David Chan Smith, "Useful Knowledge, Improvement, and the Logic of Capital in Richard Ligon's *True and Exact History of Barbados*," *Journal of the History of Ideas* 78, no. 4 (2017): 549–70.

25. Koji Yamamoto, *Taming Capitalism Before Its Triumph: Public Service, Distrust, and "Projecting" in Early Modern England* (Oxford University Press, 2018), chapter 3.

26. "William Goodson and Robert Sedgewick to the officers in Jamaica, Jan 4 1655/56," *A Collection of the State Papers of John Thurloe*, 7 vols. (London, 1742) (hereafter *Thurloe*), 4:389–90. On Indigenous and African knowledge in agricultural innovation, see Eric Otremba, "Inventing Ingenios: Experimental Philosophy and the Secret Sugar-Makers of the Seventeenth-Century Atlantic," *History and Technology* 28, no. 2 (2012): 119–47; Michael J. Jarvis, "Bermuda and the Beginnings of Black Anglo-America" in *Virginia 1619: Slavery and Freedom in the Making of English America* (University of North Carolina Press, 2019). On efforts to translate "plantation" knowledge, see Michael D. Bennett, "Caribbean Plantation Economies as Colonial Models: The Case of the English East India Company and St. Helena in the Late Seventeenth Century," *Atlantic Studies* 20, no. 4 (2023): 508–39.

27. "Ralph Austen to Samuel Hartlib, 10 Feb. 1654," 41/1/66B, in *The Hartlib Papers* (The Digital Humanities Institute, University of Sheffield, 2013), http://www.dhi.ac.uk/hartlib (hereafter *HP*); "Ephemerides 1653 Part 4, Hartlib, 1653," *HP* 28/2/72B; "Ralph Austen to Samuel Hartlib, undated," *HP* 41/1/145A.

28. For commercial empire, see Brenner, *Merchants and Revolution: Commercial Change, Political Conflict, and London's Overseas Traders, 1550–1653* (Princeton University Press, 1993), chapter 12; L. H. Roper, *Advancing Empire: English Interests and Overseas Expansion, 1613–1688* (Cambridge University Press, 2017). For ideas about state unity and empire, see David Armitage, "The Cromwellian Protectorate and the Languages of Empire," *Historical Journal* 35, no. 3 (1992): 531–55; Alexander B. Haskell, *For God, King, and People: Forging Commonwealth Bonds in Renaissance Virginia* (University of North Carolina Press, 2017), chapters 4–5; Andrew Fitzmaurice, "The Commercial Ideology of Colonization in Jacobean England: Robert Johnson, Giovanni Botero, and the Pursuit of Greatness," *William and Mary Quarterly* 64, no. 4 (2007): 791–820; Thomas Leng, "Commercial Conflict and Regulation in the Discourse of Trade in Seventeenth-Century England," *Historical Journal* 48, no. 4 (2005): 933–54.

29. Ligon, *True and Exact History*, 66.

30. Ted McCormick, *William Petty and the Ambitions of Political Arithmetic* (Oxford University Press, 2009), 183 (quote); William Petty, *History of the Survey of Ireland*, ed. Thomas Aiskew Larcom (Dublin, 1851); Vincent Gookin, *The Great Case of Transplantation in Ireland Discussed* (London, 1655); Barnard, *Cromwellian Ireland*, chapter 8; William J. Smyth, *Map-making, Landscapes and Memory: A Geography of Colonial and Early Modern Ireland, c. 1530–1750* (University of Notre Dame Press, 2006), chapter 5 (esp. 181–82). For a digital collection of the Down Survey maps correlated, see Trinity College Dublin's Down Survey of Ireland project: http://downsurvey.tcd.ie/index.html.

31. For disputes over landownership, see Puckrein, *Little England*, chapter 2. For surveying and property lines, see B. W. Higman, *Jamaica Surveyed: Plantation Maps and Plans of the Eighteenth and Nineteenth Centuries* (Institute of Jamaica Publications, 1988); R. H. Leech, "'In what manner did they devide the land': The Early Colonial Estate Landscape of Nevis and St Kitts," in *Estate Landscapes: Design, Improvement and Power in the Post-Medieval Landscape*, ed. J. Finch & K. Giles (Boydell Press, 2008), 191–204.

32. "Report Concerning Affairs in America," f. 123, MSS Egerton 2395, BL; "Edward D'Oyley to Thomas Povey, July 12 1658," f. 169, MSS Egerton 2395, BL.

33. *Interesting Tracts Relating to the Island of Jamaica* (St. Jago de la Vega, Jam., 1800), 1–2.

34. *Interesting Tracts*, 61; "William Goodson and Robert Sedgewick to the officers in Jamaica, Jan 4 1655/56," *Thurloe* 4:389–90. Modyford's plan for Guiana had actually played a role in encouraging Cromwell to pursue the Western Design.

35. Various descriptions of the geographic organization include: "A brief description of Jamaica," ff. 608–12, MSS Egerton 2395, BL; "The Scittuacon of the Island," ff. 3–5, CO1/14/57, TNA; "Thomas Modyford's Description of Jamaica," ff. 1–9, Add. MS 11410, BL; "Jamaica census 1661," CO1/15/98, TNA. For savannas as commons at this stage, see "Sedgwicke to Thurloe—5 Nov 1655," *Thurloe* 4:151–53. On the relationship of spatial distribution to previous Spanish settlement, see Pestana, *English Conquest*, 130–31; James A. Delle, *The Colonial Caribbean: Landscapes of Power in the Plantation System* (Cambridge University Press, 2014).

36. "William Brayne to Thomas Povey and Martin Noell, July 1657," ff. 129–30, MSS Egerton 2395, BL.

37. "Instructions to the Commissioners, 1656," *Thurloe* 4:634; William Noël Sainsbury et al., *Calendar of State Papers, Colonial Series* (London, 1860–1926), 9:100–101; Daniel Gookin, *To all persons whom these may concern, in the several townes, and plantations of the United Colonies in New-England* (Boston, 1656); "Brayne to Cromwell, Jan 9 1656/7," *Thurloe* 5:770–71; "Gookin to Thurloe, 21 Jan 1655/6," *Thurloe* 4:440; "Gookin to Thurloe, May 10 1656," *Thurloe* 5:6–7. On the parallels between this effort and the encouragement of puritan migration to Ireland, see Games, *Web of Empire*, 268–69.

38. "Stokes to Sedgewick, May 22 1656," *Thurloe* 5:48; "Luke Stokes to Cromwell, May 29 1656," in *Thurloe* 5:66–67; "Instructions for those going to Jamaica," *Thurloe* 4:635–6; "Instructions to Commanders in Jamaica, 1656," *Thurloe* 4:634–35. For the African settlements as "plantations," see "The Scittuacon of the Island," 5.

39. "Col. Barrington to Thurloe, 1 July 1657," *Thurloe* 6:376–77; "Goodson to Thurloe, 25 June 1656," *Thurloe* 5:151–53.

40. "The Scittuacon of the Island," 28; "D'Oyley to Cromwell, Sept 12 1657," *Thurloe* 6:512. For an example of Barrington's letters, see "Francis Barrington to Thurloe, July 1 1657," *Thurloe* 6: 376–77.

41. On religion and the Western Design, see Pestana, *English Conquest*, chapter 1; Block, *Ordinary Lives*.

42. "Col. Barrington to Thurloe, 1 July 1657," *Thurloe* 6:376–77. For previous Spanish settlements as "less considerable" plantations see Ogilby, *America*, 343. For the halting maturation of plantation capitalism on Jamaica, see Trevor Burnard, *Planters, Merchants, and Slaves: Plantation Societies in British America, 1650–1820* (University of Chicago Press, 2015), 49–52, 66–71. For conflicted attitudes toward prior Spanish settlement, see James Robertson, "'Stories' and 'Histories' in Late-Seventeenth-Century Jamaica," in *Jamaica in Slavery and Freedom: History, Heritage, and Culture*, ed. Kathleen E. A. Monteith and Glen Richards (University of the West Indies Press, 2002), 25–51.

43. "Goodson to Thurloe, June 25 1656," *Thurloe* 5:151–53.

44. "Col. Edward D'Oyley's Description of Jamaica," ff. 16–25 (quote, 19), Add. MS 11410, BL. For the slave trade, see Donohue, *Fire Under the Ashes: An Atlantic History of the English Revolution* (University of Chicago Press, 2013), chapter 6; Swingen, *Competing Visions*, chapter 2.

45. "Considerations about the Government of Jamaica," ff. 283–86, MSS Egerton 2395, BL; "Instructions to Thomas, Lord Windsor, Governor of Our Island of Jamaica, 21 March 1661/2," ff. 15–16, CO138/1, TNA. See also "Considerations about the Peopling and Settling of Jamaica," CO1/14/54, TNA; "Considerations" (quote, f. 120v), CO1/14/54, TNA. For a more public focused version of the plan see "Proposal Concerning Jamaica," CO1/14/56, TNA. For the way in which the royal plantation fit with Restoration ideas about gardening and political order, see Kate Mulry, *An Empire Transformed: Remolding Bodies and Landscapes in the Restoration Atlantic* (NYU Press, 2021).

46. "The King's proclamation for the encouraging of planters in Jamaica. Whitehall, 14 Dec. 1661," #94, CO 1/15, TNA; "Instructions to Governor D'Oyley," ff. 6–8, CO138/1, TNA; "Instructions to Lord Windsor," ff. 13–20, CO138/1, TNA.

47. Alexander Aikman, *Journals of the Assembly of Jamaica* (Kingston: 1811–1829), 1: 1; Agnes Mary Whitson, *The Constitutional Development of Jamaica, 1660–1729* (University of Manchester Press, 1929), 20–23; Pestana, *English Conquest*, 210–12.

48. Council Minutes, 11 Dec. 1662, f. 21r, CO139/1, TNA. For orders, see ff. 6–8, CO138/1, TNA. For separate jurisdiction at Port Morant, see *Calendar of State Papers, Colonial Series*, 5:#215. For savannas, see Edmund Hickeringill, *Jamaica Viewed . . .* (London, 1661), 16.

49. Jamaica Entry Book, ff. 23, 29, CO138/1, TNA; "Thomas Modyford to the Committee for Foreign Plantations, 10 May 1664," ff. 135–38 (quote, 136v), CO1/18, TNA; "Considerations Touching Jamaica, 21 July 1664," #94, CO1/8, TNA. It is possible that Modyford initially proposed a plan for Jamaica in January 1664 that more directly reproduced (almost verbatim) his earlier township proposal for Guiana (#2, CO1/18, TNA). For "Planter Governor" see A. P. Thornton, *West-India Policy Under the Restoration* (Oxford University Press, 1956), 25. On the role of the new Africa Company, see Swingen, *Competing Visions*, chapter 3.

50. Lord Windsor had taken clergymen with him when he traveled to the island, and orders had been given for their support; see *Calendar of State Papers, Colonial Series* 5: #264; Council Minutes, 20 Apr. 1663, f. 22r, CO139/1, TNA. The General Assembly that met after Windsor's departure established a small number of parishes; see ff. 39v–40r, CO139/1, TNA. Parishes had been established in Barbados in the midst of disputes around landownership in the 1640s; see Puckrein, *Little England*, chapter 2; Larry Gragg, *Englishmen Transplanted: The English Colonization of Barbados 1627–1660* (Oxford University Press, 2003), chapter 4.

51. "Thomas Modyford to the Committee for Foreign Plantations, 10 May 1664," ff. 135–38, CO1/18, TNA.

52. For laws passed by Modyford's assembly, see ff. 55r–81v, CO139/1, TNA. For the 1664 Jamaica slave code and its relationship to the evolution of slave law in the English Atlantic, see Edward B. Rugemer, "The Development of Mastery and Race in the Comprehensive Slave Codes of the Greater Caribbean During the Seventeenth Century," *William and Mary Quarterly* 70, no. 3 (2013): 429–58. For a comparative view of local government and slavery in Carolina, see Ryan A. Quintana, *Making a Slave State: Political Development in Early South Carolina* (University of North Carolina Press, 2018). On the Restoration episcopate see Grant Tapsell, "A British Patriarchy? Ecclesiastical Imperialism Under the Later Stuarts," in *The Nature of the English Revolution Revisited: Essays in Honour of John Morrill*, ed. Stephen Taylor and Grant Tapsell (Boydell, 2013), 261–84. For their political thought, see Mark Goldie, "John Locke and Anglican Royalism," *Political Studies* 31, no. 1 (1983): 61–85.

53. John Man, *Novissima et Accuratissima Jamaicae Descriptio* (1671), Cabinet Blathwayt, 35, JCB; "Thomas Modyford to the Duke of Albemarle, 28 December 1667," #100, CO1/21, TNA.

54. Ogilby, *America*, 342–43; *Tabula Iamaicae Insulae* (1678), Cabinet Blathwayt, 35, JCB. On the memory of Spanish locations, see Robertson, "'Stories' and 'Histories.'"

55. See, in particular, Koot, "Imagining the West Indies"; "Map of the River Oyapok, from the Journal of Jesse de Forrest" reproduced in Harlow, *Colonising Expeditions*, 150.

56. Richard Ford, *A New Map of the Island of Barbadoes* (1675), Cabinet Blathwayt, 35, JCB. The lack of parish boundaries may be a reflection of Ford's Quakerism and consequent aversion to formal parochial jurisdiction.

57. Higman, *Jamaica Surveyed*; Dunn, *Sugar and Slaves*; Burnard, *Merchants, Planters, and Slaves*, chapter 2.

58. Allan Greer has recently emphasized the way in which sugar producers in both the English and French empires were at the forefront of the shift to absolute forms of private landownership; see Greer, "Owning Bodies, Owning Lands."

59. John Cordy Jeaffreson, *A Young Squire of the Seventeenth Century*, 2 vols. (Hurst & Blackett, 1878), 1:198, 199.

60. Harlow, *Colonising Expeditions*, 252.

61. Ford, *A New Map of the Island of Barbadoes.*

62. Harlow, *Colonising Expeditions*, 179.

CHAPTER 6

When a River Is a Border

Rivalries and Commercial Networks in the Riverine West

KIM M. GRUENWALD

In the eighteenth century the Mississippi and Ohio Rivers rose in late spring thanks to melting snow and again in the fall thanks to seasonal rain. As an army officer, Thomas Hutchins made a detailed study of the western waters and sent a report to Thomas Jefferson in February 1784. Hutchins wrote that the Wabash and Illinois Rivers rose in early February, and the Ohio in March. The Missouri also rose in March, and the Upper Mississippi soon thereafter. The flood of the Ohio came on suddenly or gradually depending on the weather, and lasted six to eight weeks. As for the Mississippi, high water lasted on the upper river even longer than that, and the waters could rise on the lower stretches from Natchez to New Orleans from January through June. The Mississippi could flow south at a rate of five miles an hour in the spring but slowed to two miles an hour in the summer and early fall. Because boats encountered sandbanks, islands, shoals, and downed trees year-round, the faster sections could actually be more dangerous to navigation in the spring. Though the French had sent boats with goods upriver to Illinois from New Orleans once a year in the past, it was a journey that took so long and so much manpower at the oars that almost all people and cargo went downstream only. People walked home one way or another—either up the Natchez Trace to settlements south of the Ohio, or by sailing from New Orleans to ports on the Atlantic Coast and then heading west.

The popular image of North American colonies growing in isolation from Europe has long been debunked by scholars of the Atlantic world, and now a new generation have tied the history of the Ohio and Mississippi Rivers to that story. Paul Mapp focused on the interaction between empires, studying the interior of North America and adding the Pacific Ocean to the mix while exploring what Europeans misunderstood about the continent and how it affected their calculations during the decades leading up to the Seven Years' War. By the end of the conflict, France had given up on their plans, while England's grew. Spain focused on using the lower Mississippi to defend other holdings, and Native Americans lost a crucial ally they had used to make demands of the British. François Furstenberg took up the story of the trans-Appalachian West as Atlantic history for the second half of the eighteenth century. In studying "the Long War for the West," Furstenberg sifted through the various viewpoints and strategies of the English, French, Spanish, and Native Americans as all sought to control the region between 1754 and 1815. He recognized New Orleans as the focal point of many plans because of its location as the only outlet for trade beyond the mountains at that time. Furstenberg showed that far from being an isolated frontier, the Ohio and Mississippi Valleys were a key stage upon which the "sweeping forces of imperialism and global warfare" played out.[1] When historians talk about the Long Eighteenth Century and the Long War for the West, they are placing war and imperial intrigue at the center of the narrative. But what of the people who actually made use of the western rivers during those years? For merchants, rivers were more than borders—they were arteries of trade. Their letter books, journals, ledgers, and other business records allow us to track the movement and networks of people and goods that would integrate east and west along waterways between Pittsburgh and New Orleans.[2]

The rivers themselves are at the center of the story. During the second half of the eighteenth century, the Ohio and Mississippi Valleys were borderlands—"the contested boundaries between colonial domains."[3] Each empire had its own concerns—the French wanted to use portages from the Great Lakes to the Mississippi to connect Atlantic trade from the St. Lawrence to the Gulf of Mexico, while the Spanish focused their attention on protecting their assets to the south.[4] Initially the British wanted to capture the fur trade by creating their own alliances with Native Americans, while Native Americans wanted to trade with whoever gave them the best prices without encroaching on their territory. The opposite banks of the Ohio and Mississippi Rivers were often claimed by rivals during this period. After the British exiled the French from the Ohio Valley, the Shawnees wanted to protect their hunting grounds south of the Ohio River. The Spanish claimed territory west of the

Mississippi River, while the English claimed control of the eastern bank. Many of the French remained, making the best of it under either Spanish or English rule. Europeans, colonists, and Native Americans eyed each other warily and traded with each other even while trying to press their advantage.

Nobody could travel faster than the current of a river, and as they made their way downstream, they were out in the open for all to see. Trees lined the banks, making river corridors the clearest path to destinations. It was an intimate setting with face-to-face meetings in boats, as well as stops on shore to exchange news or a meal when a settlement came into view. When people descended rivers, they couldn't take a detour to avoid others as they might on the ocean or when they crossed the plains. The rivers were the artery of travel for everyone west of the Appalachian Mountains and avoiding fellow travelers was impossible.

The Atlantic world of colonial America had encompassed merchants extending their networks to the fall line of rivers that reached the coast from the Lower South to New England. Philadelphia merchants wanted to harness the downstream power of the Ohio and Mississippi Rivers and allow the Atlantic world to make landfall by extending that world of commerce across the Appalachian Mountains. Their work began with attempts by firms like Baynton, Wharton & Morgan and others before the American Revolution to corner the fur trade from one direction in Illinois and provide flour to a hungry port from another by way of the Gulf of Mexico. After the war, firms like Reed & Forde forged riverine networks all the way to New Orleans, sending goods downriver from Pittsburgh to a wave of settlers heading west. They completed the integration of east and west by sending oceangoing vessels to the southern port to export western produce to the Atlantic world in addition to selling goods. In the process, Philadelphia merchants took part in extinguishing Native American land titles, expanding slavery up the Mississippi River, and claiming a continental empire for the new United States. The purchase of Louisiana in 1803 meant that the Mississippi was a border no longer. One empire at last controlled both banks of the river, and Americans would expand to the Pacific via overland routes thereafter.

* * *

When a young man named George Morgan apprenticed with two well-respected merchants named John Baynton and Thomas Wharton, Philadelphia merchants did most of their business with England, southern Europe,

Madeira, and the West Indies, as well as participating in a lively trade with other ports along the coast. Then war broke out, and everything changed. The Seven Years' War brought a flood of bills of exchange and a host of new trading opportunities including supplying the army and acquiring land in the Illinois country. By war's end, the three men had established the firm of Baynton, Wharton & Morgan. As British colonists, they and other merchants transformed the port of Philadelphia, expanding its hinterland west and seeking land grants from the king beyond the Appalachian Mountains.[5] The youngest partner found himself dispatched to Illinois, and there he developed a contentious relationship with French merchants and rivals from his own colony, as well as English officials. Despite the fact that the venture failed, Morgan would find himself entangled in the fate of the western waters to the end of his days.

Travel accounts from early commercial ventures on behalf of Baynton, Wharton & Morgan reveal the international riverine world of western rivers as well as travelers' awareness of Atlantic rivalries that spread to the interior. John Jennings made the journey on the Ohio from Fort Pitt to Fort Chartres in the spring of 1766 with the first five boats loaded with the firm's goods. He kept a daily log of distances traveled, weather, and a description of what they saw on shore and the people they encountered. Other than one encounter that worried his party early on, they had a remarkably peaceful journey and met a wide variety of the region's inhabitants, especially Native Americans. The boat crews took precautions, watching carefully for signs of trouble, but their fears never materialized. A party of Shawnees in four canoes traded fresh meat with them in exchange for biscuit and tobacco near the Scioto River. Jennings's party traded for meat in a similar fashion with another group as they headed down the Ohio. Jennings mentioned small groups of what he labeled Indian cabins, the cabins of warriors, hunters' cabins, and large Indian villages. Below the Falls of the Ohio they heard gunfire, and Jennings assumed Indians were hunting nearby. Heading up the Mississippi at last, when Jennings encountered "a large French boat," his party put up their flag and rowed to shore to meet them. Nearly a month later, Jennings neared his destination in Illinois where he chauvinistically characterized the French he saw as "very bad" farmers. That summer George Morgan followed, and in letters to his wife, he echoed many of Jennings's observations, tending to label everyone he encountered as either French, English, or Indian.[6]

As for descending the Mississippi River, at first Jennings belittled French farmers. As he traveled further south, both he and Morgan later described them as rivals producing wheat and utilizing slave labor to grow tobacco and other crops. When Jennings headed for the Gulf he encountered French hunting parties and more unidentified Native Americans in canoes. At Natchez he wrote of the presence of a new Spanish fort and an Acadian settlement beyond. The French settlement of Pointe Coupée impressed him, and he described a fort and church on the Spanish side of the river opposite. The local economy supported exports of indigo, rice, and maize.[7] George Morgan also kept a journal of a trip from Kaskaskia to the Gulf of Mexico down the Mississippi River. He wrote of a Spanish meadow and made note of "French hunting parties" on both the English and Spanish shores. Morgan and his party ate buffalo meat on the trip, sometimes provided by others they met, and other times procured by a member of their own party. Once past Natchez and in the final stretch of his journey south, George Morgan encountered many more Frenchmen and Native Americans. He stated that there were a couple of Native American settlements and a "considerable French settlement with Fort & Garrison" on the Red River. There were more settlements along the Mississippi, some with "new French huts," and Morgan saw fifty slaves at Pointe Coupée harvesting cypress trees to be floated to New Orleans the next spring. Several nearby planters owned three dozen slaves, more owned twenty, and newcomers owned a handful each. They grew cotton, rice, indigo, and tobacco. Other words used in reference to the French include villages, huts, and families. They were rivals for the territory as much as the fur trade.[8]

As for doing business on the western waters, the official Spanish position on the Mississippi was fairly rigid, especially once defined by the new governor in New Orleans as he tried to wrest trade away from the British and integrate the lower river more fully into the Spanish empire. Native Americans and French inhabitants were not interested in trading with the Spanish, who could not provide the manufactured goods that came from England and France. In 1769, the Spanish sent Alejandro O'Reilly, who had been born in Ireland but had risen through the ranks of the Spanish Army, to New Orleans to establish their rule. When O'Reilly arrived, he wrote, "I found the English entirely in possession of the commerce of the colony." They traded with the German settlements and sold furs to France in exchange for goods. When it came to selling goods for specie, the English had the advantage because they sold their goods more cheaply than anyone else.[9] In September O'Reilly issued a proclamation that outlawed the practice of meeting with or hosting any foreigners without passports, or providing transport or guides. Inhabitants were not allowed to buy merchandise from anyone on the Mississippi or any of the lakes without formal authorization from O'Reilly, and anyone

Figure 6.1. Traveling south from Natchez, Morgan passed by Pointe Coupée and Baton Rouge, and then reported on crossing into French (actually Spanish) territory at the Iberville River. The Iberville connected the Mississippi to Lake Maurepas, with smaller rivers providing the boundary between that point, Lake Pontchartrain, and finally, the coast. The British hoped to use that route to Natchez from the Gulf of Mexico, bypassing the delta and New Orleans, though those plans never came to pass. Lieut. John Ross, *Course of the River Mississipi, from the Balise to Fort Chartres*. London: Rbt. Sayter, 1775. David Rumsey Map Collection, David Rumsey Map Center, Stanford Libraries.

violating that order would be fined with one-third of the money going to anyone who informed on such persons.[10]

Despite O'Reilly's public proclamation, New Orleans needed flour, and O'Reilly would turn to another Irishman named Oliver Pollock to provide it. With the backing of Philadelphia merchants, Pollock took consignments of flour, rum, and lumber to the West Indies and sent back spices, molasses, and sugar. Over the course of five years, he plowed his profits into more ships and learned Spanish fluently enough to trade without a translator. While in Cuba, Pollock struck up a friendship with O'Reilly at his post in Havana before following him to New Orleans. O'Reilly allowed him to bring in flour at special rates in the late 1760s and early 1770s, and Pollock worked with the firm of Willing & Morris back in Philadelphia to fulfill the port's needs. Historian Kathleen DuVal writes that Pollock managed to be English and Spanish simultaneously through the "Spanish trade privileges" he leveraged by way of personal relationships. Years later, Pollock characterized his early business as resting on the foundation of London dry goods, African slaves, and Philadelphia flour. O'Reilly's successor opened up the lower Mississippi to even more English trade to keep the colony fed, and the flour trade continued through the end of the century.[11]

Borders were fluid in Illinois, as well, with Spanish territory on the opposite bank of the Mississippi River. Some French farmers had crossed over, unhappy that the British wanted to pay them with notes on New York or London that they couldn't redeem. One knowledgeable observer wrote that the only "means of inducing many of those who have removed to the Spanish side, to return to our's with their Cattle &tc." was to offer them cash for their crops and meat. Others would follow if Morgan and others failed to "break their connextions with the people of New Orleans." A few years later under even stricter British military rules, George Morgan noted that "no person whatever officer merchant or inhabitant can cross the river from either side without a written passport," and said that the "poor inhabitants" lived in "a state of slavery." But sometimes those who lived in the west found ways to use the boundaries to their advantage. Morgan wrote that after the death of a friend outside his cabin one night, one of the murderers "escapd over the river."[12]

John Jennings stored that first shipment of goods mostly at Kaskaskia in the spring of 1766, with a small portion sent to Fort Chartres. When George Morgan arrived a few months later he opened a store at Cahokia and one in Vincennes two years after that before finally settling down to run the business at Kaskaskia. Morgan intended to trade English goods to Native Americans for furs and dry goods and groceries to his French neighbors in exchange for cattle and wheat. Selling provisions such as meat and flour to soldiers brought in reasonable profits, as did sales to the Royal Department of Indian Affairs.[13] Atlantic rivalries meant forts and soldiers, and Philadelphia merchants found them to be a ready market.

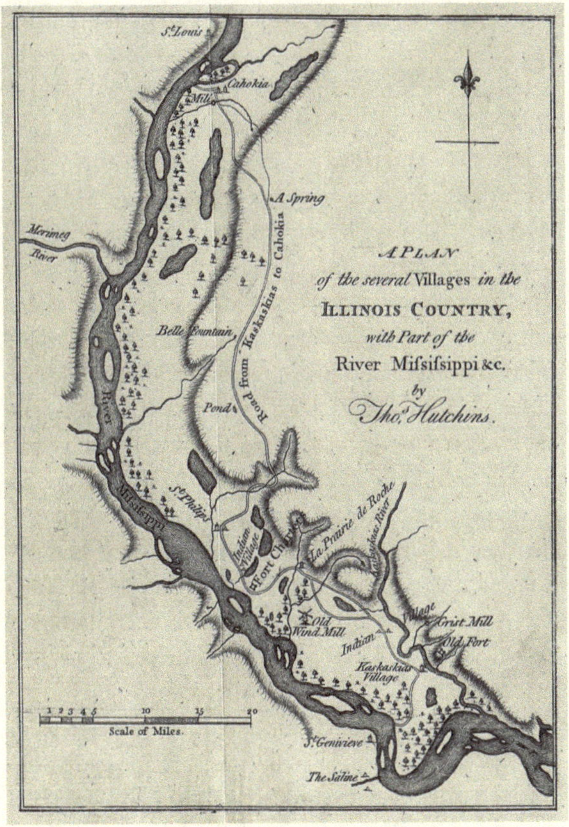

Figure 6.2. Though it was officially Spanish territory west of the Mississippi, those of French descent retained control of most of the trade there. Thomas Hutchins, *A Plan of the several Villages in the Illinois Country*, in Hutchins, *A Topographical Description Of Virginia, Pennsylvania, Maryland, And North Carolina*. London, 1778. David Rumsey Map Collection, David Rumsey Map Center, Stanford Libraries.

Atlantic rivalries also emerged in the interior in the form of competition over the purchase of forced migrants. George Morgan sold slaves to the French in Illinois. An associate named James Rumsey purchased more than three dozen Africans for the firm in Jamaica and brought them down the Ohio River from Fort Pitt to Illinois in 1767. In December Morgan wrote to his partners that Rumsey would arrive just as a group of Frenchmen planned to leave for New Orleans in search of labor. Morgan offered them slaves in exchange for pelts, specie, bills of exchange, or produce to be paid in May, and they stayed to make deals with him instead. Morgan intended to sell slaves for 400 dollars or more, and wrote that "we have it in our power to make a monopoly of all the flour in the country by purchasing it with negroes." But it soon became apparent that there were more slave-traders in the market. In February, Morgan wrote to his partners that sixty slaves were expected from New Orleans in the near future, apparently undercutting his sales.[14]

The fur trade also turned out to be a disappointment when Morgan consistently lost out to French traders, partly because British regulations stipulated that he could not go to the Native American tribes in search of furs—they had to come to him. But historian Jacob Lee concludes that George Morgan's failure in Illinois went beyond British regulations. Morgan and his French neighbors had actually united against the first English officer in charge, but that relationship deteriorated over time. Accompanied by a Seneca named Silver Heels with connections back east, Morgan found himself excluded from Native American kinship networks in Illinois. Illinois peoples had been marrying French traders and planters and each other across tribal lines for decades, and because they had retained control over their lands, they had resisted Pontiac's call for rebellion in 1763. They didn't like the English much, finding them arrogant, and there was no way for Morgan to break into Illinois society with a wife and other ties back in Philadelphia. He just didn't fit in.[15]

When it came to trading with French inhabitants, Morgan was at a disadvantage there, too, for the observations noted above. The French preferred to trade with New Orleans as they always had—Morgan didn't seem to have anything special to offer in the end. In New Orleans they could get cheaper goods and use their produce to pay for them. Thus, business did not go as planned. Baynton, Wharton & Morgan had heavily invested in goods, but large enough quantities of furs had not yet arrived from the west to pay their debts. In the fall of 1767, they threw up their hands and invited their creditors to look at their books; the creditors decided that the situation was not as bad as they feared. Eight of their creditors came together as a Board of Trustees to supervise their business over the next three years. All expected great profits to come from the Illinois venture.[16] Instead, Philadelphia merchants' economic affairs fluctuated for several more years. In 1768 as tensions on the seaboard rose, the British moved their troops east, abandoning military contracts, and British merchants did deals with the French, cutting out colonial merchants and keeping the fur trade out of American hands. The British decommissioned a number of forts and sold Fort Pitt to Alexander Ross in 1772.[17]

Though Baynton, Wharton & Morgan failed, George Morgan continued to increase his knowledge of western waters. On his very first trip down the Ohio, he struck up a friendship with Ensign Thomas Hutchins, who would eventually be named the first (and only) Geographer of the United States in 1781 and go on to leave his mark on land surveys in the northwest thereafter.

Morgan and Hutchins both recognized that understanding the courses of the western rivers would be the key to settling the interior. In the early 1770s, while Hutchins worked on a map of western rivers, George Morgan wrote to a third party seeking information about the upper Mississippi: "Engineer Hutchins a particular friend of mine is about publishing a map of the interior rivers and lakes in American." Morgan sent a proposal from Hutchins to Wharton in London hoping to find a publisher. Hutchins had been surveying all the Ohio's tributaries with an eye toward settlement. Morgan wrote to Hutchins that "great dependence is placed pointing out the best water communication between the west branch of the Susquehanna & the Allegheny River & the Potomack & the Yahagany or Cheat River." Morgan wrote Hutchins that he had been promised a map from a Father Harding that was "a very exact draft of the River Mississippi from Fort Chartres to its sources." He also wrote about a map of the Monongahela and Little Kanawha, and later, about adding other rivers.[18]

During the American Revolution George Morgan served as Indian agent at Pittsburgh. Afterwards, still convinced that he could make a fortune in the west, he got together a group of investors to petition Congress to sell them two million acres of land in the Illinois Country. His continuing interest in the west would bring him into contact with two men Spain charged with the task of protecting the empire's interests on the Mississippi a year earlier: the Spanish minister to the United States, Don Diego de Gardoqui, and the governor of Louisiana, Esteban Miró. They closed the river to American navigation, but let it be known that American settlers would be welcome on the west bank if they pledged loyalty to Spain. They promised free trade, land grants, and religious freedom to Americans willing to settle in their territory, and a number of Americans applied. In the summer of 1788, Gardoqui found someone, perhaps their mutual friend Thomas Hutchins, to tell Morgan to apply to the king to establish a colony on the banks of the Mississippi. Morgan did so, backing out of his deal with Congress.[19]

Before heading down the Ohio, Morgan sent a report to Gardoqui in November. He warned that the western population was growing, and the people resented taxes from an American government that couldn't guarantee them access to markets. Spain needed to act fast, Morgan insisted, for the Ohio Valley had all the raw materials needed to manufacture their own goods if they failed to gain access to New Orleans. On the journey downriver, Morgan noted that Kentuckians were on the fence about allying with Spain and were interested in a British spy's propaganda currently making the rounds about the legality of seizing New Orleans. Morgan rated freedom of religion and free trade to be the most important factors that would dictate the west's loyalty in the end.[20]

George Morgan promised to bring many families with him and take an oath of allegiance to the King of Spain if his demands were met. Morgan wanted to be in control of the colony—appointing officials, distributing land, and heading two militia units. He wanted a rank in the Spanish Army equivalent to his rank in the United States, one thousand acres of land for his wife and each of his children, a salary, and word of the king's protection over his family. His land grant was to be opposite the mouth of the Ohio, and he insisted that the king promise free trade, support for schools, and freedom of religion. The colony would provide a port of entry for Americans who wanted to trade with the Spanish in New Orleans, he argued. Give him all that, he proposed to the king, and Spain would control trade west of the Allegheny Mountains. Alas, various intrigues scuttled those plans, and Morgan abandoned the colony of New Madrid after only a year. But the venture showed that George Morgan both looked backward, turning to the tried-and-true method of petitioning a king for land, and forward, articulating the role that the rivers would play in the years to come. Merchants and settlers alike wanted access to markets, and only the western waters could provide it. Morgan belonged to a group of Atlantic world merchants willing to give their loyalty to either a British or a Spanish king, choosing whoever would serve their purpose at the time.[21]

* * *

When thirteen of the British colonies declared independence, Philadelphia became a national port, and then for a time the new nation's capital. As citizens rather than subjects, a new generation of merchants sought to connect east and west along the Ohio and Mississippi Rivers and corner the market on land warrants issued by states in an attempt to raise money to pay war debts. As they looked beyond the Appalachians, newly independent Americans believed that seizing land from Native Americans and selling it to settlers would kill two birds with one stone: First, it would provide funds to pay the new nation's debts and invest in the future, and second, it would provide land to help citizens in the new republic start off right in their quest to create "kingless self-government" based on a landed citizenry.[22] Thomas Doerflinger characterized the merchants of revolutionary Philadelphia as men driven by "a vigorous spirit of enterprise." They took risks and developed innovative business methods and networks to get ahead. In addition to serving in the army, John Reed got his start as a trader during the American Revolution. An army official out to make a profit from his position helped Reed to plan a venture that included taking a load of goods to Boston by public wagon and exchanging them for saddles and medicine, among other things, making contact with various quartermasters along the way. Reed later spent time in the West Indies and tried his hand at opening a store in Wilmington, Delaware—he was still trying to collect on those accounts years later. Reed took Standish Forde as a partner after the war, and the two men imported dry goods from England and speculated in western lands, hoping to sell both quickly at a profit. Instead, the postwar economic depression nearly did them in, and by the late 1780s, Reed & Forde began casting their net wider in an effort to pay their debts.[23] Reed & Forde's business would look quite different from that of Baynton, Wharton & Morgan as Americans flooded west.

Like many others, including some of the wealthiest and most influential men on the eastern seaboard, John Reed and Standish Forde speculated in hundreds of thousands of acres of land in western Virginia, Pennsylvania, and Kentucky. Initial surveys indicated that their lands off a branch of the Monongahela contained iron ore, and they sent out feelers trying to find an experienced man to establish an ironworks, hoping to supply "all the settlements on the western waters." Whenever Reed and Forde wrote of land, they emphasized easy river access to markets. Once they boasted that farmers would be able to drive cattle to the headwaters of the Potomac, as well as shipping flax, hemp, and tobacco by the same route. Other lands lay closer to Philadelphia or Baltimore trade routes, and settlers passing through on their way to Kentucky were said to provide another ready market.[24] The partners purchased a flour mill on Ten Mile Creek, a branch of the Monongahela River, near Redstone, nearly 40 miles south of the Ohio River's headwaters at Pittsburgh to process the settlers' wheat into flour and ship it south to New Orleans. After the American Revolution, Pittsburgh grew as men of commerce used their prime location to outfit settlers coming from the south and the east, heading for new homes in the Ohio Valley.[25] As for Philadelphia goods destined for western markets, warehouses were needed to store goods when the Ohio ran low in the summer and winter, awaiting the river's rise in the spring and fall.

The partners had debts to pay and land to sell, so they continued expanding their network to New Orleans, providing markets for settlers by linking the headwaters of the Ohio to the Gulf of Mexico and beyond. Reed and Forde used their partnership to take a very hands-on approach to business.

Figure 6.3. Red Stone lies directly south of Fort Pitt on the Monongahela River; roads are indicated by parallel lines from Fort Pitt and other settlements. Thomas Hutchins, *The Western Parts of Virginia, Pennsylvania, Maryland, and North Carolina*. London, 1778. David Rumsey Map Collection, David Rumsey Map Center, Stanford Libraries.

In the fall, one or the other—they usually took turns—made their way across the mountains from Philadelphia to tend to business in Pittsburgh and its hinterland, and then moved on to Kentucky before heading down the Mississippi to Natchez and New Orleans. Forde made the journey in late 1788 or early 1789, Reed went west in the fall of 1789, and then Forde a year later. The partner who stayed behind sent letters with updates about prices in Philadelphia, Natchez, New Orleans, and Europe with any acquaintance who happened to be heading across the mountains, hoping to advise the traveler on the best way to invest their cash, goods, or flour, or perhaps the best way to collect their debts with news of notes that had been protested. During Reed's trip over the winter of 1789–90, intelligence did not flow back east quickly, which is evident because Forde sent letter after letter asking for news so that he could arrange insurance for the trip.[26]

Trade within the Ohio Valley itself had grown since Morgan's sojourn in Illinois. Forde instructed Reed to visit Lancaster, Hagerstown, Clarksburg, and Morgantown to collect debts and seek out goods, then to take charge of the flour at their own mill and visit others along the Monongahela so that they could send as large a cargo as possible south. He mentioned a "clever fellow" he'd met the year before who could give his partner advice on which mills were the most trustworthy. Reed was to purchase oil at Redstone to go with the kegs of paint the partners intended to send downriver and to purchase butter and whiskey, as well as skins and furs. Forde wrote with news of a tinman heading Reed's way, and asked him to find tin in Pittsburgh for the artisan and take payment in cups that would sell well in Natchez.[27]

Personal networks were all well and good, but the river always had its say in the end—Forde's journal a year later makes that clear. During most of November 1790 he traveled from town to town, conducting business along the various rivers and streams of the Upper Ohio Valley. He spent a couple of days waiting for wagons with goods to arrive from the east at Redstone on the Monongahela, then loaded the goods and flour onto boats and headed for the Ohio's headwaters at Pittsburgh. Before reaching Wheeling, Forde and his men had to unload flour, drag the boats forward, then reload as they tried to move downstream. The river's course proved to be deeper thereafter, and they made good time to Louisville at the Falls of the Ohio.[28] The falls consisted of a two-mile stretch of rapids nearly 600 miles below Pittsburgh and represented the one major navigational hazard on the river's journey to the Mississippi. George Rogers Clark settled there on the southern bank with others during the Revolution and attacked Native Americans to the north. Just as the growth of a commercial center was inevitable at the Ohio's headwaters, so too was the growth of such a town at the point where goods would either have to be moved by portage around the falls from one set of boats above to another below, or be stored, waiting for the river to rise enough for boats to make it over the rocky narrows.[29] Once over the falls, Forde purchased flatboats and a Kentucky boat at Shippingport below, but bad weather forced them to stop on shore as they floated toward the Mississippi. They broke through ice on the lower reaches of the Ohio, and then had to contend with low water and sandbars on the Mississippi. They arrived at New Madrid about seven weeks after leaving Pittsburgh.[30]

The partners didn't intend for all of the flour shipped from the Ohio's headwaters to reach the mouth of the Mississippi. From Natchez in January 1792, John Reed directed William Shannon, one of their agents following behind him with the cargo, to sell a barrel here and there when he could, if there was a profit to be found. Reed wrote him to be on the lookout for any chance to trade flour for indigo, too, which was bringing in good prices in Europe. A man couldn't be too long en route, however, for the flour that arrived down south first brought the best prices.[31] Because everyone had to ship their cargo during the time of high water, gluts in the market and falling prices in New Orleans were common.

European rivalries continued to require attention as Forde headed for the Gulf Coast. Once Reed & Forde's goods reached the lower Mississippi, a new array of commercial tasks awaited the partners. The British had taken possession of West Florida and Natchez, Mississippi, after the Seven Years' War, and the settlement grew during the Revolution as a refuge for Loyalists. Settlers from the region "provided a lasting larder for New Orleans." Flour was in high demand throughout Louisiana, and in Cuba as well. By the early 1780s, Natchez was the largest town in the region north of New Orleans with a population of 1,000 free men and 300 slaves. Residents sent regular shipments of indigo and cotton, as well as corn and tobacco, south to the Gulf of Mexico. Newcomers from the United States brought large herds of cattle in the 1780s. The Spanish encouraged further growth for a couple years by guaranteeing a market for American tobacco in 1788. The Natchez District became best known for cotton production a decade later, after the invention of the cotton gin.[32]

A man named David Ferguson ran their store there. In the summer of 1790, he wrote that the hot weather might lead to spoiled flour. Flour prices were high, but there was little cash to be had, and most sales had to be done by credit. Ferguson also bartered flour for furs. Prospects were gloomy, however, and he advised against another shipment of goods. Planters had had their tobacco refused the past spring, and Ferguson used caution when extending credit. The next year, Ferguson proved much more optimistic—cash sales had been brisk, he wrote, and he expected to sell a big tobacco crop. A Mr. Cochran seemed to be their main competition, and Ferguson told the partners that their rival's selection of goods was "much better calculated for this place." He expected to have $700 or $800 on hand in cash for Reed when he arrived, and there would be a big crop coming in from contracts he'd made. That fall Ferguson reported that plenty of men from Kentucky were content to sell their tobacco for goods in Natchez rather than making the trip all the way to New Orleans. The tobacco crop was so good that year that many used it in order to settle their debts with the firm. In 1792, Ferguson wrote to Reed downriver that "the sales have been much better this spring than they have been since your commencement of business here." Unfortunately, Reed's reply from New Orleans the next week contained the bad news that the tobacco inspections had not gone well, which would curtail payments.

The river's navigational challenges weren't the only setbacks merchants faced when doing business on the western waters.[33]

In addition to their river ventures, Reed & Forde was one of the first firms to send oceangoing vessels regularly to the port of New Orleans starting in 1789. They used well-connected men in Philadelphia to obtain passports allowing them to trade in New Orleans when the Spanish closed the port to American trade. Scholars have weighed in on early New Orleans often: Daniel Usner labeled the lower Mississippi Valley a frontier exchange economy, François Furstenberg called it a hot spot, John Clark called it a pawn, and Arthur Whitaker wrote of its significance nearly a century ago—it was clearly the key to getting goods to market from the interior.[34] But holding New Orleans wasn't easy because both the French and Spanish found that the soil around the city just wouldn't support food crops—they needed flour from Illinois or Pennsylvania, so exceptions to strict trade policies were made time and again just to keep those who lived there fed. The industries that did grow up along the Mississippi relied on slaves for labor. Reed & Forde first sent cigars, tea, and sugar, and took payment early on in the form of beaver, otter, bear, and deer skins. A couple years later the payments took the form of tobacco and commercial notes. Tobacco came through New Orleans from both Kentucky and Natchez. Natchez plantations also supplied indigo. In 1791, five of the nine ships that arrived in New Orleans from Philadelphia came with "cargoes that were consigned wholly or in large part to Reed and Forde." The partners hoped to get better prices for their western cargoes that year in France rather than England, and they began shipping flour and cheese to New Orleans.[35]

The firm they did the most business with was Coxe & Clark, who also provided the most direct link between the ports. Daniel W. Coxe brought his good Philadelphia name to the partnership and Daniel W. Clark brought his facility for languages and working the angles in New Orleans—he had gotten his start there under Spanish rule years before. In March 1793, Daniel W. Coxe and Daniel Clark Jr. wrote to Reed & Forde that they had formed a new firm, hoping that together they would be able to facilitate more business for "our Friends." They believed that trade in New Orleans would grow, especially if war broke out between England and France, and Spain would be drawn in. They mentioned trade with the Spanish Empire specifically, stating that the government "will doubtless continue to wink at the mode in which the business is carried out." It would be Daniel Clark's connections that got the job done. Clark wrote personally to the Philadelphia partners a couple weeks later that competition with Coxe had been harming both of their businesses, and he expected to have even more trade for the firm now that Coxe had become his partner.[36]

Unfortunately for Reed & Forde, they failed rather spectacularly shortly after the turn of the century: Land sales stagnated, they lost brigs to privateers in the 1790s, new Spanish policies forced them to close their store in Natchez, the Whiskey Rebellion disrupted trade in Philadelphia at a crucial juncture when "the shoemakers with whom we have contracted have all been on campaign against the Western Insurgents," and they failed to fulfill a contract in New Orleans.[37] Others followed in their footsteps, however, as rivals from not only Philadelphia, but Baltimore and New York, as well, flooded into New Orleans after Pinckney's Treaty in 1795 in anticipation of opening up trade on the Mississippi River. Spain relinquished New Orleans back to France in 1800, and President Thomas Jefferson engineered the purchase of Louisiana in 1803 to ensure western farmers would retain access to Atlantic trade. The purchase doubled the size of the new nation and extended its claims west to the Rocky Mountains. Rivals claimed opposite banks of the western rivers no longer.

Philadelphia merchants like Reed & Forde made it seem possible to claim the trans-Mississippi West for the United States by creating strong ties between east and west through commercial ventures along the Ohio and Mississippi Rivers. They supported both industry and trade rather than simply selling land and recording deeds as fast as they could. Indeed, one partner or the other was often at hand when there was business to be done in the west—they used their partnership to best advantage with an almost constant flow of information across the mountains. John Reed and Standish Forde completed the circuit from Philadelphia to New Orleans along the rivers and from New Orleans back to Philadelphia by way of the Atlantic Ocean. Though Philadelphia lost its advantage to New York in the port of New Orleans itself after the signing of Pinckney's Treaty, it would be Philadelphia merchants who continued to supply residents of the Ohio and Mississippi Valleys with eastern goods by shipping them overland to Pittsburgh and then downriver before steamboats arrived to take goods *upriver* from New Orleans after the War of 1812.[38]

* * *

By chance, the year 1806 would be a momentous one for George Morgan, John Reed, and Standish Forde. Though Morgan had sought the approval of the King of Spain just over a decade and a half before, when Aaron Burr stopped by his home in western Pennsylvania on his way west to pursue plans to separate it from the east, George Morgan acted the only way the loyal citizen of a republic could: He informed his president of the danger. Cleaving the continent in two had seemed possible in the years immediately following the Revolution, but such schemes were bound to fail after the Louisiana Purchase. That same year Forde signed a document ending his partnership with Reed, probably to protect assets when he fell ill. Writing that they would start anew if he recovered, Forde noted that they had been in business for twenty-five years. Unfortunately, he died two days later.[39] Though not quite a full generation apart in age, George Morgan, John Reed, and Standish Forde pursued their businesses in different eras: one as the merchant of a colonial port, and the other pair as merchants of a national port. What they had in common was their pursuit of integrating the eastern half of the continent into the trade world of the Atlantic seaboard along western waters not fully under the control of any one empire.

The year 1806 also saw the return of the Lewis and Clark expedition. After their half-century of relying on the Ohio and Mississippi Rivers, perhaps it only seemed natural for Americans to head up another river—the Missouri—to explore the Louisiana Purchase and beyond. Though Lewis and Clark failed to find a Northwest Passage, they succeeded in focusing the new nation's attention on the trans-Mississippi West, ushering in a new century of settlement by overland journeys and railroads. Philadelphia merchants laid the groundwork for those continental ambitions by laying claim to the trans-Appalachian West through the creation of commercial networks that extended from Pittsburgh to New Orleans along the Ohio and Mississippi Rivers during the second half of the eighteenth century. As for the western rivers, in the nineteenth century, Americans would erase the centuries-old Native American routes and portages with canals and land reclamation projects from the shores of Lake Michigan to the Gulf Coast, trying to assert dominance by destroying the landscape.[40]

What shall we call the critical half-century between the eras of ocean-going ships and the overland journeys west from the Mississippi? United States historiography recognizes the North, the South, the mid-Atlantic, the Midwest, the West, and the borderlands. Previous scholars of "the frontier in the formative years" simply called it "the trans-Appalachian West."[41] It's interesting that the history of the trans-Mississippi West doesn't have much to do with the Mississippi River, and the most salient geographical features of the trans-Appalachian West are in fact the Ohio and Mississippi rivers rather than the Appalachian Mountains. Of course, early historiography traced frontier history from east to west, until borderlands historians came along to shift the focus to the movement of peoples who started from the southwest. But the Riverine West deserves a category of its own. It's all about moving downstream along the most efficient route for goods to travel and having to share that journey with peoples of all nations, including Native American confederacies, because rivals claimed opposite banks.

Notes

Adapted from Gruenwald, Kim M. *Philadelphia Merchants on Western Waters: Commerce and Empire in the Riverine West, 1750–1803*. © 2025 Johns Hopkins University Press. Reprinted by permission.

1. Paul W. Mapp, *The Elusive West and the Contest for Empire, 1713–1763* (University of North Carolina Pres, 2011); François Furstenberg, "The Significance of the Trans-Appalachian Frontier in Atlantic History," *American Historical Review* 113, no. 3 (June 2008): 647–77.

2. See for example, Elizabeth Urban Alexander, "Daniel Clark: Merchant Prince of New Orleans," in *Nexus of Empire: Negotiating Loyalty and Identity in the Revolutionary Borderlands, 1760s–1820s*, ed. Gene Allen Smith and Sylvia L. Hilton (University Press of Florida, 2010), 242–43; James Alton James, *Oliver Pollock: The Life and Times of an Unknown Patriot* (D. Appleton-Century, 1937), 1–3; Kathleen DuVal, *Independence Lost: Lives on the Edge of the American Revolution* (Random House, 2015), 35–38.

3. Jeremy Adelman and Stephen Aron, "From Borderlands to Borders: Empires, Nation-States, and the Peoples in Between in North American History," *American Historical Review* 104, no. 3 (1999): 814–41, esp. 816. See also David E. Narrett, *Adventurism and Empire: The Struggle for Mastery in the Louisiana-Florida Borderlands, 1762–1803* (University of North Carolina Press, 2015); Matthew Salafia, *Slavery's Borderland: Freedom and Bondage Along the Ohio River* (University of Pennsylvania Press, 2013).

4. For French plans, see John William Nelson, *Muddy Ground: Native Peoples, Chicago's Portage, and the Transformation of a Continent* (University of North Carolina Press, 2023).

5. Max Savelle, *George Morgan, Colony Builder* (Columbia University Press, 1932), chapter 1; Harry D. Berg, "Economic Consequences of the French and Indian War for Philadelphia Merchants," *Pennsylvania History* 13, no. 3 (July 1946), 185–93; Jacob M. Price, "Economic Function and the Growth of American Port Towns in the Eighteenth Century," *Perspectives in American History* 8 (1974): 123–86; Mark Abbott Stern, *David Franks: Colonial Merchant* (Pennsylvania State University Press, 2010).

6. John Jennings, "Journal from Fort Pitt to Fort Chartres in the Illinois Country," *Pennsylvania Magazine of History and Biography* 31, no. 2 (1907), 145–56; Peter Marshall, "Lord Hillsborough, Samuel Wharton and the Ohio Grant, 1769–1775," *English Historical Review* 80, no. 317 (October 1965): 717–39; Savelle, *George Morgan*; George Morgan letters in *The New Regime, 1765–1767*, ed. Clarence Walworth Alvord and Clarence Edward Carter (Illinois State Historical Library, 1916), 311–16.

7. John Jennings, "John Jennings' Journal at Fort Chartres, and Trip to New Orleans, 1768," *Pennsylvania Magazine of History and Biography* 31, no. 3 (1907): 304–10; Sylvia L. Hilton, "Spanish Louisiana in Atlantic Contexts," in *Louisiana: Crossroads of the Atlantic World*, ed. Cecile Vidal (University of Pennsylvania Press, 2014), 69–74.

8. Alvord and Carter, eds., *The New Regime*, 438–48.

9. Margaret Fisher Dalrymple, *The Merchant of Manchac: The Letterbooks of John Fitzpatrick, 1768–1790* (Louisiana State University Press, 1978), 9; Hilton, "Spanish Louisiana," 76–77.

10. O'Reilly Proclamation, 1769, New Orleans, http://cdm16313.contentdm.oclc.org/cdm/ref/collection/LPC/id/281, accessed December 20, 2016.

11. James, *Oliver Pollock*, 1–21, 53–56; DuVal, *Independence Lost*, 35–43.

12. George Croghan to General Gage, 12 January 1767, in Alvord and Carter, *New Regime*, 479–81; Memo 9 May, 1770, George Morgan Letterbook, Baynton, Wharton, and Morgan Papers, Pennsylvania State Library, Harrisburg (hereafter BWMP).

13. Savelle, *George Morgan*, 31–36.

14. Clarence Walworth Alvord and Clarence Edwin Carter, eds., *Trade and Politics, 1767–1769* (Illinois State Historical Library, 1921), 126, 128, 135, 140, 161–62, 181, 225, 331, 348.

15. Jacob Lee, *Masters of the Middle Waters: Indian Nations and Colonial Ambitions Along the Mississippi* (Harvard University Press, 2019), 148–50; Robert Michael Morrissey, *Empire by Collaboration: Indians, Colonists, and Governments in Colonial Illinois Country* (University of Pennsylvania Press, 2015) 206–211; Savelle, *George Morgan*, 39–46.

16. Savelle, *George Morgan*, 46–7.

17. Walter S. Dunn Jr., *Opening New Markets: The British Army and the Old Northwest* (Praeger, 2002), 132–36.

18. Morgan to Hutchins, 19 December 1770; Morgan to Wharton, 21 December 1771; Morgan to Hutchins, 18 February 1772; Morgan to Hutchins, 27 February 1772, George Morgan Letterbook, BWMP.

19. Savelle, *George Morgan*, 176–228.

20. Ibid., 209–210; Bruce Tyler, "The Mississippi River Trade, 1784–1788," *Louisiana History* 12, no. 3 (Summer 1971): 255–67; Paul A. Gilje, *Free Trade and Sailors' Rights in the War of 1812* (Cambridge: Cambridge University Press, 2013), 131.

21. Savelle, *George Morgan*, 205–6. For an excellent account of Wilkinson and Miró squaring off against Gardoqui and Morgan, see Narrett, *Adventurism and Empire*, 169–86. Narrett labels Morgan, Pollock, and others in Louisiana as adventurers: anyone who risked money in a commercial venture, and who used intrigue, rumor, and deception in the borderlands where imperial control was weak, shifting allegiance at will. Ibid., 1–7.

22. Michael A. Blaakman, *Speculation Nation: Land Mania in the Revolutionary American Republic* (University of Pennsylvania Press, 2023), 7–19, 15 (quote).

23. Thomas Doerflinger, *A Vigorous Spirit of Enterprise: Merchants and Economic Development in Revolutionary Philadelphia* (W. W. Norton, 1986), 157–64, 216, 234, 295; Price, "Economic Function and the Growth of American Port Towns in the Eighteenth Century," 123–86; Arthur P. Whitaker, "Reed and Forde, Merchant Adventurers of Philadelphia: Their Trade with Spanish New Orleans," *Pennsylvania Magazine of History and Biography* 61, no. 3 (July 1937): 237–62.

24. Redstone was the site of an old fort at a good spot to ford the Monongahela River—the site is now Brownsville, Pennsylvania. Reed & Forde to Ephraim Grant, 3 June 1787, Letterbook;

Reed & Forde to Charles Steward, 3 June 1787, Letterbook; Reed & Forde to James Groome, 8 September 1788, Letterbook; Reed & Forde to Joseph Thompson, 14 September 1788, Letterbook; Reed and Forde Papers, 1759–1823, Historical Society of Pennsylvania, Philadelphia, Pennsylvania (hereafter referred to as R&F).

25. Richard C. Wade, *The Urban Frontier: Pioneer Life in Early Pittsburgh, Cincinnati, Lexington, Louisville, and St. Louis* (University of Chicago Press, 1964), 7–13, 44–45. For the mill, see letters between Reed & Forde and Caleb Copeland in Letterbook and Correspondence, various dates between November 1788 and February 1792, R&F.

26. John Reed to Standish Forde, 11 June 1789, 6 July 1789, Letterbook; Standish Forde to John Reed, undated November 1789, 25 November 1789, 7 December 1789, 10 December 1789, 7 January 1790; Standish Forde to John Reed, 28 October 1791, 5 December 1791, Correspondence, R&F; Standish Forde to John Reed, 7 June 1793, 19 June 1793, Letterbook, R&F.

27. Standish Forde to John Reed, undated November 1789, 25 November 1789, Letterbook, R&F.

28. John W. Jordan, "Notes of a Journey to New Madrid, Tennessee, 1790," *Pennsylvania Magazine of History and Biography* 36, no. 2 (1912): 209–16, esp. 209–12.

29. Wade, *The Urban Frontier*, 13–18, 64–66; Wallace Carson, "Transportation and Traffic on the Ohio and Mississippi Before the Steamboat," *Mississippi Valley Historical Review* 7, no. 1 (June 1920): 26–38, esp. 29, 34n32.

30. Jordan, "Notes of a Journey to New Madrid," 212–6; Reginald Horsman, *The Frontier in the Formative Years, 1783–1815* (Holt, Rinehart and Winston, 1970), 64; Ralph Lee Woodward Jr., "Spanish Commercial Policy in Louisiana, 1763–1803," *Louisiana History* 44, no. 2 (Spring 2003): 133–64, esp. 151–52. One of their partners in a venture to New Orleans, John Lewis, had similar troubles with cold weather and ice the year before: see John Lewis to Reed & Forde, 8 February 1789, Correspondence, R&F.

31. John Reed to William Shannon, 22 January 1792; David Ferguson to Reed & Forde, 5 August 1791, Correspondence, R&F. For more on trade for provisions from Pittsburgh along the Ohio, see Randolph C. Downes, "Trade in Frontier Ohio," *Mississippi Valley Historical Review* 16, no. 4 (1930): 467–94, esp. 473.

32. Woodward, "Spanish Commercial Policy," 145–46 (quote), 151–53; Susan Gaunt Stearns, *Empire of Commerce: The Closing of the Mississippi and the Opening of Atlantic Trade* (University of Virginia Press, 2024), 136–52; Horsman, *The Frontier in the Formative Years*, 56–58.

33. David Ferguson to Reed & Forde, 10 June 1790, 4 July 1790, 5 August 1791 (first quote), 20 November 1791, 28 March 1792 (second quote), 3 April 1792, Correspondence, R&F.

34. Whitaker, "Reed and Forde," 245–58; John G. Clark, *New Orleans, 1718–1812: An Economic History* (Louisiana State University Press, 1970); Arthur Preston Whitaker, *The Mississippi Question, 1796–1803: A Study in Trade, Politics, and Diplomacy* (D. Appleton-Century, 1934) and *The Spanish-American Frontier, 1783–1795: The Westward Movement and the Spanish Retreat in the Mississippi Valley* (Houghton Mifflin Company, 1927); Daniel H. Usner, *Indians, Settlers, and Slaves in a Frontier Exchange Economy: The Lower Mississippi Valley Before 1783* (University of North Carolina Press, 1992).

35. Whitaker, "Reed and Forde," 251 (quote); Day Book, 1785–1791, R&F; Daniel Hodge to Reed & Forde, 29 May, 11 June, 7 August, and 10 September 1789; James Jones to Reed & Forde, 5 October 1789; Thomas Dunsford to Reed & Forde, 19 July 1790; Standish Forde to John Reed, 5 December 1791, all in Correspondence, R&F; Ship Records, Ulali, R&F; Linda K.

Salvucci, "Atlantic Intersections: Early American Commerce and the Rise of the Spanish West Indies (Cuba)," *Business History Review* 79, no. 4 (Winter 2005), 781–809.

36. Alexander, "Daniel Clark," 241–45; Coxe & Clark to Reed & Forde, 6 March 1793, and Daniel Clark, Jr. to Reed & Forde, 16 March 1793, R&F.

37. Whitaker, "Reed and Forde," 255–60 (Reed & Forde quote 255).

38. Whitaker, *The Mississippi Question*, 142, and Kim M. Gruenwald, *River of Enterprise: The Commercial Origins of Regional Identity in the Ohio Valley, 1790–1850* (Indiana University Press, 2002).

39. Savelle, *George Morgan*, 234–35; Standish Forde Memo, April 26, 1806, box 10, R&F.

40. Christopher Morris, *The Big Muddy: An Environmental History of the Mississippi and Its Peoples from Hernando de Soto to Hurricane Katrina* (Oxford University Press, 2012); and Nelson, *Muddy Ground*.

41. Horsman, *Frontier in the Formative Years*; Malcolm J. Rohrbough, *The Trans-Appalachian Frontier: Peoples, Societies, and Institutions, 1775–1850* (Oxford University Press, 1978).

CHAPTER 7

Military Lines

How the Introduction of Contours Affected Maps and Movement

KAREN RANN

For nineteenth-century military men, the map was a powerful tool with which to plan campaigns. With a good map they could thwart the movements of enemies and enable the transportation of their own troops and provisions. Where a campaign met with success, land could be relabeled, apportioned, and appropriated with boundary lines redrawn. Traditionally, making maps required movement through terrain regardless of borders or boundary lines; an ideal had been for cadets to learn to both traverse and visualize land as would a hunter, before committing his vision to paper. A poor map could result in lost ground, defeat, and death. During the nineteenth century, a new technology would transform mapping and reinvent the visualization of terrain. This revolution in mapmaking is explored here through the maps of three military surveyors. Their descriptions of mapping processes throw light on both the advantages and issues that delayed the immediate adoption of a game-changing innovation: the contour line.

Contour lines delineate heights of equal elevation. They are a simple (if somewhat abstract) graphic convention used to convey the positions, heights, and gradients of hills on maps.[1] Their adoption by the military for maps of fortifications was both swift and uncontentious. Less successful were their trials with contouring for reconnaissance mapping.

The reconnaissance map entailed cadets and officers envisaging land from above and transforming this conceived viewpoint—at speed and under threat of enemy fire—into lines on paper to facilitate the movement of battalions, provisions, and artillery. Prior to aerial photography, reconnaissance mapmaking required movement through potentially hostile terrain. More adept artist-soldiers painted vivid, pictorial renditions of hills and valleys for the map reader's eye, but danger lurked within the allure of these deft results.[2] Distortions in scale and inaccuracies for the positions, gradients, and altitude of hills risked the map reader and strategist being horribly—and in battle scenarios, fatally—deceived.[3]

During the early years of the nineteenth century, advances in the range and precision of artillery meant that armies could shoot, and be shot at, over greater distances. To increase commanders' confidence in moving and positioning troops, military mapmakers sought ways to acquire and convey more accurate data for the positions, heights, and gradients of hills.[4] The French had used contours for fortification projects as early as the 1760s,[5] and by the mid-1840s contouring had come into regular use by the Ordnance Survey.[6] Improvements in, and the standardization of measuring instruments enabled the production of more uniform results, whereas previously the military relied on individualistic pictorial interpretations made by artist-soldiers.[7]

Historically, military strategists viewed isolines as purveyors of truth, and earlier scholarship echoed this sentiment. The geographer and cartographer Arthur Robinson defined isolines as lines "that show by their absolute and relative positions the locations and gradients within a set of numbers." Robinson espoused their use "because they can be produced by machine mapping methods," portraying the location of "gradients, troughs, hills and valleys . . . [as] abstract geographical quantitative distributions."[8] Rarely employed early in the nineteenth century, by the end of that century their use was ubiquitous throughout Europe. As pictorial maps came to be seen as unreliable and subjective, so contoured maps appeared both more accurate and objective on paper. But isolines were not always truthful imparters of fact, particularly when—as with the last map discussed in this chapter—lines were drawn by eye and with unstandardized measurements.

Robinson portrayed cartography as objective science, a position contested by others writing on the power of maps, colonialism, isolines, and contours later in the twentieth century. By the 1980s, the perceived empirical objectivity of maps had been thoroughly dismantled; some scholars compared the study of maps to colonial acquisition practices, and others concluded that the power of (lines on) maps was "inherently political."[9] This work underscores the point that contour lines do not convey whether a hill is wooded, or a valley boggy. Like diagrams and charts, they are an abstraction, "a form of sampling: the space between two contour lines tells us nothing of the nature of the terrain."[10] All the same, historical military campaigns required cartographic assistance. As map historian J. H. Andrews noted: "By dehumanizing a landscape, maps make it easier for military commanders . . . when inflicting cruelty or hardship on the inhabitants of a mapped area."[11] According to historical geographer David Livingstone, isolines "convey every impression of impartiality and neutrality, and thereby inspire trust in their objectivity," but went on to observe that through "the power of cartographic discourse [they] present as natural what is culturally constructed."[12] For their perceived objectivity and truthfulness isolines are appealing, but our trust in them is partially maintained through failing to dig below the surface and examine the processes used to determine them. The emotive subjectiveness of pictorial maps encouraged viewers to question their reliability, particularly when compared with the supposed scientific precision conjured by isoline and contour. From the 1850s, the Ordnance Survey used contour lines to structure pictorially drawn relief on their maps, acknowledging both the potential for improving precision *and* making maps easier to read at a glance. In *Cartographic Relief Presentation*, Eduard Imhof contrasted the visual utility of unsupported contours on large-scale maps (such as for fortifications) with the difficulties of making sense of them at smaller scales such as for an atlas. Imhof's observations help us understand why many nineteenth-century military commanders argued against the adoption of contour lines for reconnaissance maps. A pictorial map painted a picture, whereas a contour map required patience to understand what occasionally appeared to be "formless confusions of lines."[13] The contoured map appeared objective and instilled a belief in its accuracy and trustworthiness. But with features omitted between the lines, contours present land as abstraction. They enable both a dispassionate viewing and the potential to dehumanize the land and its inhabitants.

Two issues dogged the military's adoption of contour lines. First, for the contoured map to be trustworthy, accurate surveying was necessary—though not always feasible. Second, contours made it difficult to visualize the mapped landscape quickly. The first map in this chapter describes mapping options available to the military prior to contour lines, and the second two explore the functional advantages and disadvantages of contour lines on military maps. All three highlight how drawing different types of maps affected the movements of military mapmakers in the landscape. Military maps were unlikely to be printed, and few manuscript maps survive.[14] However, those discussed here were printed in military publications and used to illustrate arguments for improved mapmaking practices.

The first mapmaker considered is Colonel Philip Bainbrigge (1786–1862). During the Peninsular War he drew *Field Sketching before the Enemy* on an intelligence gathering mission.[15] Bainbrigge was trained to be attentive to salient features while moving through the landscape, to understand rudiments of geology, flora and fauna, and to confer with (or question) inhabitants. He sketched his reconnaissance map on horseback in two and a half hours. Speed was of the essence. He did not use contour lines; any precise drawing technique would have been considered almost impossible under the circumstances. He was pleased enough with the results—drawn directly into a sketchbook—to keep it. Thirty-four years later, it was printed in an article titled "Field Sketching," appearing alongside a commentary on its making in a subsection "Sketching against time, and in the neighbourhood of the enemy."[16]

The second map was for fortifications. Drawn by Henry Drury Harness (1804–1883), it was published in 1838 to illustrate a paper he wrote titled "On Contoured Plans and Defilade."[17] It clearly demonstrated the cutting-edge advantages contour lines conferred on defense plans and contrasts with the first and third maps in both scale and function. Harness's depicts a small section of hillside below a wall of fortification. He drew contour lines at five-foot intervals to ascertain which areas were vulnerable to enemy observation or attack. Harness's mapping process was aided by accurate surveying equipment, plentiful manpower, and time. His illustration and description displayed a solid theoretical comprehension of the advantages that isolines offered. From the mid-nineteenth century, diagrammatic isolines grew in popularity as compelling visual explicators of geographical data in atlases and on maps, yet the wholesale adoption of contour lines by the military was not assured. Part of the reason for this is demonstrated by the third map.

Illustrating a paper by Charles Edmund Webber (1838–1904), the third map was published in 1866, nearly thirty years after Harness's map. Webber's paper was part of a series on "Military Sketching" published between 1863 and 1874.[18] All papers in the series were written to address the problem that "for the tactical operations of armies" there was "no uniform and systematic plan for expressing ground."[19] Like Harness's map, the map that illustrated Webber's arguments was contoured; however, as it was for reconnaissance (like Bainbrigge's map), it was drawn with minimal instruments and using the body as a scale. A significant difference between the reconnaissance maps was that Bainbrigge's was drawn during battle and Webber's to train cadets.

Also, the cadets learning to contour with Webber were not expected to map autonomously as Bainbrigge had done. Instead, they were required to systematically survey sample strips of land.[20] Webber attempted to demonstrate the potential advantages of contours for reconnaissance purposes, yet his cadets' slow pace left many in the military unconvinced. Comparison of the first and third maps is important because Webber himself compared them. The first map, made swiftly from the saddle, was prone to mistakes and the heights and gradients of hills could only be assessed in relation to each other. The introduction of contouring enabled mapmakers to convey heights and gradients more precisely, but still imperfectly, because cadets augmented the maps after surveying in ways that could introduce discrepancies.

As artillery became more powerful, military men desperately required the advances in mapmaking that could more accurately demonstrate height, shape, and gradient of terrain; as one officer noted, "What we expect to get from contours is, that knowledge of the ground, that accuracy in estimating slopes, which shall make the artillery officer right about his ranges."[21] Through the scrutiny and comparison of these three military men's maps a picture emerges of relative functionality, including the ability to enable and thwart movement by the military and the potential to facilitate the domination of land through firepower over ever greater distances. Even by the century's end, this process was incomplete.

* * *

Maps did not miraculously materialize upon campaign tables; they required men to enter the field to survey and draw them. This section contextualizes the early conditions under which military men drew maps. In the 1790s, a "Corps of Royal Military Surveyors and Draftsmen" was formed. The creation of the corps heralded an acknowledgment that some uniformity and standardization of training would facilitate the making of military maps that were likely "to be employed on active service in the war with France."[22] At this time, drawing practice was as essential a component of a cadet's training as the ability to clean a rifle or fire a cannon; it was a compulsory element of classroom training in all military schools.[23] In the classroom, cadets learned to draw from models. A *Treatise* of the time recommended that "an officer, who wishes to be employed in reconnoitrings, should studiously apply himself to figure drawing, and particularly to the drawing of antique statues," and he should "work after specimens of the most graceful and difficult outlines which it may be attempted to imitate."[24]

Alongside classroom drawing, cadets trained outdoors to draw at speed and use the body as a measuring instrument, which was termed "sketching without instruments."[25] It required adjudging altitude and distance by eye, sometimes with a "pencil held at arms length," or by pacing: counting steps, by foot or hoof.[26] Timepieces, even a man's pulse, were also made use of: "Slopes and declivities may be judged of by the greater or less time which it may take a man, or a horse . . . to ascend or descend them."[27] Military men on both sides of the Atlantic prized a style of speedy reconnoiter that required the ability to move through, and notice, the landscape as would a huntsman who is "everywhere at home," or by applying to "*wood-cutters*, *hunters*, *trappers* and *Indians*, [as the best guides for] examining the face of the country."[28] Once attained, these skills were applied on forays into unknown and uncharted terrain, and to reconnoiter, read, and draw the landscape without becoming lost. Through an education of the eye, officers and cadets learned to memorize and draw territory while on the move; their sketches required a "kind of *coup d'œil* which arises from practice, and that acute discernment which the mere habit of seeing and observing give."[29] This teaching style was only suited to gifted cadets, and as the century progressed, military instructors sought more formulaic methods for teaching the accurate shading of heights and gradients in the landscape.

In 1812, George Gordon and J. H. Bedford Smith reflected in their *Essay on Military Drawing*, "It must be a subject of regret . . . that no uniform and defined method of expressing the inequalities of ground, has hitherto been adopted by the British Army." They sought to define a series of rules "founded on incontrovertible principles, [that] admit of no variation" and argued that accurate data for gradient and altitude could avert dangers in the movement of troops and artillery.[30] Their essay included a table that divided gradients into those shallow enough to "admit of Manœuvres," those between 20 and 30 degrees to be "ascended and descended singly," and finally, the steepest, which only "Chasseurs and Riflemen" could climb up.[31] Prior to the adoption of contours, the question that vexed these and other writers was whether a pictorial drawing method for shading hills—a scale of shade—could be devised that would differentiate between gradients of 5 and 10 degrees.[32]

Another issue for military men was ascertaining and marking the heights of hill summits on maps. From 1801, a quick shorthand for "numbering

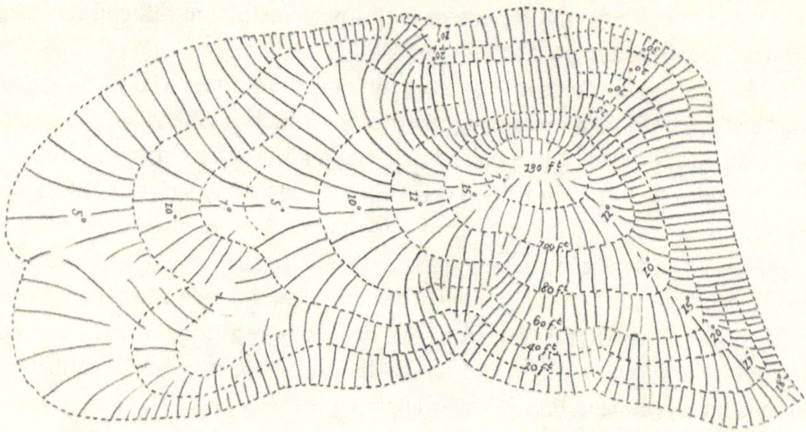

Figure 7.1. Karen Rann, *A drawing of William Siborn's "Military Sketch of a Hill" from 1822*, pencil on paper, 2025. The drawing demonstrates how contour lines were typically used to guide and facilitate the placement of vertical hachures. © Karen Rann.

hills in ascending order of height" came into use.[33] Originally developed by the French, it was most likely introduced to military training institutions in England by General François Jarry (1733–1807).[34] Called Relative Command, it entailed assessing the relative heights of summits in the field by eye. It enabled lines of sight to be quickly ascertained, so that for fortification works it was clear which "site dominated or commanded another."[35] It was an efficient shorthand that lacked precision. Of two hills marked with a 2, one might be much nearer a 3 in height, yet the system did not enable the map reader to distinguish between them. The digits added to hilltops related neither to feet nor meters (although it was later adapted by the Ordnance Survey to gauge measured heights).[36] Initially the numbers merely indicated that a hill marked as a 3 was understood to be higher than one marked as 2.[37] It was this system that Bainbrigge used on his reconnaissance map.

Bainbrigge's map was published thirty-four years after its making, alongside his recollections of it, for an article on "Field Sketching." He wrote that on July 21, 1812, in preparation for the battle of Salamanca, he stood beside the Duke of Wellington on high ground from which they could observe the French army crossing the river Tormes in the valley below.[38] Bainbrigge recalled that the duke turned to him and requested he "cross the river, and see what sort of a position there was . . . for stopping the advance of the

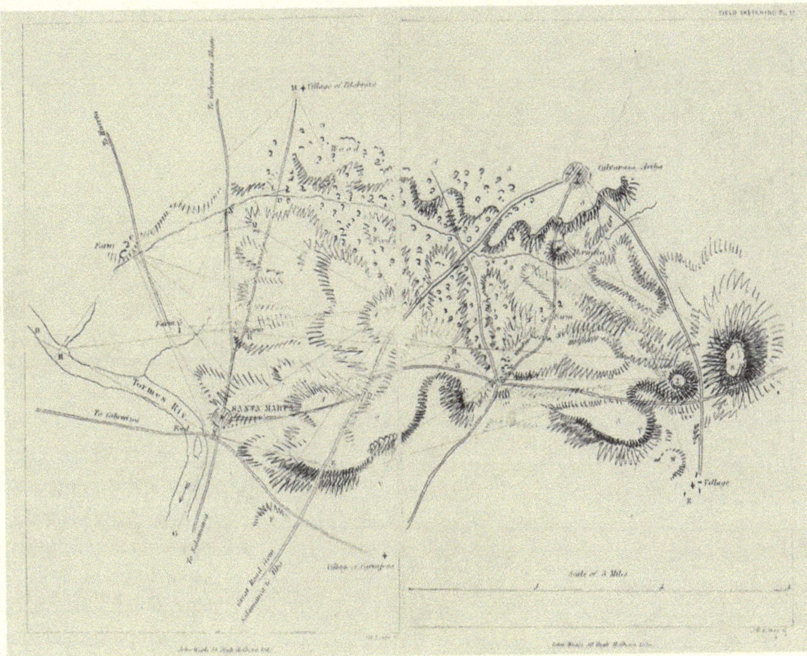

Figure 7.2. Colonel Bainbrigge, from "Field Sketching," in *Aide-mémoire to the military sciences* (London: Lockwood & Co, 1853). This was a reproduction of a sketch made by Colonel Bainbrigge in 1812.

French, and to make a sketch of it as quickly as possible."[39] On fording the river, Bainbrigge sought out "the intersections of remarkable objects" and marked these upon his map using a pocket sextant and compass to triangulate between these and hilltops. He judged distances by pacing so as to show "All ground within cannon-shot (1,200 yards)," and numbered heights of hills by eye using Relative Command.[40] His progress up some inclines was thwarted: "The rocks were so rugged I could not ride up," he wrote.[41] Riding within earshot of enemy skirmishers' fire, he shaded the sketch with rough and rapid vertical strokes, closer and darker where the slopes were steeper. In the article that accompanied his published map, the advantages of horizontal strokes—that appear to resemble densely drawn contour lines—were discussed in a subsection titled "sketching at leisure." However, the unidentified author went on to say that the "vertical mode" (that Bainbrigge had used) was "more easily understood in hasty sketches, where but little detail is admissible."[42]

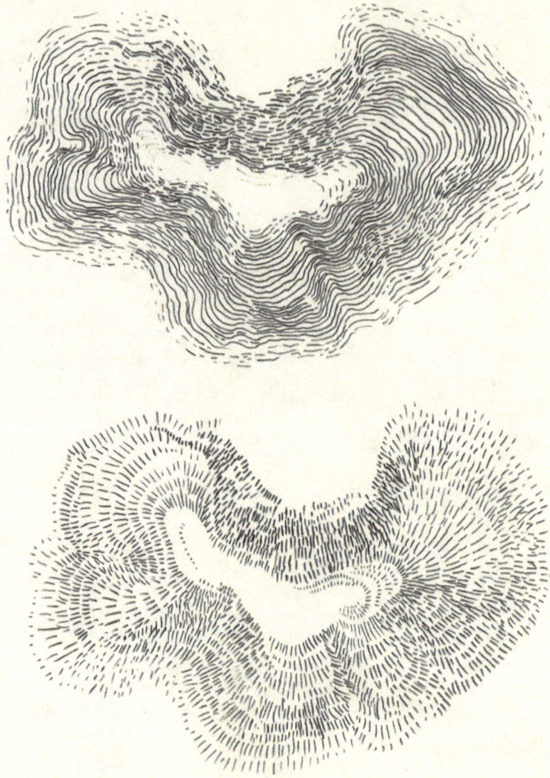

Figure 7.3. Karen Rann, *Drawings based on those from plate x in A Course of Military Surveying*, pencil on paper, 2025. The drawings compare an archetypal hill drawn in the horizontal and vertical styles, as demonstrated by Basil Jackson in 1838. © Karen Rann.

The rough vertically sketched strokes on Bainbrigge's map were good enough for the duke to reposition his army so that the French initially withdrew theirs. Even though military topographer Charles Edmund Webber thought Bainbrigge was "probably the best topographer in the army" at that time, he was disparaging of the map.[43] He thought Bainbrigge's map was too small and hastily made to give Wellington accurate information about the terrain.[44]

However well or poorly the reconnaissance sketch had functioned for Wellington (he eventually won the battle), for Bainbrigge it became ingrained in his memory. Decades later he could recount precisely the route he followed and his intentions. As a gifted artist-soldier, this aliveness to place and immersion in the landscape resulted in a visually literate map. Yet the drawbacks were considerable. Bainbrigge's techniques evoked the type of heroic endeavor that, in 1829, a military magazine urged "young officers" to guard against. It considered officers were "too apt to imbibe the implicit credence in the miraculous powers of the *coup-d'œil*, and to imagine that a pencil and paper, a quick eye, and a hand gallop, are the only instruments needed for 'heaven-born genius.'"[45] Heroics aside, to the military men who came after Bainbrigge, the role of the artist-soldier appeared outmoded. By the 1830s, the preference was already for more systematic, mechanistic surveying and drawing methods such as contouring.

* * *

During the period that Arthur Robinson dubbed a "golden age" in "the development of geographic cartography" between 1835 and 1855, the British began to adopt French mapmaking techniques including contour lines.[46] For fortification projects, contours were promoted by Henry Drury Harness as they gave more comprehensive information than traditional leveling lines. Leveling entailed the lengthy process of taking cross-sections of the rise and fall of the land along ruler-straight lines run through the terrain.[47] These measurements only afforded a series of snapshots rather than a complete picture.

By the 1830s, the range and precision of armaments had improved dramatically. The British military now needed more comprehensive information about terrain than leveling could ever offer and sought out French innovations such as contouring.[48] The French were ahead of the curve in the introduction of contour lines for military purposes.[49] In 1749 Louis Milet de Mureau, a fortifications officer in the French army, developed a system of planes suspended one above the other by which he desired to demonstrate "the degree to which one fortification overlooked another."[50] Map historian Josef Konvitz considered his method a "first step toward the use of contour lines." From the 1760s, students at Mézières (a famous school for military engineers) learned Mureau's system. In the 1780s, Jean-Baptiste Meusnier de la Place (1754–1793) refined it by drawing lines on maps as "curves connecting points of equal value of elevation."[51] Then, circa 1801, a man described as "battalion commander Haxo" was said to have used contours at two-meter intervals for fortifications required by Napoleon in Italy.[52] By 1809, Pierre-Antoine Clerc (1770–1843) was using contour lines extensively for fortifications projects for Napoleon.[53] Clerc probably also taught Guillaume Henri Dufour (1787–1875).[54] Dufour was a Swiss general who, from 1812, created what has been described as *scientific* contour line plans of fortifications on Corfu.[55] It is clear that the French were drawing contour lines for fortification projects at the time of the Battle of Salamanca (in 1812), but as reconnaissance maps were less likely to be printed, there is no evidence of the French using them for this purpose at that time.

As an instructor in fortification at Woolwich Royal Military Academy, Henry Drury Harness taught contouring and advocated for cadets to study "French military authors."[56] He was said to possess a "strictly logical" mind, turned "chiefly in the direction of mathematical and cognate problems."[57] Rather than artistic drawing skills, he placed emphasis on what he termed "excellence in Mechanical Drawing."[58] In his 1838 paper "On Contoured Plans and Defilade," Harness explained how contour lines—rather than leveling—could be applied practically in the field (See Figure 7.4). His paper explained how to apply contouring, through instrumental surveying, so that "the horizontal and vertical position of every point can then be at once discovered."[59] His written argument, however, was both densely detailed and ambiguous. He did not specify which measuring instruments he recommended (we must assume a theodolite) but covered in great detail how the men should hold pickets, move with leveling staffs, and signal to each other. He recommended a scale of 200 yards to an inch "with contours at vertical intervals of about 5 feet."[60] These intervals would enable a thorough assessment of "planes of fire" with the "consideration of the power of firearms of all kinds [which was] rendered necessary by the more powerful guns and small arms."[61] Harness promoted a standardized, methodological, and scientific approach to contouring that elevated the collection of data and statistics over individuality and art.

While Harness struggled to convey his ideas in words, his graphic visualizations were impressively farsighted. The planimetric accuracy and tactical advantages he conceived in the use of contour lines was innovatively adapted for an entirely different series of isolines for a civil project in statistical mapping. In 1837, the Irish Railway Commissioners invited him to develop a series of statistical maps to facilitate the "movement of people" across Ireland.[62] Harness designed and delivered three maps: a passenger map "of remarkably modern appearance," a traffic map, and a map "Showing by the varieties of shading the comparative Density of the Population."[63] Each map presented a wealth of statistical information in a visual format without precedent.[64] To glance at Harness's third map, the impression is of Ireland apportioned by contour lines that depict regions differentiated through shading. These regions, however, are not grouped by altitude—as a shaded contour map is. Instead, it appears he adapted the concept of contour lines and used their visual qualities to group together regions differentiated by population

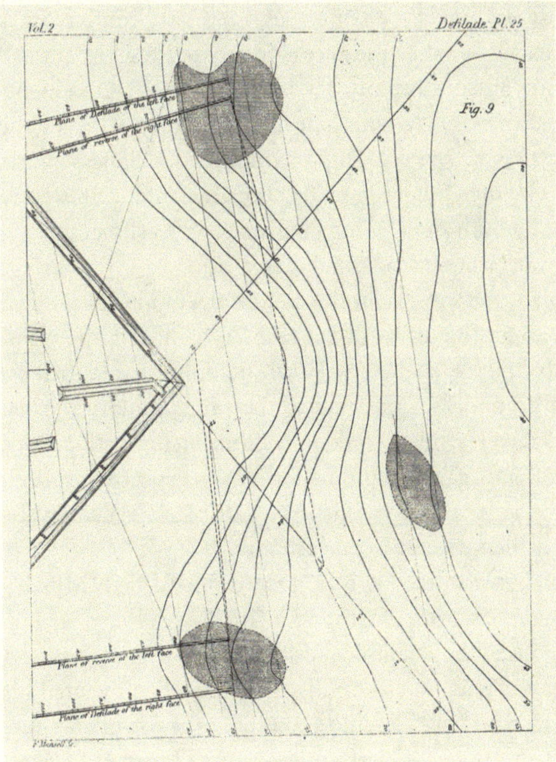

Figure 7.4. Henry Drury Harness, plate 25 from "Contoured Plans and Defilade," in *Papers on subjects connected with the duties of the Corps of the Royal Engineers*, vol. II. London: J. Barker, 1838.

density. Harness never wrote up his novel methodology. After the famous cartographer Augustus Heinrich Petermann (1822–1878) saw Harness's map, he deployed the concept for his own cartographic enterprises, wrote and published extensively on the subject, and came to be "erroneously credited with this innovation."[65]

* * *

Harness's maps had proved that isolines were useful at vastly different scales, from siting fortifications to statistically mapping whole countries. Between the 1840s and 1870s, the question remained whether contouring—for reconnaissance surveys—could be achieved, and Webber took up this query. In Britain, contour lines appeared on Ordnance Survey (OS) maps from the 1840s. Yet some people argued that they were unreadable. In reports and parliamentary debates between 1846 and 1856, their adoption by the OS was discouraged.[66] Those arguing for them recognized that contour lines alone "informed, at a glance, whether the access to any particular pass will be direct or circuitous in course, steep or easy in inclination."[67] Opponents claimed that "mechanical drawing systems" created visual obstacles for the uninitiated map reader; they risked creating "characterless," "lifeless" maps of unreadable ugliness.[68] By the end of the 1850s, though, contour lines were in regular use in Britain by both the OS and civil surveyors, who employed them to site industries, waterworks, roads, and railways.

Within what was called "the theatre for the tactical operations of armies" military men disagreed about the utility of contour lines; indeed, the problem was exacerbated as by the 1860s, there was still "no uniform and systematic plan for expressing ground."[69] Appearing in the same military journal that had previously published Harness's work on contouring for fortifications were a series of papers for and against contours for reconnaissance purposes. As well as the papers, the journal also published a "Discussion" held at the War Office in May 1866, which was recorded verbatim.[70]

The papers published between 1863 and 1874 were either titled "Military Sketching" or "On the Representation of Ground."[71] The first of these, written by Henry Y. D. Scott, espoused the scientific merits of contour lines drawn in the field. Over twenty pages, plus nineteen full-page plates, Scott vented his frustration at staff officers, who have "had no training in the use of contours . . . [which] can alone give accuracy of work and certainty of reading."[72] Scott was convinced that contours could be drawn "with a minimum of equipment" in the field and that traditional vertical strokes (like Bainbrigge's) were "unscientific" and drawn "according to the genius of the draftsman."[73] Subsequent papers presented alternative drawing methods, as well as various systems by which a man's body could be used as a surveying instrument.[74] These were written by E. R. James and Charles Edmund Webber.

Webber aired his concerns in two essays: "On the Representation of Ground, Especially in Military Reconnoissance" (1865) and "Military Sketching, Representation of Ground, &c" (1866). He denigrated the individualistic methods Bainbrigge had employed, arguing that "giving a rough idea of ground" was insufficient for the modern military to invade, occupy, or hold ground.[75] Bainbrigge's reliance on crude sketching and Relative Command had been flawed, producing a "pictorial effect" that was "*most* calculated to deceive the commander."[76] Webber recommended that instead a scale of shade should be "entirely based on a skeleton of contours," proposing that "the scale of shade in the one case, and use of contours on the other, put the multitude on a par with the few."[77]

To demonstrate his argument, Webber conceived a scenario whereby he took the place of Bainbrigge, all those years ago, to fulfill the Duke of Wellington's request for a sketch. Webber thought Bainbrigge should not have worked alone. He deemed a party of sixteen men, and a further three hours, better suited to the mission. Webber described how he would have divided the ground and apportioned men to work in pairs. With minimal equipment, his men would pace to draw contours "to intervals of 75 feet, with intermediate 'half-contours,' or shaded to scale where necessary." A priority was to ascertain the gradient of slopes, particularly "points of steepest inclination" in order to select "the safest parts" to position troops.[78]

Webber emphasized steps to be followed rather than taking initiative or independent thinking. In his first paper, he framed the argument for contour lines in terms of "the practicability of movement across an open or enclosed country, *irrespective of routes*," that entailed a mode of sketching "calculated on the searching accuracy of rifled arms."[79] He thought Scott's drawing method likely to "produce uniformity in results of the work of indifferent draughtsmen," though considered his own method superior as it was faster to execute.[80] Plate I, "Reconnoissance between Plumstead and Eltham," from his second Essay, illustrated his system for using the body to produce contours (see Figure 7.5). He conceived of an average pace and height for men's bodies by which "each step *is so mechanical* that any one could learn it in a short course of instruction."[81] Webber stressed the mechanical, replicable nature of his surveying and drawing methods. A year later, E. R. James agreed that "sketching must be strictly mechanical, and the sketcher need not necessarily be an artist."[82] Yet, rather than mapping under enemy fire, both he and Webber were employed as instructors in military academies and typically used models of hills to teach contour lines.

James argued that contour lines were too difficult for commanding generals to read, and too abstract.[83] In 1866, he recommended that for "relief to the eye," slopes should be given a "rounded appearance" and drawn "modified" by "what artists term the line of beauty, that is to say the ground at first falls easily, then rapidly, and lastly easily again."[84] It is as though he thought military officers would prefer idealized gradients on their maps, rather than what would actually confront them in the field! He couched his emphasis on "pictorial effect" as a compromise, a "means of agreement between the mechanical and artistic schools."[85] James was writing from a theoretical standpoint and for those who still desired soldiers to be artists.

Arguments over these papers took place during the "Discussion" chaired by Sir John Fox Burgoyne. During this, Webber railed at James's pictorial rendering of contours, even accusing him of "basing his sketching on contour lines, without actually contouring."[86] More significantly, both Webber and James received further flak from Captain Marsh—another instructor in military surveying—for their standardization of men's bodies. Marsh considered their "height of the eye" and pace ratios too arbitrary and stated, "We are not prepared to go into the field and suppose that our eyes are necessarily 5 feet 4 inches in height, or that our pace is necessarily 32 inches in length."[87] Both Webber and James built their arguments on theoretical practices; they had trained cadets at military establishments, rather than on active duty.

Webber chose a curious series of maps to illustrate his argument. They depicted ground in the vicinity of the Royal Military Academy divided into six strips that had been contoured by "mere beginners" under his instruction.[88] The first plate showed all six strips; a second illustrated in detail "the mode of contouring while surveying one of the strips in plate I."[89] Yet a comparison of the two plates reveals that contours and shading do not match up. It looks as though the cadet, once returned to the classroom, doubted his original contouring of the hill and doctored his work.[90] These discrepancies are of little consequence in the classroom but could have been disastrous for the movement of troops in the line of fire.[91]

Scott summed up the risk for a commander who relied on a sketched contour map: "Continuous contour lines in rough sketches . . . might lead to the supposition . . . that the sketcher felt more certainty in their truthfulness than his means of attaining accuracy justified."[92] Scott, Webber, James, and the other military men who took part in the "Discussion" were unable to reach a consensus on how best to depict relief on reconnaissance maps. Instructors in military drawing, such as Webber, based their ideas on what could be achieved in training, rather than on active duty. Their comprehension of what could be managed in the field, using the body to contour, was too optimistic and "unsuited to military sketches."[93] During proceedings, Captain Farrell found both James's and Webber's scales of shade "nothing more nor less than the multiplication of contours," and concluding that contours were "a decided strain on the imaginative powers," his preference was for the reconnaissance methods Bainbrigge had employed in 1812.[94] "This scale of shade will destroy all freedom of hand," he argued.[95] In agreement, Colonel

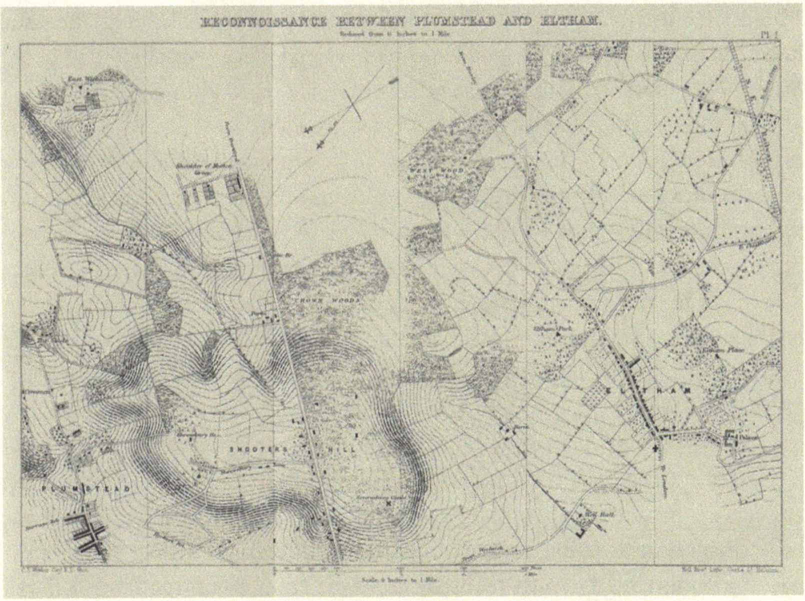

Figure 7.5. Charles Edmund Webber, plate I, "Reconnoissance between Plumstead and Eltham," from "II Military Sketching, Representation of Ground, &c.," in *Papers on subjects connected with the duties of the Corps of the Royal Engineers, New Series Vol XV*. Woolwich: Jackson & Son, 1866. Reproduced by kind permission of the Syndics of Cambridge University Library.

Simmons and Lieutenant Colonel Fisher thought most cadets were capable of more than "a simple, mechanical sketch" and could be relied upon to learn to draw faithful, pictorial maps.[96] They thought mechanistic drawing methods minimized the active intelligence required of a cadet to choose for himself how to move and what to notice. It was not until 1880 that the War Office uniformly adopted contouring.[97] And even then, their recommendation was that "the scale of shade [between contour lines] is only used *indoors*, and not much there, excepting when the sketcher is new to his work."[98]

* * *

Maps were considered "an indispensable item in war," and the information they contained was essential for military leaders.[99] In particular, the representation of relief was vital "to the tactical movement of armies." In 1869 this was the view of Auguste Frédéric Lendy, treatise writer and director of a "Practical

Military College." The importance of military maps was undisputed; the issue had been how best to draw them. Lendy considered much "mischief [had been] done by the diversity of methods employed . . . for representing ground."[100]

Early nineteenth-century maps created by soldier-artists, such as Bainbrigge, were vague in their descriptions of height and gradient, particularly when contrasted with the precision a uniform scale of shade—as applied to a contoured map—could impart. Bainbrigge's reconnaissance sketch showed how numbering the relative heights of hills, coupled with speedy instinctive drawing (rather than a system) could convey enemy terrain. His sketch enabled his commander to picture the land before moving his troops. Once Bainbrigge had returned, Wellington could further question him on particulars such as which slopes might be too difficult for cavalry to traverse.

The contour lines proposed by Harness were designed for the defense of a position rather than reconnaissance. His fortification plans were intended to secure a position from which enemy attacks could be repelled. There was little dissension to using contour lines for fortification purposes, and compared with leveling—which required the same equipment—contours offered a more comprehensive overview of the height, volume, and gradients of every hill surveyed. Because speed was a less critical factor, and the sites for fortifications were generally smaller, military surveyors could draw contour lines with relative accuracy. Already in use by the French, from the 1830s onward the advantages of contour lines for fortifications became obvious to the British.

We do not know if Webber's system was ever trialed on a battlefield. He expected cadets to labor mechanically, contouring narrow strips of terrain. This restricted their movements and did not permit them to build a comprehensive picture of enemy terrain. Another flaw connected with using the body for contouring was the (mis)perceived truthfulness of contours. Through their use on fortification plans, contour lines were associated with objective accuracy, yet contouring by eye—as Webber advocated—created lines that looked more accurate than they actually were. Neither Harness nor Webber thought that the military should rely on the artistic abilities of individuals. The trade-off was a loss of autonomy and freedom of initiative for surveyors.

The three military mapmakers discussed were selected because their views appeared in publications and because how they wrote about their maps could be compared with their illustrative examples. Each was passionate about the methods he used to draw his maps. Qualities such as accuracy, readability, and utility, contested then, can be debated anew now—particularly with regard to the advantages of contouring by eye that Webber advocated. Contouring with the body as instrument would never produce the same level of accuracy that measuring equipment such as theodolite and spirit level would. However, there were occasions when equipment—other than the body—remained unviable for reconnaissance, and it was advisable to have cadets trained for such instances. Even today, mountain leader courses teach students to "read contours" through their feet.[101]

Contour lines, unlike those for walls or barricades, are designed to be crossed as "they are but imaginary lines, that . . . have no natural existence."[102] They aid traversal of land without prescribing specific routes. Contour lines provided comprehensive information on gradients and height and facilitated the military to both build defenses and to move troops through hostile territory and across enemy lines. An advocate of contouring, writing in 1854, stated, "The system of contouring substitutes positive information for pictorial effect, and, *once understood* must convey the same general impression."[103] His words neatly summed up the contour map's evolving ability to convey data precisely.

In the nineteenth century, both making a map and using it required bodies moving through terrain. In the twenty-first century, paper maps—with or without contour lines—have become an obsolete military technology. Printed contour maps are still produced for hikers; they enable anyone to imagine or plan forays—not limited to roads and paths, and which might cross boundary lines (as the military once did)—from the comfort of home. Through twentieth-century advances in aerial photography and satellite technology, and subsequently unmanned aerial vehicles (UAVs) in the twenty-first, there is no longer a requirement for mapmakers to be physically present in the terrain they are surveying. A military campaign can be "guided autonomously . . . [by UAVs that carry] offensive ordnance . . . to interfere with or destroy enemy targets. Unencumbered by crew."[104] It is a technology with the potential to take life and territory with alarming speed and precision. Once considered a cutting-edge technology, contour lines are now as useless to the military as the cannonballs they once guided. Now, military personnel may have a presence *virtually* anywhere. From the comfort of a control center they can cast an eye over villages and track across fields a continent away. Unlike Bainbrigge, moving *physically* through enemy terrain two centuries before, military presence on the ground now would never entail carrying a sketchbook. Instead, through possession of real-time live-feed and remotely controlled firepower, the military may now hold and conquer terrain, rewrite borders and boundary lines, and literally take the ground from under our feet.

Notes

1. I use the term "map" broadly, covering everything from reconnaissance sketches made under enemy fire to facilitate movement of troops; general surveys to identify areas prone to attack; and fortification plans designed to hinder enemy movement. A plan depicts a specific location, and a map generally covers a region. For simplicity, I use the term map to describe both reconnaissance sketches and plans. The contour line is a "line of equal altitude" within the larger category of isolines, or "the sets of lines . . . that show by their absolute and relative positions" depictions on paper that ranged "from temperatures and elevations to population densities." Arthur H. Robinson, "The Genealogy of the Isopleth," *Cartographic Journal* 8, no. 1 (1971): 49–53 (quote 49).

2. Yolande Jones, "Aspects of Relief Portrayal on 19th Century British Military Maps," *Cartographic Journal* 11, no. 1 (1974): 19–33.

3. Charles Edmund Webber, "Military Sketching, Representation of Ground, &c.," *Papers on subjects connected with the duties of the Corps of the Royal Engineers, New Series Vol XV*, (Woolwich: Jackson & Son, 1866), 37; Anon., "Correspondence," *The United Service Journal and Naval and Military Magazine*, Part I (Henry Colburn and Richard Bentley, 1830), 38–40.

4. Jones, "Aspects of Relief Portrayal," 19–33. This article has informed and inspired my arguments throughout.

5. Josef Konvitz, *Cartography in France 1660–1848: Science, Engineering, and Statecraft* (University of Chicago Press, 1987), 96–7.

6. Although the Ordnance Survey was an arm of the military, its role was separate. See W. A. Seymour, ed., *A History of the Ordnance Survey* (W. M. Dawson & Sons, 1980); Rachel Hewitt, *Map of a Nation: A Biography of the Ordnance Survey* (Granta, 2011).

7. Jones, "Aspects of Relief Portrayal," 20.

8. Robinson, "The Genealogy of the Isopleth," 49, 50.

9. Jeremy W. Crampton, *Mapping: A Critical Introduction, to Cartography and GIS* (Wiley-Blackwell, 2010), 3–6 (the acquisition of maps); M. Dodge, R. Kitchen, and C. Perkins, eds., *The Map Reader: Theories of Mapping Practice and Cartographic Representation* (John Wiley and Sons, 2011), 5 (quote). Much has been written on the power of maps. A small sample includes Denis Wood, *The Power of Maps* (Guilford Press, 1992); Peter Barber, *Magnificent Maps: Power, Propaganda and Art* (British Library, 2010); Chris Perkins, "Cartography—Cultures of Mapping: Power in Practice," *Progress in Human Geography* 28, no. 3 (2004): 381–91; Jeremy Black, "Government, State, and Cartography: Mapping, Power, and Politics in Europe, 1650–1800," *Cartographica* 43, no. 2 (Summer 2008): 95–105.

10. Karen Rann, *Horizontal Hills: A Creative Historical Geography of the Emergence of Contour Lines in Nineteenth-Century Britain and Ireland* (PhD thesis, Queen's University Belfast, 2022), 19.

11. J. H. Andrews, "Introduction: Meaning, Knowledge, and Power in the Map Philosophy of J. B. Harley," in *The New Nature of Maps: Essays in the History of Cartography*, by J. B. Harley (Johns Hopkins University Press, 2001), 1–32, esp. 24.

12. David N. Livingstone, *Putting Science in Its Place: Geographies of Scientific Knowledge* (University of Chicago Press, 2003), 160.

13. Eduard Imhof, *Cartographic Relief Presentation* (De Gruyter, 1982), 157.

14. The National Archives in London, however, hold twenty-three MS maps made during the Peninsular War; see MR 1/167 and MPH 1/1010, TNA.

15. R. J. N., "Field Sketching," in *Aide-mémoire to the military sciences: framed from contributions of officers of the different services and edited by a Committee of the Corps of Royal Engineers, in Dublin, 1845–46*, ed. G. G. Lewis, H. D. Jones, T. A. Larcom, and John Williams (London: John Weale, 1846), 1:534–8 and Plates XI, XII.

16. The essay consisted of four subsections: I. Notices of the Apprehension of Ground; II. Preliminary Arrangements and Instruments, &c. necessary; III. Sketching at leisure, and under no restriction from the neighbourhood of the enemy; IV. Sketching against time, and in the neighbourhood of the enemy.

17. Henry Drury Harness, "On Contoured Plans and Defilade," in *Papers on subjects connected with the duties of the Corps of the Royal Engineers* (London: J. Barker, 1838), 2:75–90 and Plate 3.

18. Including Henry Y. D. Scott, "On the Representation of Ground" (1863); Webber, "II Military Sketching" (1866); E. H. James, "III Military Sketching" (1866); and J. A. Millar, "The Scale of Shade Simplified" (1874).

19. Henry Y. D. Scott, "On the Representation of Ground," in Officers of the Royal Engineers, *Papers on subjects connected with the duties of the Corps of the Royal Engineers* (Woolwich: Jackson & Son, 1863), 12:144–64, esp. 144.

20. Webber, "II Military Sketching" (1866), 139–40 ("each step *is so mechanical* that any one could learn it in a short course of instruction").

21. W. D. Marsh, in "VII. The Scale of Shade Simplified, by Capt J. A. Millar . . . Discussion on Ditto," in Officers of the Royal Engineers, *Papers on subjects connected with the Duties of the Corps of the Royal Engineers* (Jackson & Son, 1874), 22:162.

22. Seymour, *A History of the Ordnance Survey*, 48–49.

23. Jones, "Aspects of Relief Portrayal," 19.

24. C. S. de Malortie, *A Treatise on Topography, for both Civil and Military Purposes* (T. Egerton, 1815), 2:198–99.

25. Alfred Wilks Drayson, *Practical Military Surveying and Sketching, with the use of the compass and sextant, theodolite, mountain barometer, etc* (Chapman & Hall, 1861), 77–84; William Paterson, *Notes on Military Surveying and Reconnaissance*, 6th ed. (Trübner and Co., 1882), 82.

26. Drayson, *Practical Military Surveying and Sketching*, 77–84.

27. Auguste Frédéric Lendy, *A Practical Course of Military Surveying, including the Principles of Topographical Drawing* (Atchley and Co., 1869), 31 ("If no watch is to be had, the time is obtained by counting the pulsations of an artery"); D. H. Mahan, *A Treatise on Field Fortification containing instructions on the methods of laying out, constructing, defending, and attacking Intrenchments, with the general outlines also of the arrangement, the attack and defence of permanent fortifications*, 3rd ed. (John Wiley, 1861), 134.

28. Malortie, *A Treatise on Topography*, 2:183 (first quotation); D. H. Mahan, *A Treatise on Field Fortification*, 132, his italics.

29. Malortie, *A Treatise on Topography*, 2:104. Coup-d'œil is to take in the scene with one glance (from French, "a stroke of the eye").

30. George Ham. Gordon and J. H. Bedford Smith, *An Essay on Military Drawing. From the German of Schienert* (T. Egerton, 1812), i.

31. Gordon and Bedford Smith, *An Essay on Military Drawing*, 26.

32. William Siborn devised scales of light and shade to differentiate ground that "1. admits of manoeuvres 2. can be ascended and descended obliquely by light cavalry 3. rifleman and light infantry are most suitable . . . to be employed." William Siborn, *A Practical Treatise on Topographical Surveying and Drawing* . . . (C. and J. Rivington, 1827), 80.

33. Jones, "Aspects of Relief Portrayal," 23.

34. J. E. O. Screen, "Jarry, Francis (1733–1807), army officer and military educationist," *Oxford Dictionary of National Biography*, Sept. 23, 2004; accessed Aug. 9, 2022. https://www.oxforddnb.com/view/10.1093/ref:odnb/9780198614128.001.0001/odnb-9780198614128-e-14669. As well as Relative Command, it is likely Jarry introduced horizontal hachuring, and even contouring, to the British military training establishments he worked in.

35. Jones, "Aspects of Relief Portrayal," 22–24; Royal Engineers, *Aide-mémoire to the military sciences* (London: Lockwood & Co., 1853), plate 12. Two pages reprinted from Colonel Bainbrigge's sketchbook, drawn for the Duke of Wellington during the 1812 Peninsular War. They include Relative Command numbering from 1 to 5 for the hills.

36. Rann, *Horizontal Hills*, 129–32.

37. Jones, "Aspects of Relief Portrayal," 22–24.

38. Salamanca is roughly halfway between Madrid in Spain and the Portuguese border.

39. R. J. N., "Field Sketching," 536.

40. R. J. N., "Field Sketching," 535.

41. R. J. N., "Field Sketching," 537.

42. R. J. N., "Field Sketching," 528.

43. Webber, "II Military Sketching" (1866), 37. Webber drew the third map that this chapter considers, so it is interesting that he considered Bainbrigge thus because elsewhere in his article he pointed out many of the deficiencies in Bainbrigge's 1812 map.

44. Webber, "II Military Sketching" (1866), 42. According to Webber, the sketch was "not large enough to describe the tactical points, the aids to defence, or the capabilities of the ground," and (in a footnote on page 43) even the "relative command" numbers were wrong: "None here could tell by the drawing shown that the Bussaco ridge is 250 feet above that opposed to it."

45. Anon, "Military Surveying," in *The United Service Journal and Naval and Military Magazine*, part 1 (London: Henry Colburn, 1829), 340.

46. Arthur H. Robinson, "The 1837 Maps of Henry Drury Harness," *Geographical Journal* 121, no. 4 (1955): 440–50 (quote 440).

47. In addition to the military's use of leveling for fortifications, before contouring they were essential for deciding the routes of railway lines and canals.

48. "The French cartographic style was conveyed to England by men such as Isaac Landmann who had taught at the École Royale Militaire in Paris before becoming professor of artillery and fortification at the Royal Military Academy at Woolwich, from 1777 to 1816." Rose Mitchell, "Contours of Conflict: The Highs and Lows of Military Mapping at the National Archives of the United Kingdom," in *History of Military Cartography*, Lecture Notes in Geoinformation and Cartography, E. Liebenberg et al., eds. (Springer International Publishing, 2016), 394.

49. Konvitz, *Cartography in France*, 97. A plan of a fort was made using a contour method in 1761.

50. François de Dainville, "From the Depths to the Heights: Concerning the Marine Origins of the Cartographic Expression of Terrestrial Relief by Numbers and Contour Lines," trans. Arthur H. Robinson and M. Carlier, *Surveying and Mapping* xxx, 3 (Sept. 1970): 398.

51. Konvitz, *Cartography in France*, 97. Though the lines Mureau drew were isobaths, on a hydrographic chart, rather than for fortifications.

52. Dainville, "From the Depths to the Heights," 400.

53. Luisa Rossi, ed., *Napoleone e il golfo della Spezia. Topografi francesi in Liguria tra il 1809 e il 1811* (Silvana Editoriale, 2008). Pierre-Antoine Clerc made contour maps of the Gulf of La Spezia, a military port within the French Empire.

54. John David Fenniman, "The Olsen-Bredsdorff Map and the Adoption of Contours for Relief Depiction on Atlas Maps in the Early Nineteenth Century" (Master's diss., University of Wisconsin–Madison, 1974), 33. Clerc "taught at the École Polytechnique and at the Ecole de Metz." Dufour studied at both these institutions: at the Polytechnique in 1807, and Metz in 1809.

55. Fenniman, "The Olsen-Bredsdorff Map," 33; Guillaume Henri Dufour, *Plan general de la place et des forts de Corfou et du terrain extérieur à la distance d'environ deux kilomètres*, MS map, 136 x 100cm, scale 1:5000 (1811–12). Original in the Dépôt de la guerre à Paris (BGE, fad e 105); image in Hans-Uli Feldmann and Hans-Peter Höhener, "Biographie Dufours," *Cartographica Helvetica*, 54, (2017): 7 fig. 9, https://www.e-periodica.ch/digbib/view?pid=chl-001:2017:0::12#9.

56. Harness to Denison, letter dated 1836, in Charles Edmund Webber, ed., *General Sir Henry Drury Harness, K. C. B. Colonel Commandant Royal Engineers* (Royal Engineers Institute Committee, 1903), 16–17, https://archive.org/details/bub_gb_gEIoAAAAYAAJ/page/n3/mode/2up, "a knowledge of French . . . would enable the cadet to study with perfect facility French military authors."

57. Webber, *General Sir Henry Drury Harness*, 10.

58. Webber, *General Sir Henry Drury Harness*, 17.

59. Harness, "Contoured Plans and Defilade," 75.

60. Harness, "Contoured Plans and Defilade," 89–90.

61. Webber, *General Sir Henry Drury Harness*, 15.

62. "Second Report of the Irish Railway Commissioners," *Journal of the Statistical Society of London*, volume 1 (September 1838): 257–77, https://archive.org/details/jstor-2337983/page/n1/mode/2up.

63. Robinson, "The 1837 Maps of Henry Drury Harness," 442. The map is available online: UCD digital library, "Atlas to accompany second report of the Railway Commissioners Ireland," https://digital.ucd.ie/view-media/ivrla:45724/canvas/ivrla:45728.

64. Robinson, "The 1837 Maps of Henry Drury Harness," 447. Robinson stated, "To my knowledge nothing remotely like these maps preceded them."

65. *Petermanns Geographische Mitteilungen* (the annually produced journal for which Petermann served as editor); Robinson, "The 1837 Maps of Henry Drury Harness," 449 (quote).

66. The reports were House of Commons, *Report from the Select Committee on Ordnance Survey (Ireland)*, 664, vol. xv. 1846; HC, *Report from the Select Committee on Ordnance Survey (Scotland)*, 519, vol. x, July 10, 1851; HC, *Report from the Select Committee on the Map of Ireland*, 921, vol. xxxix, 1853; HC, *Report from the Select Committee on Ordnance Survey of Scotland*, 198, vol. xiv, May 6, 1856. Collated correspondence was published as *Correspondence between the Treasury and the Board of Ordnance* in 1852 and *Correspondence Respecting the Scale of the*

Ordnance Survey and upon Contouring and Hill Delineation was collated and published in 1854. These debates centered on the scale and function of maps, though these issues impinged on whether contour lines were desired, or not.

67. W. J. Macquorn Rankine (Civil Engineer) in *Correspondence Respecting the Scale of the Ordnance Survey and upon Contouring and Hill Delineation. Presented to both Houses of Parliament by Command of Her Majesty* (George Edward Eyre and William Spottiswoode, 1854), 148.

68. Lewis D. B. Gordon in *Correspondence Respecting the Scale of the Ordnance Survey and upon Contouring and Hill Delineation. Presented to both Houses of Parliament by Command of Her Majesty* (George Edward Eyre and William Spottiswoode, 1854), 160, British Parliamentary Papers (House of Commons) online.

69. Henry Y. D. Scott, "On the Representation of Ground," in *Papers on subjects connected with the duties of the Corps of the Royal Engineers* (W. P. Jackson, 1863), 12:144.

70. J. F. Burgoyne in the chair, "Discussion on papers II and III," in *Papers on subjects connected with the duties of the Corps of the Royal Engineers* (Jackson & Son, 1866), 15:57–78. The papers by E. R. James and Charles Edmund Webber were read at this meeting.

71. The papers and discussion were printed in volumes of *Papers on subjects connected with the duties of the Corps of the Royal Engineers* in 1863, 1865, and 1866; as no consensus was reached, the journal printed three further papers on the subject in 1868, 1869, and 1874.

72. Scott, "On the Representation of Ground," 145.

73. Scott, "On the Representation of Ground," 155.

74. Burgoyne was also one of the Irish Railway Commissioners that employed Harness in 1837.

75. Webber, "II Military Sketching" (1866), 37.

76. Webber, "II Military Sketching" (1866), 44. All quotes. (His italics.)

77. Webber, "II Military Sketching" (1866), 34, 39.

78. Webber, "II Military Sketching" (1866), 42, 44.

79. Charles Edmund Webber, "On the Representation of Ground, especially in Military Reconnaissance," in *Papers on subjects connected with the duties of the Corps of the Royal Engineers*, XIV (Woolwich: Jackson & Son, 1865) 14:137–47, esp. 137. (My italics.)

80. Webber, "On the Representation of Ground," 138.

81. Webber, "On the Representation of Ground," 139–140. (My italics.)

82. E. H. James, "III Military Sketching," in *Papers on subjects connected with the duties of the Corps of the Royal Engineers* (Jackson & Son, 1866), 15:46.

83. James, "III Military Sketching," 46.

84. James, "III Military Sketching," 46. All quotes

85. James, "III Military Sketching," 47.

86. Webber in "Discussion on papers II and III," 76.

87. Marsh in "Discussion on papers II and III," 68.

88. Webber, "II Military Sketching," 41.

89. Webber, "II Military Sketching," Plates I, V, and page 45, "Description of Plates."

90. J. B. Harley and R. R. Oliver, *The old series Ordnance Survey maps of England and Wales, scale: 1 inch to 1 mile* (Harry Margary, 1991), 8:x. In the Ordnance Survey, the exaggeration of hachures between contour lines was a frequent mistake made by trainee draftsmen.

91. Webber, "II Military Sketching," 45. In a footnote he acknowledged some issues with scale; the drawing was reduced by photography and "consequently the contours are not at intervals adopted."

92. Scott, "On the Representation of Ground," 157.

93. F. J. Hutchison and H. G. MacGregor, *Military Sketching and Reconnaissance* (C. Kegan Paul & Co., 1878), 34.

94. Farrell, in "Discussion on papers II and III," 61 (both quotes). His stance is surprising as he introduces himself as previously working for "the contour department of the Ordnance Survey" (page 62).

95. Farrell, in "Discussion on papers II and III," 63.

96. Fisher, in "Discussion on papers II and III," 67.

97. Jones, "Aspects of Relief Portrayal," 19.

98. War Office, "Ordnance Survey—processes of hill sketching," in *Report of Committee on TextBook, and System of Instruction for Military Sketching and Surveying etc.* (Harrison and Son, 1880), 49. Quotation from W. St. George Burke. GBR/3296/Maps/CCSA.IM_509_1, Charles Close Society Archives, Cambridge University Map Library.

99. Lendy, *A Practical Course of Military Surveying*, 2.

100. Lendy, *A Practical Course of Military Surveying*, preface (no page number), 17.

101. Anna Fleming, "Skills for the Hills," *The Guardian*, January 25, 2020, 9.

102. Farrell, in "Discussion on papers II and III," 62.

103. Gordon, in *Correspondence Respecting the Scale of the Ordnance Survey*, 160.

104. John F. Guilmartin, "Unmanned Aerial Vehicle: Military Aircraft," Britannica.com, accessed May 19, 2022, https://www.britannica.com/technology/unmanned-aerial-vehicle.

PART III

Movement

CHAPTER 8

Indian Centers, Colonial Peripheries

Locating the International in Early America

CHARLES W. A. PRIOR

What defined the international society of early America?[1] This essay seeks to frame a preliminary answer to this question by using diplomacy and spatial representations of Indigenous power as lenses to consider interpolity relations in the Northeastern Woodlands. A series of interlinked historiographical moves help us to bring the international character of this region into focus. The first is the re-centering of Indigenous power, a development that has cut colonialism down to size and redrawn the map of power dynamics within, across, and between zones of cross-cultural interaction. The reconceptualization of Indigenous power in terms of location, movement, linkages, and networks foregrounds interactions within common worlds across multiple spatial and jurisdictional contexts.[2] Second, a range of studies has revealed these zones—grounds, lands, frontiers, inposts, meridians—as places where "local affairs" were "world revealing," linked to and illustrative of wider movements in international affairs.[3] Third, historians have sought to bring order to this eclectic and borderless set of places by searching for common and commensurate experiences, languages, and meanings.[4] Europeans were obliged to both understand Indigenous epistemologies of power and to work within them by adopting the sets of customs through which they were expressed.[5] In the process, they became treaty people.

Where colonial, imperial, and state power were concentrated around enclaves of settlement and a loose arc of posts and fortifications, diplomatic links with Indigenous polities pulled Europeans into a multinodal world of kinship, alliance, and exchange. Anglophone officials sought to create a colonial world, but they were repeatedly obliged to operate in an international one. This reality is captured in contemporary maps that juxtapose Indigenous and colonial polities in spaces where neither had a monopoly of power. John Mitchell's map of British and French domains (1755) depicts spaces where violence, movement, and diplomatic accommodation shaped multilayered relationships among Indigenous and Anglophone social formations. On the ground, treaties—particularly the sequence of Covenant Chain councils held in New York and Pennsylvania—constituted points of intersection in a broader geography of international relations that were conducted according to a hybrid set of political and cultural norms.

This essay locates the international in two senses. First it places Indigenous polities within a framework of relations that is inflected by the epistemologies of colonialism and the state. Recent work on imperial locations helps to position diplomatic relationships in contexts that are shaped by multiple interactions, shared concepts, historical ties, and cultural knowledge.[6] Second, it locates the international in points of intersection between Indigenous and (for the purposes of this essay) Anglophone polities. To do so, it uses Mitchell's map as an illustration of Indigenous territoriality and history, centered on Iroquoia and its precincts, that places colonies as one power grouping sited in a landscape of multiple nations and diffused sovereignties. The deep textures of relations within these spaces are evident, albeit transmitted through European eyes, in the records of diplomacy that were produced by imperial officials as they negotiated the complex protocols of the Covenant Chain. What New York's secretary of Indian affairs Peter Wraxall described as "this peculiar Specie of Politics" was conducted in locales of the international and defined by cross-cultural diplomacy that was characterized by distinct languages, norms, and customs.[7]

In Marshall Beier's concise formulation, Indigenous diplomacies are not new but "newly noticed," having been obscured by European epistemologies and concepts of international relations.[8] Historians of international thought have suggested that we focus on the locations where the norms that structured interpolity relations were formed and applied.[9] Iroquoia and the broader spaces illustrated by Mitchell form one such locale. Interactions within these locations were not exclusively statist, but included individuals, small nations, peoples, and companies; they constituted an "array of actors" who observed negotiated principles of "amity."[10] Most importantly, sovereignty—the defining concept of modern states and the states system—emerged as contingent and malleable, rather than as a "fixed" and perfect category. Negotiated and plural concepts of sovereignty defined the politics of interpolity zones, in which diplomatic alliances and treaty protocols created a particular space and set of nation-to-nation relations.[11] The Covenant Chain exemplifies this pattern of relations. It was a mechanism of consensus, sovereign negotiation, and the regulation of commerce and settlement. It transformed Britain's imperial domain into an international system that was regulated by cross-cultural norms of amity, alliance, and negotiated sovereignty.

Locating the International

Historians of early American diplomacy have long debated where Indigenous peoples fit both in relation to colonial and imperial powers, and to the foreign policy of the early federal government.[12] To consider Indigenous nations as purely domestic entities has the effect of "normalizing" colonialism, subtracting them from the sphere of international relations.[13] Additionally, international relations as a discipline is difficult to disentangle from the conceptual frameworks that surround both the state and the emergence of international law, which vindicated "advanced colonial practice."[14] Much of the historiography on this topic is focused on the early national period, where the supremacy of the domestic sovereignty of the United States went in parallel with its emergence as a "treaty-worthy" international power. A different kind of internationalism can be applied to the colonial northeast, a place that was defined by social worlds, village worlds, and middle grounds that contained what Richard White called "aspects of a world system." The Covenant Chain—multinational and multilocal—was an approach to alliance that created a "peculiar political arena" of movement, exchange, and alliance whose sinews were Iroquoian languages and concepts of law and diplomacy.[15]

Diplomatic, military, and commercial engagement with Indigenous nations contributed to the formation of international law.[16] These concepts contained embedded cultural assumptions that defined Indigenous peoples as stateless nomads, whose rights to land were recognized for the purely instrumental purpose of rendering their cessions legally sound.[17] Native Americans were classified as "subjects," as peoples requiring protection, incorporation, and ultimately—in the formulation of the Marshall court—as "domestic dependent nations."[18] Likewise, in common with the diffused sovereignty of the British imperial constitution, the character of the federal order that emerged in the 1780s was Janus-faced; its relations with individual states, as well as the Indigenous nations of the continental interior, tilted between domestic and foreign affairs. It is little wonder, then, that so much early work on foreign policy in colonial and early national settings made only tentative steps toward positioning Indigenous polities as diplomatic sovereigns.[19] One state of the field essay that appeared in the wake of *The Middle Ground* noted "how profitably the tribes can be made full actors in the story of white rivalries."[20]

By contrast, historians of early America broadly agreed that internationalism was implicit in the broader processes of state formation and exchange that shaped the geographical proliferation of the colonial system.[21] This is not a new insight. As early as 1934 Max Savelle argued that diplomacy was shaped by impulses that were "inherent in the geographical situation of colonies," and that the rise of European imperial formations in the west created a "new international situation."[22] Emily Rosenberg noted that the edges of the expanding American domain were not protean national spaces, but the "borders of global power." It followed that colonialism generated "international relationships" where frontiers dissolved into "international zones," and where colonial wars, the American Revolution, and westward expansion were all "episodes in *international* history."[23] For Karen Kupperman, "America was international before it became national" and in the process of the consolidation of national state power, "margins were crucial locations."[24] Even the "west" was internationalized as a multinational "meeting ground" and a "global frontier."[25] This frontier comprised a set of locales where settlerism operated within "spaces of transnational history," for example in the Great Lakes region that linked the Indigenous west to the colonial outposts of the Atlantic coast.[26]

Such transnational spaces of interaction required hybridized and cross-cultural norms of diplomacy. On one view, colonial-era diplomacy was an extension of the Westphalian system of states into Indigenous domains, where European rules and norms were introduced and adopted by a process of "global dissemination."[27] With colonialism came public law and diplomatic conventions, with the consequence that Europeans did not consider treaties concluded with Indigenous peoples to be equal to state-to-state agreements.[28] As Leonard Sadosky argued, while Europeans were initially obliged to work within the framework of borderlands diplomacy, these unformed spaces were ultimately incorporated into the Westphalian system through a process of territorial and juridical state-formation. Here, borderlands were peripheries that would over time be transformed by European-style statecraft and, ultimately, by the imperatives of republican state-formation. It followed that if borderlands were impermanent, then so too were the diplomatic cultures that structured relations within them.[29] Yet the nascent United States found itself confronted with Indigenous confederations that "encircle the Union from Maine to Georgia" and assumed their previous role as foils to the expansion of a land-based imperial formation.[30] The pivot of the United States away from middle ground–style diplomacy toward a single sovereignty may have closed off internationalism east of the Mississippi, but the Haudenosaunee reforged Covenant Chain links with Britain in the Grand River, establishing new international locales in the process.[31]

One premise of the legal framework of colonialism was that European claims to *de jure* sovereignty had to be worked out in a context of "intersocietal law."[32] "Treaties"—mutually ratified agreements with Indigenous sovereigns—were instruments of colonialism in that they provided evidence of the rightful possession of "foreign" territory and signaled Indigenous acceptance of European presence.[33] In this sense, colonial North America was incorporated into the "extended treaty system" of Europe.[34] The treaty-making revolution of the nineteenth century led to the extension of international society, as new American states entered the international system as "highly active treaty makers."[35] International law served to legitimize colonialism, but colonialism was also defined in a juridical sense by treaties that revealed the "international quality of relations" between states and Indigenous sovereigns. Notwithstanding, diplomatic agreements with Indigenous powers were not constrained by the universalist principles of equity and justice within international law.[36] Colonialism helped to universalize aspects of international law, but it did not foreclose the sovereignty of Indigenous powers within that system—a premise that underpins current calls for the recognition of Indigenous rights in the international sphere.[37] As I will suggest below, the Covenant Chain reveals the prevalence of locally determined norms of interpolity relations, which functioned as parts of an international society that was formed by and within processes of imperialism.[38]

Mapping the International

The international locales of early America were defined by strong diplomacy and weak borders. In such a context, an account of sovereignty predicated on absolute power within defined territories was less prevalent than legitimacy, recognition, and the capacity for diplomacy. Additionally, sovereignty

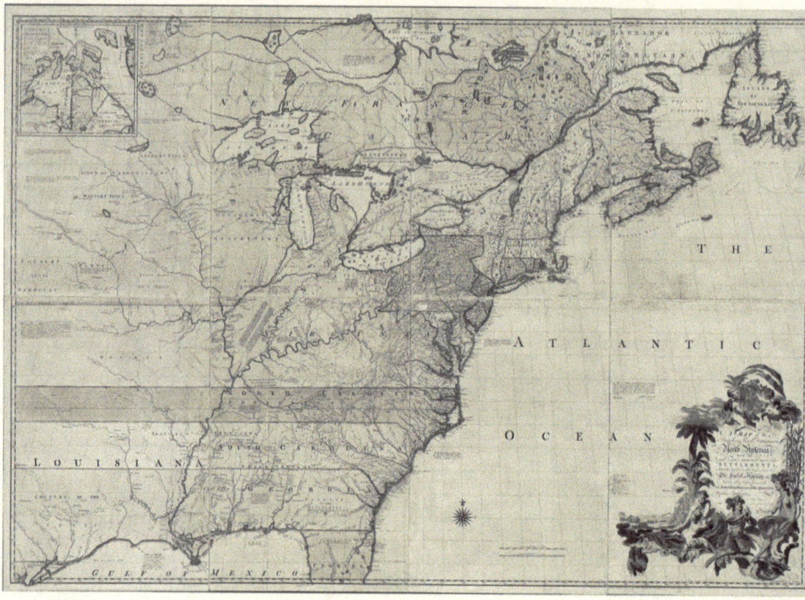

Figure 8.1. John Mitchell, *A Map of the British and French Dominions in North America* (1755). Library of Congress, Geography and Map Division, Louisiana: European Explorations and the Louisiana Purchase. http://hdl.loc.gov/loc.gmd /g3300.np000009.

was relational rather than purely spatial: People moved and interacted and in doing so formed alliances that stretched across space and cut across borders.[39] As European states sought to project their power into Indigenous spaces, that power "weakened at the periphery." Perfect spaces of sovereignty under the jurisdiction of states and bounded by borders were, in fact, a "fabric full of holes," confined to "islands of occupation," an "archipelago" situated inside and among vividly mapped concentrations of Indigenous powers.[40] Europeans created maps to exhibit their territorial claims to themselves and rival states, but in the process, they also mapped the domains of Indigenous polities and the relationship of imperial powers to them.

John Mitchell's 1755 *A Map of the British and French Dominions in North America* is a well-known point of reference in negotiations over territory and borders in the American interior.[41] The map of eastern North America presented a layered landscape of imperial, colonial, and Indigenous domains—a conflation of homelands and empires.[42] Its imperial elements are indicated most obviously in the title, and the cartouche, dominated by the royal arms and depicting the Indigenous as subjects under the protection of the Crown.[43] Out in the field of the map, Mitchell ran lines that demarcated the "bounds" of Virginia and New England as they were set out in early seventeenth-century charters.[44] Positioned around the map are blocks of text that narrate episodes in the diplomatic history of the interior. For instance, Mitchell indicated the territory "formally surrendered" by the Cherokee to the Crown under the terms of the Treaty of Westminster.

The colonial northeast is conspicuous in its density of place names, partially obscuring an Indigenous toponymy, while at the same time historicizing a long process of conflict and the acquisition of territory.[45] Intercolonial borders are dimly drawn, whereas putative claims to the trans-Appalachians are indicated by prominent labels, particularly Virginia's, reflecting a long-sought "prospect into unlimited Empires."[46] Imperial officials remarked on the raw potential of colonial dominion, all the more apparent "for viewing a map made on a large scale"—which map, we do not know—"we found them seated on an extensive continent . . . and which we dreaded, least time should [be] established into an empire of dangerous consequence."[47]

The most compelling feature of Mitchell's map is its treatment of Indigenous territoriality. At least since 1751, agents of the British empire had corresponded with Mitchell to discuss the relationship between Indigenous polities and British territorial claims. In July 1751, Cadwallader Colden reminded him that "the Indians have no registers whereby they can prove their right to what they claim. It wholly depends on Tradition and hearsay evidence [. . .] therefore their claims must be subjected to disputes."[48] Haudenosaunee territoriality is present in the diplomatic record, where council speakers characterized the Confederacy as an expansive regional presence with extended domains, influence, and a strategy of using conquest and alliances as mechanisms for consolidating power. Mitchell's map indicated those Indigenous claims to space through mediums that included Indigenous toponyms and explanations of migrations and territorial transactions.

Mitchell's map portrayed the Haudenosaunee as an expansive confederacy whose political history was inscribed on the land and whose power was amplified by water and projected through key nodes of communication and exchange. On the map, the domain of the Six Nations is, next to New France and Louisiana, the largest territorial formation. These prominent homelands stretch from Virginia to northeast New York and are labeled with each of the constituent nations' territories and networks of villages. Viewers can see locations of strategic importance. Canahogue—the site of modern Cleveland—is labeled as a "seat of War, the mart of Trade & the Chief Hunting Grounds" of the Six Nations. Mitchell also depicted the history of the territoriality of the Haudenosaunee, using the French name given to the peoples of the Longhouse: Irocoisia, "the original country of the Iroquois" is positioned in the vicinity of Lake Champlain. By 1755, the St Lawrence River was the "river of the Iroquois," above which were the lands of the "northern Iroquois."

Historians have tended to view Haudenosaunee claims to "conquest" with skepticism, interpreting them as part of a symbiotic relationship with Britain that added territories gained by conquest to the Crown's imperial domains.[49] Certainly, the Crown sought power in borderland spaces that were not captured by the rational geography of surveyed bounds and lines of property; cession of land from Confederacy to Crown pushed British domains west. At Quadoghe (modern Chicago), Mitchell indicated the "Bounds of their deed of Sale to ye Crown of Britain," agreed first at Albany in 1701, and then renewed in treaties in 1720 and 1744. Around the Great Lakes, Mitchell noted the "ancient" villages and homelands of the Illinois, Eries, Outaouais (Odawa), Hurons, and Algonquians, which were absorbed by "Right of Conquest," the details of which were narrated in two blocks of script that also detailed the history of those nations who were under the "alliance and subjection" of the British Crown.[50] "Conquest" and the sites of abandoned villages are markers of layers of movement, driven by mourning wars that were themselves spurred by the biological effects of colonial intrusion in the form of smallpox. Haudenosaunee, carrying superior weapons and moving swiftly by water, pushed into the villages of the western Great Lakes in the search for furs and captives.[51] Mitchell's labels convey processes of change within and between Indigenous polities.

The Mitchell map presents us with a conflation of imperial and Indigenous territoriality. First and foremost, it is a map defined by Indigenous presence rather than erasure. Colonial knowledge of Indigenous nations was shaped by proximity, and so the coastal and trans-Appalachian regions are more detailed than the speculative and conjectural trans-Mississippian spaces. Iroquoia is the literal center of the map: Cutting across colonial borders, its toponyms intermingled with new names in a patchwork of villages. Reaching out from this center were unseen connections that Mitchell hinted at in brief commentaries: kinship, exchange, tribute, and other forms of linkage between Indigenous nations in lands well beyond what the map labeled as the "extent of English settlements"—located at the headwaters of the Cumberland River in present-day Kentucky. A particularly striking example of this can be found in a general congress held by William Johnson with representatives of multiple Native nations in March 1768, in which an unnamed speaker reminded the audience of the ability of the Haudenosaunee to extend their senses across wide spaces:

> We have large Wide Ears and we can hear that you are going to Settle Great numbers in the heart of our Country, and our Necks are stretched out, and our faces set to the Sea Shore to watch their motions . . . our Legs are long, and our sight so good that we can see a great way thro' the Woods.[52]

The map reveals the extent to which contemporaries defined space in terms of activities, a set of dynamic processes defined by the "ability to control movement and access" to specific locations and resources, and where pathways were spaces of "balance" that leavened "uneven levels of power."[53] We need to think beyond our assumptions of inevitable settler expansion that the open western borders of the colonies suggest and consider a second story, defined by deep cultural interaction, diplomatic negotiation, and movement along negotiated routes of trade.

The Covenant Chain as an International System

Where many historians have been guarded in their characterizations of Iroquois power, work on international relations has taken a very different tack. The Confederacy has been characterized as a "security regime," an "international organization" whose power was extended through a realist foreign policy of alliances.[54] The Great Law of Peace was a normative system that governed "inter-nation(al) relations," while the oral tradition that contained the founding of the Confederacy was an account of "nation-formation," the development of foreign policy, and positions on the "rights of foreign nations."[55] Iroquoian power "rested on location, numbers, and political organization," and in the capacity to influence diplomacy, both in terms of conduct and outcomes.[56] The Iroquois sent diplomatic missions to articulate their foreign policy to locations throughout the Great Lakes, the Southeast, and to the British imperial capital.[57] But perhaps the most obvious expression of their power lay in the ability to set and shape the terms on which the diplomacy of the American Northeast was conducted, which resulted in relationships framed by the use of wampum, condolence ceremonies, and councils held in key locations in Iroquoia. The Covenant Chain was the vehicle of this diplomatic power and reflected Iroquoian self-determination and what Iris Marion Young has aptly described as "procedural unity."[58] These procedures were components of a diplomatic relationship that was historically rooted, culturally complex, and that bound and constrained Anglophone officials within a set of conventions and norms.[59]

Covenant Chain diplomacy was a system that the British joined rather than fashioned and only partially understood. Its character became a topic of discussion in metropolitan diplomatic circles and in the wider public sphere of print. Accounts of treaties and "Indian affairs" formed part of the expansion of the political culture of empire and the imperial claims to territory that accompanied it. The Iroquois appeared in these texts as both fearsome adversaries and shrewd diplomats, what one observer described as "thorough bred politicians." Cadwallader Colden noted that they "use many Metaphors in their Discourse," which "under the Pen of a skilful Interpreter may strongly move our Passions by their lively Images."[60] William Johnson, writing in the 1750s and describing a collection of treaties he had negotiated, explained that the (male) reader would learn about "the sentiments of nations, as far distant from him on the globe, as, perhaps, he imagines them to be, with respect to his superior attainments as a rational creature." However, the student of global affairs needed to understand that while a "Hyde-park education" was a requirement for grasping European politics, it was "very useless, indeed" when "transported to the banks of the Ohio."[61] Abridged sections of Colden's text became a standard preface for treaties printed in London. For instance, the record of a council held at Philadelphia (1742) was prefixed with an account of the "first Confederacy" and a list of the principal nations, tributaries, and peoples in alliance with the Iroquois. The power of the Confederacy reached—as Mitchell's map showed in detail—deep into the interior to "the remotest Indian Nations of North America, and forc'd them to court the Friendship and Protection of such a formidable Power." Once again, Iroquoian diplomatic practice attracted comment, with the author observing that the Confederacy possessed a "nice Address in the Conduct of their Affairs."[62] It is notable that these accounts approached the Haudenosaunee in terms of power politics and contained no glimpses of their inner workings, whether that included the foundational nature of the Great Law, the centrality of Clan Matrons, or the importance of clans or oral traditions.

The Covenant Chain was a potent mixture of history, symbol, and metaphor, a flexible approach to alliance that revolved around rituals of greeting and renewal; it bound people together and it expanded to the scope of the community that the alliance contained.[63] Its treaty protocols are well known. Council speeches provided narrations of sovereignty and were a medium for the representation of temporal and spatial connections and paths, an Indigenous epistemology of unity, amity, and fictive kinship; they affirmed alliances, served as calls to war, and as declarations of peace, trust, and friendship.[64]

In elements of council speech, wampum was critical, and belts produced for Covenant Chain councils presented multilateralism in pictorial form. At Philadelphia, the Seneca orator Kanickhungo presented the British with a "large belt of white Wampum of eleven Rows, with four black St George's Crosses in it." This belt combined key symbols of the political culture of each nation. Speaking on it, he explained that it bound "all the *English* Governments and all the Indians," and that those "who have the Direction of Affairs in our own nations" employ wampum as "an Account of all these Things . . . faithfully preserved." The Covenant Chain was not only "between the Chiefs of our Nations, and the Chiefs or Principal Men of this Government, but likewise between all our People, and all your People" and "several other Nations of Indians around us."[65] As an object, the belt had been fashioned over a period of time in order to provide a map of alliances, a networked structure that bound together the nations of the Confederacy, the Crown, colonial governments, and the wider international community of Indigenous and Anglophone peoples.

Wampum was also a central element of the treaty council that gathered at Easton in 1758. This was a large conference that convened during war, and its importance is indicated by the fact that colonial clerks—unusually—noted the presence and number of Iroquois women, likely Clan Matrons.[66] A number of belts are mentioned in the record: those exchanged at earlier councils and kept as part of the Iroquoian repository of the material and oral culture of diplomacy; belts sent from place to place to summon nations to councils; belts presented by William Johnson; individual strings of wampum, and requests for the return of belts sent to relay messages. Colonial officials arrived with their own belt, which was fashioned to symbolize the agreements reached around the "ancient" council fire at Philadelphia, presenting what is described as "A large white Belt, with the Figure of a Man at each End, and Streaks of Black, representing the Road from the Ohio to Philadelphia."[67]

The metaphors of path and road predominate in council speech and were employed by both Indigenous and European speakers, not to reference hard boundaries, but to underscore mobility, limber networks, and lines of communication. Yet behind the ritual and metaphorical elements of diplomatic culture lay clear objectives that the Confederacy sought to achieve in diplomatic transactions. Metaphors expressed through speeches and wampum conveyed Haudenosaunee policy that Europeans who entered Indigenous territory exercised no direct supremacy there, instead becoming the newest among neighboring nations. In the words of the Cayuga sachem Toka'en:yon, "A Road has been made from our Country to this Council Fire, that we might treat about Friendship; and as we came down the Road, we saw, that by some Misfortune or other, Blood has lately been spilt on it."[68] The reply, from Thomas King (Saghughsuniunt, Oneida), reveals how the central issues of trade, land, and strategic alliance were put in motion:

> When we first heard of the French coming to *Ohio*, we immediately sent Word to the Governors of *Virginia* and *Pennsylvania*; we desired them to come, and likewise to supply us with such Things as were proper for War, intending to defend our Lands, and hinder the *French* from taking Possession of them: but these Governors did not attend to our Message; perhaps they thought there was no Foundation for our Intelligence. The *French* [] came, and become our Neighbors; and you neither coming yourselves, nor assisting us with warlike Stores, our People of Necessity were obliged to trade with them for what we wanted, as your Traders had left the Country. The Governor of *Virginia* took Care to settle on our Lands for his own Benefit; but when we wanted his Assistance against the *French*, he disregarded us.[69]

The speech reveals not just connections, but their internal dynamics. The orator's words indicate agency in the sense that the nations of the Haudenosaunee determine the nature of their relationship with Europeans and specifically seek trade in desired goods that could be relayed along Indigenous pathways of exchange. Trade, access to Indigenous networks of alliance, and military assistance were vital to the Confederacy's pragmatic political stance that sought to preserve the cohesion and influence of the Haudenosaunee amid interimperial rivalry in the Great Lakes. Alliances with Europeans were effective levers to provide security of territory and to establish buffer zones to limit foreign incursion. Underpinning the whole is a seam of pragmatism combined with an insistence on reciprocity, but this did not mean that the Iroquois were irrevocably bound in obligation to the English.

In the decade preceding the Seven Years' War, British officials used councils to gather information about the reach of the alliance structure of the Covenant Chain. At a council held at Philadelphia in 1742, Governor George Thomas sought information about the location of the warriors of each nation, where zones of peace were established, and whether the details of treaties concluded at Albany had made their way west. His principal interest was the "State and Condition of the Nations to the Westward of the Great Lakes." The answer, by an unattributed speaker, was that the Confederacy "had always Abundance of their Men out amongst the Nations situate to the West of their Lakes. That they had kindled a Fire with a vast many Nations, some whereof were Tributaries, and they had a good Understanding with all." They described embassies to the "southern Indians" and another sent to the Governor of New York. The reply shows the Iroquois fanning out to all points of the compass, extending diplomacy into the Great Lakes Middle Ground to bring nations there into the commercial and political orbit of the British polities of the Northeast, an association that gave the Six Nations additional political power.[70]

The extent of Iroquoian networks and conduits of diplomacy both penetrated and encircled the domains claimed by colonial governments. At the pivotal council at Lancaster (1744), the Onondaga diplomat Canassatego described a landscape of Indigenous alliances and connections.[71] This framework of Indigenous power relations is one that does not explicitly include colonial powers or peoples and concerns passage through Virginia to the Tuscarora:

> with whom, we hold Correspondence; but the Road between us and them, has been stop'd for some Time, on account of the Misbehaviour of some of our Warriours, and they shall keep to that Road; but as that would be very inconvenient, for Messengers, going to the *Tuscaroro's*, we desire, they may go the old Road.[72]

The passage is important not for the substance of the issue but rather for the way in which it reveals the networks of diplomacy between spaces that lacked exact borders. There are two roads, one recently taken by war parties making their way south and the other agreed at Albany in 1722. At a later Albany council (1745), George Clinton reminded colonial officials that the

Covenant Chain bound them to "all the far Nations of *Indians* in League with our Great King," while Canassatego confirmed the reach of Iroquoian alliances "with a Great Number of far *Indians*."[73]

British officials studied and coveted the power that was conveyed by these alliances. From the standpoint of metropolitan officials, the Iroquois opened the door to the west and had the potential to draw other nations into relationships with the colonial inposts of the Anglophone east. According to a commentary attached to the 1746 treaty at Albany, "The Reputation they have gain'd among all the *Indian* Nations in North *America*, gives them an Influence in the Councils of every Nation." The same observer recorded that the Mohawk delegation at Albany was not concerned with "Affairs on the other Side of the Sea" and preferred to attend to "what is properly their Business."[74] That business, as William Johnson noted in a memo to the Board of Trade, centered on relations with other Native Americans. He wrote, "The Eyes of all the Western Tribes of Indians are upon the behaviour of the Six Nations." He continued with the shrewd observation that their "fame of power" lay not in the actual use of force but in the ability to skillfully play the "neutral part." This, he concluded, "I apprehend to be their modern State."[75]

In memos to the Board of Trade, Johnson was emphatic that international diplomacy was a vital element of imperial reason of state: To tap into the circuits of power that shaped the Indigenous interior, the British needed the Iroquois. After the French capitulation in 1761, the English found themselves facing a Confederacy that could set aside neutrality for a renewed phase of imperial policy, and each power looked at the western reaches of colonies and the Ohio as a space to widen access to markets, to consolidate buffer zones, and to cultivate new allies. Johnson was keenly aware of this shift and sought to find a balance between colonial security (forts, frontier settlement) and the expansion of diplomacy using the Covenant Chain as a mechanism: "the Six Nations in particular having a great influence over the rest and being the best barrier for [New York] and the neighboring provinces."[76]

Given the strategic importance of the Haudenosaunee, metropolitan officials and their colonial counterparts devoted a great deal of money and attention to diplomacy, which became a significant burden on the fiscal and military resources of an imperial system spread thinly on American and Asiatic Indigenous continents. The British found themselves under heavy obligations to maintain their standing with the Iroquois, which is suggestive of the cultural "soft" power that the Confederacy could deploy to cement alliances and their accompanying obligations—access to trade goods, gifts of tribute, recognition of Iroquoian territoriality. Johnson sought to maintain clear and swift lines of communication and used the spatial metaphor of the Longhouse to stress the importance of relaying news from locales of the interior back to his base at the eastern door in the Mohawk Valley:

> As you have always been looked upon as the door of the Six Nations where all news, especially from the Westward and Southward must enter and go out, we don't hear this door open as we used to do formerly, and believe it to be worn out, & think it necessary to hang on a new one of such wood as will never decay; the noise of which when it opens should alarm all the Confederacy.[77]

The removal of the French and the decommissioning of forts and garrisons in the Ohio left the British with an imperial infrastructure of costly outposts that served as hubs for trade. In a memo to George III an official admitted that forts were costly, and neither "awe or distress the Indians, tho' they may contribute to our information of their transactions." The question that imperial officials faced in the 1760s was whether the necessary costs of diplomacy could be borne by an imperial economy that emerged from the war with a colossal debt. Writing to Thomas Gage in the wake of the Treaty of Niagara, where over sixty thousand pounds was spent on gifts, Johnson admitted, "I am greatly concerned at the amount of the Indian expences." He continued to insist that diplomatic affairs were worth the costs of "supporting an interest with many nations."[78] Although the Haudenosaunee were willing to negotiate with many other Indigenous nations, they expected that British diplomacy would require the attention of a single sovereign power, national as opposed to local. A memorandum prepared by Lords of Trade on "Indian Affairs" referred to the "alteration" in the "State" of the Crown's dominions, which requires an "alteration in the System," particularly concerning "matters which the Indians would not submit to the discussion of particular colonies." The Confederacy's concern was not with marking out imperial possession but rather with creating a zone of international interaction. This zone, the Lords of Trade acknowledged, was "necessary for the preservation of the British interest with those Indians and for the preventing all foreign influence and Connection" in a setting where the British role is "increased by alliances with those numerous Nations heretofore under the

Dominion of France."[79] Here again, Mitchell's careful placement of Indigenous nations in spaces where borders were either tentative or absent suggests that contemporaries recognized broad zones of interaction.

Interaction with the Haudenosaunee also influenced the shape of the political system that developed in the colonies. Benjamin Franklin's well-known remarks about the political organization of the Six Nations, which has "subsisted Ages, and appears indissoluable," is one of many instances where colonial officials recognized that the model of confederacy was well-suited to the diffusion of nations throughout the northeast.[80] Colonial proximity to Indigenous sovereigns was a strong prompt to the articulation of distinct sovereign powers, all of which pertained to affairs outside sketchy colonial borders. The United Colonies in 1643 declared that "we live encompassed with people of several nations and strange languages," and this fact required a federal approach to "all affairs of our war or peace leagues" necessary for "such a confederation of amity." Similarly, the Albany Plan of Union gave the proposed president general the power "to hold or direct all Indian treaties in which the general interest of the colonies may be concerned; and make peace or declare war with Indian nations."[81] And, at Albany in 1775, the Twelve United Colonies presented themselves to the Iroquois as a diplomatic federation. In draft Articles of Confederation completed the previous month, Franklin recommended "A perpetual Alliance offensive and defensive, is to be enter'd into as soon as may be with the Six Nations; their Limits to be ascertain'd and secur'd to them."[82] The United Colonies' speaker at Albany recalled Canassatego's words (spoken at Lancaster in 1744) on the advantages of confederal government, and on this basis compared the union of colonies to "twelve arrows" lashed together—a common Iroquoian usage.[83] They presented a union belt that signified the historic Covenant Chain and asked the attending nations to remain neutral in the "family quarrel"—a rift between kin—between colonies and Crown that broke the "covenant chain" and led the king to "enter into a new covenant against us." The commissioners declared their intention to hold firm to the alliance, keeping the path "open with all our people and yours, to pass and repass without molestation," and pledging to "seek measures for healing the breach" in the event of conflict.[84] The discussions at Albany took place before the drafting of the Model Treaty and represent a moment when, as Locke expressed it a century earlier, a newly coalescing "foederative power" exercising "the Power of War and Peace, Leagues and Alliances" engaged in structured international relations with "Communities without the Commonwealth."[85] The capacity to conduct international affairs was an intrinsic part of a long gestation of an account of sovereignty that crystallized in the Declaration of Independence, a document that positioned the United States within and beside an international community of sovereigns. Indigenous polities were also clearly present among these powers of the earth.

* * *

If we accept that the Covenant Chain was an international system, locating the international in early America places two key conceptual issues in focus. The first is the premise that the state is both the primary unit in international relations and the product of the dismantling of imperial structures at the end of the eighteenth century.[86] The Iroquois did not relate to states as they are defined by these assumptions: Colonies were dependencies sited—as Mitchell's map indicated—on a ragged periphery of thinly applied imperial power that was inevitably diffused through and shaped by local circumstances. Colonies lacked fixed borders and instead negotiated with Indigenous sovereigns to agree on their legitimacy and course; they lacked full-scale international recognition outside of the northeast; they did not exercise complete sovereignty but sat within an imperial structure of descending law and authority. Therefore, a definition of "international society" comprised of state and nonstate actors does not capture the complexity of nation-to-nation diplomacy that characterized the Covenant Chain.[87] Instead, in this context the international was characterized by embedded diplomatic customs and rituals that were intrinsic to the political culture of the Haudenosaunee. Alliances of kin were thickened by diplomacy and reached toward the far edges of Mitchell's map, drawing Europeans into the orbit of the Confederacy along multiple paths, routes, and circuits of exchange and power.

The second issue is the relationship between colonialism and diplomatic encounter. The spaces and peoples drawn into Covenant Chain diplomacy undoubtedly comprised a "peculiar political arena," a space that contained colonial and imperial political orders but was not ruled by their sovereignty. Likewise, imperial claims to territory were as thin as the lines on Mitchell's map. The sequence of Covenant Chain councils also represents a particular moment in colonial and Indigenous relations, sitting between the more aggressive settler colonial contexts of seventeenth-century Virginia and New England, and the slow but certain orientation of the United States as a power that pledged to respect Indigenous territoriality in legislative mechanisms—the Northwest Ordinance—that were designed to annex it to the public domain of the state.[88] Yet we should not allow the settler colonial framework to trick us into thinking that the pre-Revolutionary eighteenth century was a period "in between," a world of cross-cultural mixing in borderlands that were destined to become states. Instead, the Covenant Chain has a deeply rooted and living history, providing the basis of an account of Indigenous sovereignty within a living "right" relationship among sovereign nations, and in this form remains at the center of contemporary and international discussions of resource stewardship, reconciliation, land claims, and Indigenous rights.[89]

Notes

Research for this essay was supported by the Arts and Humanities Research Council of the United Kingdom (Grant ref: AH/T006099/1).

1. Erez Manela, "International Society as a Historical Subject," *Diplomatic History* 44, no. 2 (2020): 184–209, esp. 185, 186.

2. Pekka Hämäläinen, "The Shapes of Power: Indians, Europeans, and North American Worlds from the Seventeenth to the Nineteenth Century," in *Contested Spaces of Early America*, ed. Juliana Barr and Edward Countryman (University of Pennsylvania Press, 2014), 31–68.

3. Richard White, "Is There a North American History?" *Revue française d'études américaines* 79 (January 1999), 8–28; Joshua Piker, "Lying Together: The Imperial Implications of Cross-Cultural Untruths," *American Historical Review* 116, no. 4 (2011): 964–86, esp. 969, 984; Kristin Hoganson, "Inposts of Empire," *Diplomatic History* 45, no. 1 (2021): 1–22; Brooke L. Blower, "Nation of Outposts: Forts, Factories, Bases, and the Making of American Power," *Diplomatic History* 41, no. 3 (2017): 439–59.

4. Gregory Ablavsky, "Species of Sovereignty: Native Nationhood, the United States, and International Law, 1783–1795," *Journal of American History* 106, no. 3 (2019): 591–613.

5. Michael Witgen, "The Native New World and Western North America," *Western Historical Quarterly* 43, no. 3 (2012): 292–99, esp. 293.

6. Yarimar Bonilla, "Ordinary Sovereignty," *Small Axe* 17, no. 3 (2013): 151–65.

7. Peter Wraxall, "Some Thoughts upon the British Indian Interest in North America," 1756, in *Documents Relative to the Colonial History of the State of New-York*, ed. E. B. O'Callaghan (Weed, Parsons and Company, 1856), 7:26 (hereafter *NYCD*).

8. *Indigenous Diplomacies*, ed. J. Marshall Beier (Palgrave, 2009), 2, 3, 11; Kevin Bruyneel, *The Third Space of Sovereignty: The Postcolonial Politics of U.S.-Indigenous Relations* (University of Minnesota Press, 2007), 23; Roger Epp, "At the Wood's Edge: Toward a Theoretical Clearing for Indigenous Diplomacies in International Relations," in *International Relations: Still an American Social Science?*, ed. Robert Crawford and Darryl Jarvis (SUNY Press, 2001), 299–324, esp. 301, 311, 312, 313.

9. Here, there has been an emphasis on "local legal politics." See Lauren Benton, *A Search for Sovereignty: Law and Geography in European Empires, 1400–1900* (Cambridge University Press, 2010), 5; Annabel Brett, *Changes of State: Nature and the Limits of the City in Early Modern International Law* (Princeton University Press, 2011), chapter 7.

10. Manela, "International Society," 207.

11. Lauren Benton, "Made in Empire: Finding the History of International Law in Imperial Locations," *Leiden Journal of International Law* (2018): 473–78 (quote 475).

12. P. Richard Metcalf, "Who Should Rule at Home? Native American Politics and Indian-White Relations," *Journal of American History* 61, no. 3 (1974): 651–65; Arthur N. Gilbert, "The American Indian and United States Diplomatic History," *History Teacher* 8, no. 2 (1975): 229–41, esp. 230, 233; Emily S. Rosenberg, "Walking the Borders," *Diplomatic History* 14, no. 4 (1990): 565–73; Brian DeLay, "Indian Polities, Empire, and the History of American Foreign Relations," *Diplomatic History* 39, no. 5 (2015): 927–42; Gautham Rao, "The New Historiography of the Early Federal Government: Institutions, Contexts, and the Imperial State," *William and Mary Quarterly* 77, no. 1 (2020): 97–128.

13. Nick Estes, *Our History Is the Future: Standing Rock Versus the Dakota Access Pipeline, and the Long Tradition of Indigenous Resistance* (Verso, 2019), 138; Isabelle Schulte-Tenckhoff, "Reassessing the Paradigm of Domestication: The Problematic of Indigenous Treaties," *Review of Constitutional Studies* 4, no. 2 (1998): 239–89.

14. Antony Anghie, "The Evolution of International Law: Colonial and Postcolonial Realities," *Third World Quarterly* 27, no. 5 (2006): 739–53; "AHR Forum: Liberal Empire and International Law," *American Historical Review* 117, no. 1 (2012): 67–148; Jennifer Pitts, *Boundaries of the International: Law and Empire* (Harvard University Press, 2018), chapter 1; Charles S. Maier, "Dis/Relocating America: Approaches to Global History in the United States," in *Internationale Geschichte in Theorie und Praxis*, ed. Barbara Haider-Wilson, William D. Godsey, and Wolfgang Mueller (Austrian Academy of Sciences Press, 2017), 322; J. Marshall Beier, *International Relations in Uncommon Places: Indigeneity, Cosmology, and the Limits of International Theory* (Palgrave, 2005), 4.

15. Witgen, "The Native New World and Western North America," 294; Jay Gitlin, *The Bourgeois Frontier: French Towns, French Traders, and American Expansion* (Yale University Press, 2009), 31; Richard White, *The Middle Ground: Indians, Empires and Republics in the Great Lakes Region, 1650–1815* (Cambridge University Press, 2011), xxvii; David Armitage, *Foundations of Modern International Thought* (Cambridge University Press, 2013), 7; Adom Getachew, *Worldmaking After Empire: The Rise and Fall of Self-Determination* (Princeton University Press, 2019), 18–19.

16. J. H. Elliott, *The Old World and the New, 1492–1650* (Cambridge University Press, 1970); L. C. Green and Olive P. Dickason, *The Law of Nations and the New World* (University of Alberta Press, 1989); Alexander Orakhelashvili, "The Idea of European International Law," *European Journal of International Law* 17, no. 2 (2006): 315–47, esp. 318, 325.

17. Robert A. Williams, *The American Indian in Western Legal Thought: The Discourses of Conquest* (Oxford University Press, 1990), 7, 287; Robert J. Miller, *Native America, Discovered and Conquered: Thomas Jefferson, Lewis and Clark, and Manifest Destiny* (University of Nebraska Press, 2008), 1; Paul McHugh, *Aboriginal Societies and the Common Law: A History of Sovereignty, Status, and Self-Determination* (Oxford University Press, 2004), 293; Craig Yirush, "'Chief Princes and Owners of All': Native American Appeals to the Crown in the Early Modern British Atlantic," in *Native Claims: Indigenous Law Against Empire, 1500–1920*, ed. Saliha Belmessous (Oxford University Press, 2012), 129–51; Daniel K. Richter, "To 'Clear the King's and Indians Title': Seventeenth-Century Origins of North American Land Cession Treaties," in *Empire by Treaty: Negotiating European Expansion, 1600–1900*, ed. Saliha Belmessous (Oxford University Press, 2015), 5–77.

18. Stuart Banner, *How the Indians Lost Their Land: Law and Power on the Frontier* (Harvard University Press, 2005); Lindsay G. Robertson, *Conquest by Law: How the Discovery of America Dispossessed Indigenous Peoples of Their Lands* (Oxford University Press, 2005); Lauren Benton, *Law and Colonial Cultures: Legal Regimes in World History, 1400–1900* (Cambridge University Press, 2001), 3; Richard Tuck, *The Rights of War and Peace: Political Thought and the International Order from Grotius to Kant* (Oxford University Press, 1999), 111, 122, 138.

19. Kinley Brauer, "The Great American Desert Revisited: Recent Literature and Prospects for the Study of American Foreign Relations, 1815–61," *Diplomatic History* 13, no. 3 (1989): 395–417; William Weeks, "New Directions in the Study of Early American Foreign Relations," *Diplomatic History* 17, no. 1 (1993): 73–96; Kinley Brauer, "The Need for a Synthesis of American Foreign Relations, 1815–1861," *Journal of the Early Republic* 14, no. 4 (1994): 467–76.

20. Bradford Perkins, "Early American Foreign Relations: Opportunities and Challenges," *Diplomatic History* 22, no. 1 (1998): 115–20, esp. 120.

21. Paul A. Gilje, "Commerce and Conquest in Early American Foreign Relations, 1750–1850," *Journal of the Early Republic* 37, no. 4 (2017): 735–70.

22. Max Savelle, "Colonial Origins of American Diplomatic Principles," *Pacific Historical Review* 3, no. 3 (1934): 334–50, esp. 335.

23. Emily S. Rosenberg, "A Call to Revolution: A Roundtable on Early U.S. Foreign Relations," *Diplomatic History* 22, no. 1 (1998): 63–70, esp. 63, 65.

24. Karen Kupperman, "International at the Creation: Early Modern American History," in *Rethinking American History in a Global Age*, ed. Thomas Bender (University of California Press, 2002), 105, 106, 109 (margins).

25. Patricia Nelson Limerick, *The Legacy of Conquest: The Unbroken Past of the American West* (W. W. Norton, 1987), 27; Patricia Nelson Limerick, "Going West and Ending Up Global," *Western Historical Quarterly* 32, no. 1 (2001): 4–23; Nathan J. Citino, "The Global Frontier: Comparative History and the Frontier-Borderlands Approach in American Foreign Relations," *Diplomatic History* 25, no. 4 (2001): 677–93, esp. 680.

26. Ian Tyrrell, "Beyond the View from Euro-America: Environment, Settler Societies, and the Internationalization of American History," in *Rethinking American History in a Global Age*, ed. Thomas Bender (University of California Press, 2002), 171, 177, 178; Witgen, "Native New World," 293.

27. Paulina Starski and Jörn Axel Kämmerer, "Imperial Colonialism in the Genesis of International Law—Anomaly or Time of Transition?" *Journal of the History of International Law* 19 (2017): 56, 58.

28. Eliga Gould, *Among the Powers of the Earth: The American Revolution and the Making of a New World Empire* (Harvard University Press, 2012), 30.

29. Leonard J. Sadosky, *Revolutionary Negotiations: Indians, Empires, and Diplomats in the Founding of America* (University of Virginia Press, 2009), 3, 7, 8.

30. Robbie J. Totten, "Security, Two Diplomacies, and the Formation of the U.S. Constitution: Review, Interpretation, and New Directions for the Study of the Early American Period," *Diplomatic History* 36, no. 1 (2012): 77–111, esp. 79, 104; *The Federalist with Letters of "Brutus,"* ed. Terrence Ball (Cambridge University Press, 2003), 124; Max M. Edling, *Perfecting the Union: National and State Authority in the US Constitution* (Oxford University Press, 2021), 7–15; Gregory Ablavsky, *Federal Ground: Governing Property and Violence in the First U.S. Territories* (Oxford University Press, 2021), chapters 4, 5; Bethel Saler, *The Settler's Empire: Colonialism and State Formation in America's Old Northwest* (University of Pennsylvania Press, 2015), 27, 28.

31. Alan Taylor, *The Divided Ground: Indians, Settlers and the Northern Borderland of the American Revolution* (Alfred A. Knopf, 2006), chapter 9.

32. Kent McNeil, "Factual and Legal Sovereignty in North America: Indigenous Realities and Euro-American Pretensions," in *Sovereignty: Frontiers of Possibility*, ed. Julie Evans et al. (University of Hawai'i Press, 2013), 50.

33. Jeffrey Glover, *Paper Sovereigns: Anglo-Native Treaties and the Law of Nations, 1604–1664* (University of Pennsylvania Press, 2014), 3, 5.

34. Dorothy V. Jones, *License for Empire: Colonialism by Treaty in Early America* (University of Chicago Press, 1982), 1.

35. Edward Keene, "The Treaty-Making Revolution of the Nineteenth Century," *International History Review* 34, no. 3 (2012): 479–80.

36. Nancy Shoemaker, *A Strange Likeness: Becoming Red and White in Eighteenth-Century North America* (Oxford University Press, 2004), 86, 87; DeLay, "Indian Polities," 939; Donna L. Akers, "Decolonizing the Master Narrative: Treaties and Other American Myths," *Wicazo Sa Review* 29, no. 1 (2014): 58–76.

37. *Indigenous Rights in the Age of the UN Declaration*, ed. Elvira Pulitano (Cambridge University Press, 2012).

38. Jennifer Pitts, *Boundaries of the International: Law and Empire* (Harvard University Press, 2018), chapter 1. See titles cited 194n5. Anghie argues that it was colonialism that made international law "universal," but retained what he calls a "dynamic of difference," which served as the justification for denying sovereignty to certain cultures. Anghie, "Evolution of Natural Law," 742.

39. Charles Maier, *Once Within Borders: Territories of Power, Wealth and Belonging Since 1500* (Harvard University Press, 2016), 76; Benjamin Mueser, "The Nation and Property in Vattel's Theory of Territory," *Global Intellectual History* 3, no. 2 (2018): 140.

40. White, *Middle Ground*, xxvii; Benton, *Search for Sovereignty*, 2; Samuel Truett, "Settler Colonialism and the Borderlands of Early America," *William and Mary Quarterly* 76, no. 3 (2019): 438.

41. S. Max Edelson, *A New Map of Empire: How Britain Imagined America Before Independence* (Harvard University Press, 2017), chapter 4; *An Argument Delivered on the Part of New-York at the Hearing [] to Settle and Determine the Boundary Line* (London?, 1769), 28; Matthew H. Edney, "John Mitchell's Map of North America (1755): A Study in the Use and Publication of Official Maps in Eighteenth-Century Britain," *Imago Mundi* 60, pt. 1 (2008): 63–85; Chad Anderson, "Rediscovering Native North America: Settlements, Maps, and Empires in the Eastern Woodlands," *Early American Studies: An Interdisciplinary Journal* 14, no. 3 (2016): 478–505.

42. Jeffers Lennox, *Homelands and Empires: Indigenous Spaces, Imperial Fictions, and Competition for Territory in Northeastern North America, 1690–1763* (University of Toronto Press, 2017), chapters 1–2.

43. Martin Brückner, *The Geographic Revolution in Early America: Maps, Literacy, & National Identity* (University of North Carolina Press, 2006), 65.

44. Ken MacMillan, *Sovereignty and Possession in the English New World: The Legal Foundations of Empire, 1576–1640* (Cambridge University Press, 2006), chapters 3, 5.

45. J. B. Harley, "New England Cartography and the Native Americans," in *The New Nature of Maps: Essays in the History of Cartography*, ed. Paul Laxton (Johns Hopkins University Press, 2011), 170–95; Arthur J. Krim, "Acculturation of the New England Landscape: Native and English Toponymy of Eastern Massachusetts," in *New England Prospect: Maps, Place Names and*

the Historical Landscape, ed. Peter Benes (Boston University Press, 1982), 69–88; Mark Warhus, *Another America: Native American Maps and the History of Our Land* (St. Martin's Press, 1997), chapter 2; Benton, *Search for Sovereignty*, 10–12, 14, 56; Karen Haltunnen, "Grounded Histories: Land and Landscape in Early America," *William and Mary Quarterly* 68, no. 4 (2011): 513–32.

46. William Berkeley, *A discourse and view of Virginia* (1663?), 3, 4, 6, 10; William Talbot, *The discoveries of John Lederer, in three several marches from Virginia to the West of Carolina* (London, 1672), "Dedication," n. p.; 3–5.

47. [Anon], *The Late Occurrences in North America, and the Policy of Great Britain Considered* (London, 1766), 3, 10, 31.

48. *Early American Indian Documents: Treaties and Laws, 1607–1789*, ed. Barbara Graymont (1996), 9:593 (hereafter *EAID*).

49. *EAID*, ed. Barbara Graymont (2001), 10:549, 550, 552. This "myth" and "mystique" of Iroquois conquest was used as an expedient by William Johnson to neutralize the Shawnee and Delaware, but figures like Gage were skeptical: "If we are to search for truth and examine her to the Bottom, I don't imagine we shall find that any conquered Nation ever formally ceded their Country to their Conquerors, or that the latter ever required it." Quoted: White, *Middle Ground*, p. 352; see also Colin G. Calloway, *Pen and Ink Witchcraft: Treaties and Treaty Making in American Indian History* (Oxford University Press, 2013), 49.

50. Francis Jennings, *The Ambiguous Iroquois Empire: The Covenant Chain Confederation of Indian Tribes with English Colonies* (W. W. Norton, 1984), chapter 2.

51. Daniel K. Richter, *The Ordeal of the Longhouse: The Peoples of the Iroquois League in the Era of European Colonization* (University of North Carolina Press, 1992), 60–74.

52. *NYCD* (1857), 8:47.

53. Rachel St. John, "Imperial Spaces in Pekka Hämäläinen's *The Comanche Empire*," *History and Theory* 52 (2013): 76, 77; Rachel St. John, "State Power in the West in the Early American Republic," *Journal of the Early Republic* 38, no. 1 (2018): 87, 90; Alejandra Dubcovsky, *Informed Power: Communication in the Early American South* (Harvard University Press, 2016), 15, 25.

54. Neta C. Crawford, "A Security Regime Among Democracies: Cooperation Among Iroquois Nations," *International Organization* 48, no. 3 (1994): 345–85, esp. 345.

55. David Bedford and Thom Workman, "The Great Law of Peace: Alternative Internation(al) Practices and the Iroquoian Confederacy," *Alternatives: Global, Local, Political* 22, no. 1 (1997): 88; Christopher Vecsey, "The Story and Structure of the Iroquois Confederacy," *Journal of the American Academy of Religion* 54, no. 1 (1986): 79, 86, 89.

56. Jones, *License for Empire*, 21, 30; Hämäläinen, "Shapes of Power," 46–50.

57. Alexandra Harmon, "American Indians, American Law, and Modern American Foreign Relations," *Diplomatic History* 39, no. 5 (2015): 943–54, esp. 950, 954; Coll Thrush, *Indigenous London: Native Travelers at the Heart of Empire* (Yale University Press, 2016), chapter 3.

58. Iris Marion Young, "Hybrid Democracy: Iroquois Federalism and the Postcolonial Project," in *Political Theory and the Rights of Indigenous Peoples*, ed. Duncan Ivison, Paul Patton, and Will Sanders (Cambridge University Press, 2000), 241.

59. Gail MacLeitch, *Imperial Entanglements: Iroquois Change and Persistence on the Frontiers of Empire* (University of Pennsylvania Press, 2011), 2; Eric Hinderaker, "Diplomacy Between Britons and Native Americans, c. 1600–1830," in *Britain's Oceanic Empire: Atlantic and Indian Ocean Worlds, c. 1550–1850* (Cambridge University Press, 2012), 221; Nancy L. Hagedorn,

"'A Friend to Go Between Them': The Interpreter as Cultural Broker During the Anglo-Iroquois Councils, 1740–70," *Ethnohistory* 35, no. 1 (1988): 60–80, esp. 61.

60. *Memoir of Lieut. Col. Tench Tilghman, Secretary and Aid to Washington*, ed. S. A. Harrison (Albany, 1876), 99; Cadwallader Colden, *The History of the Five Indian Nations Depending on the Province of New York In America* (New York, 1727), ix.

61. *An Account of Conferences held and Treaties made, between Major-general Sir William Johnson [and the] Indian Nations of North* America (London, 1756), ii, iii.

62. *The Treaty Held with Indians of the Six Nations at Philadelphia* (London, 1742), ix, xi.

63. Charles W. A. Prior, *Settlers in Indian Country: Sovereignty and Indigenous Power in Early America* (Cambridge University Press, 2020), 12–26; Jacob T. Lee, *Masters of the Middle Waters: Indian Nations and Colonial Ambitions Along the Mississippi* (Harvard University Press, 2019), 5.

64. William N. Fenton, *The Great Law and the Longhouse: A Political History of the Iroquois Confederacy* (University of Oklahoma Press, 1998), chapter 16.

65. *A Treaty of Friendship held with the Chiefs of the Six Nations at Philadelphia* (Philadelphia, 1736), 7, 8.

66. MacLeitch, *Imperial Entanglements*, chapter 4.

67. *The Minutes of a Treaty held at Easton in Pennsylvania, in October 1758* (Woodbridge, New Jersey, 1758), 25

68. *Minutes of a Treaty*, 8.

69. *Minutes of a Treaty*, 19.

70. *Treaty Held with Indians of the Six Nations at Philadelphia*, 9.

71. Craig Yirush, "'Since We Came Out of This Ground': Iroquois Legal Arguments at the Treaty of Lancaster," in *Justice in a New World: Negotiating Legal Intelligibility in British, Iberian, and Indigenous America*, ed. Brian P. Owensby and Richard J. Ross (New York University Press, 2018), 118–150.

72. *The Treaty Held with the Indians of the Six Nations at Lancaster* (Williamsburg, 1744), 70.

73. *An Account of the Treaty held at the City of Albany, in the Province of New York . . . with the Indians of the Six Nations* (Philadelphia, 1745), 10, 13.

74. *Treaty between his Excellency the Honourable George Clinton . . . and the Six United Indian Nations* (New York, 1746), 9, 23.

75. "Colonel Johnson's suggestions for defeating the designs of the French," July 1754, *NYCD* (1855), 6:897.

76. "Sir William Johnson to the Lords of Trade," 13 November 1763, *NYCD* (1856), 7:581.

77. "Conference between the Mohawks and Colonel Johnson," 26 July 1753, *NYCD*, 6:811.

78. "Memorandum on the American Colonies," October 1766?, GEO Main 415a, Royal Archives, Windsor, fol. 5; *The Papers of Sir William Johnson*, ed. Milton W. Hamilton (University of the State of New York, 1953), 11:276, 521; William Johnson, "Review of the Trade and Affairs of the Indians in the Northern District of America," September 1767, *NYCD*, 7:973.

79. "Representation of the Lords of Trade on the State of Indian Affairs," 7 March 1768, *NYCD*, 8:20, 21, 23.

80. *The Papers of Benjamin Franklin*, vol. 4, *July 1, 1750, through June 30, 1753*, ed. Leonard W. Labaree (Yale University Press, 1961), 117–21.

81. "The New England Confederation" (1643), in *Colonial Origins of the American Constitution: A Documentary History*, ed. Donald S. Lutz (Liberty Fund, 1998), 368, 373.

82. *The Declaration of Independence in Historical Context*, comp. and ed. Barry Alan Shain (Yale University Press, 2014), 644.

83. Prior, *Settlers*, 23; *EAID*, 10:87–88; *NYCD*, 7:146.

84. "Treaty with the Six Nations, Albany, 25 August 1775," *NYCD*, 8:615, 616, 620; *Declaration of Independence*, 358.

85. John Locke, *Two Treatises of Government*, ed. Peter Laslett (Cambridge University Press, 1988), §. 146; David Armitage, *The Declaration of Independence: A Global History* (Harvard University Press, 2007), 35, 47, 83.

86. Armitage, *Declaration of Independence*, 17–18.

87. Manela, "International Society," p. 186.

88. Saler, *Settler's Empire*, ch. 1; Jeffrey Ostler, "'Just and Lawful War' as Genocidal War in the (United States) Northwest Ordinance and Northwest Territory, 1787–1832," *Journal of Genocide Research*, 18, no. 1 (2016): 1–20.

89. *The Right Relationship: Reimagining the Implementation of Historical Treaties*, ed. John Borrows and Michael Coyle (University of Toronto Press, 2017); *Keeping Promises: The Royal Proclamation of 1763, Aboriginal Rights, and Treaties in Canada*, ed. Terry Fenge and Jim Aldridge (McGill-Queen's University Press, 2015); Michael Asch, *On Being Here to Stay: Treaties and Aboriginal Rights in Canada* (University of Toronto Press, 2014); *From Recognition to Reconciliation: Essays on the Constitutional Entrenchment of Aboriginal and Treaty Rights*, ed. Patrick Macklem and Douglas Sanderson (University of Toronto Press, 2016); Mark D. Walters, "Brightening the Covenant Chain: Aboriginal Treaty Meanings in Law and History After Marshall," *Dalhousie Law Journal* 24, no. 2 (2001): 75–138, esp. 88, 89.

CHAPTER 9

"Playing the Old Game of Double"

Navigating Creek and Spanish Geopolitics in the Post-Revolutionary Gulf South

CHAD MCCUTCHEN AND JENNIFER MONROE MCCUTCHEN

In many ways, the Peace of Paris in 1783 signaled the death knell of European colonialism in North America. The British proved incapable of preventing the loss of a sizeable portion of their original colonies, holding on to Canada and a few lucrative Caribbean islands. The French, having already abandoned most of their claims in the region, ended the Revolutionary War on the verge of bankruptcy, with little to show for their efforts beyond the satisfaction of playing a part in the misfortune of a rival. The independence gained by the former colonies seemed to be more ominous than atypical, and the Spanish indeed soon followed suit, losing most of their empire in the Americas. Within four decades, the bulk of this territory would be free of European colonial rule.

Yet the outlook among Spanish officials in the Indies at the close of the war was hardly bleak. Compared to its European counterparts, Spain emerged from the American Revolution well positioned to reassert its dominance in the region. Unable to achieve a primary goal of reclaiming the island of Gibraltar from the British in the Mediterranean, Spain fared much better in the Americas as the British ceded control of East and West Florida. Coupled with their existing colonies in Cuba and Louisiana, the Spanish were now strategically prepared to control the Caribbean and the Gulf of Mexico. And unlike their allies in France, they had not completely voided themselves of their financial resources to accomplish these objectives. But the region's Native populations also remained politically and militarily powerful. The Cherokees, Choctaws, Chickasaws, Creeks (Muscogees), and Seminoles found themselves in a familiar position, wedged between competing factions intent on realizing their own designs. Experience had revealed the necessity of adjusting community-based diplomacy and local-level sociopolitical structures to meet new challenges to the members of the towns, clans, and coalitions known to Euro-Americans as the Creek Confederacy. Consequently, Muscogee Creeks drew from adaptive approaches that they utilized, first, to undermine British influence after the Seven Years' War, and then to preserve autonomy and independence in their interactions with the Spanish and the Americans after the Revolution.[1]

This chapter recenters Spain as a major political power in the region immediately following the American Revolution and analyzes the Creeks and other Native peoples as prominent political and economic agents rather than passive observers. It argues that actions and events during this time were representative of a transitional phase in which the above-mentioned groups crafted diplomatic strategies based on both colonial experience and emerging sociopolitical ideologies. Yet, while both sought to achieve relative regional sovereignty through diplomatic means, the ways in which each interpreted the meaning and outcome of that sovereignty depended on their own cultural and historical understandings in addition to their present and future designs. Enlightenment thought and Bourbon economic reforms ushered in a shift in Spanish policy over the course of the eighteenth century that emphasized trade and diplomatic alliances over the conquest of the Natives in the Gulf South. The Creeks' noncentralized, consensus-based sociopolitical structure, with its emphasis on tension and opposition, allowed them to replicate the conditions of imperial rivalry that served as a fundamental piece of their Indigenous identity earlier in the century, applying it to a new era during which neither Spain nor the fledgling United States were powerful enough to impose their will. As new political pressures arose, such as the potential of a now-unchecked US appetite for westward expansion, Spanish and Creek designs unexpectedly came to overlap, with territorial claims, access to coastlines, and control over interior waterways becoming increasingly important to Spanish officials and Indigenous community leaders.

Recent scholarship largely examines the decades that follow the American Revolution within either a broad transatlantic context by tracing the spread of liberal thought and revolutionary movements or a national perspective that emphasizes the emergence of the United States. In the North

American southeast, these earlier works focused largely on US efforts to "civilize" Native Americans for assimilation into white American society, followed by campaigns to dispossess Native peoples of their lands to facilitate territorial expansion and the growth of the "cotton kingdom."[2] Such arguments gave rise to the traditional, westward-facing narrative of US history, which highlights white American exceptionalism, democracy, and progress. But some works, like Kathleen DuVal's *Independence Lost*, explore the post-Revolutionary period through a continental lens. This approach expands historical understandings of nation-building beyond that of the United States to include Indigenous projects occurring in the same period. Alan Taylor's *American Revolutions* also takes a continental approach. It incorporates the perspectives and experiences of North America's diverse populations and uses the plural "Revolutions" to underscore that these groups all pursued their own vision of revolution and nationhood during the United States' formative years. Colin Calloway's *The Indian World of George Washington* builds upon these narratives by examining the complexities of American nation-building in relation to Native peoples and Washington's efforts to acquire Indigenous lands following the Revolution.[3]

Others have focused on Native peoples as active agents in international diplomacy, challenging the longstanding historical argument that Indigenous societies lacked a formalized foreign policy, remained largely disconnected from global events, and served no substantial role in larger narratives of US imperialism. Brian DeLay, in a historiographical essay, details these oversights, arguing that traditional concerns of late eighteenth- and nineteenth-century history are "in fact deeply, intricately entangled with Indian history."[4] DeLay urges scholars to investigate local-level power and diplomatic relationships, as well as formal diplomacy, and pushes the field toward "engage[ment] with Indian polities and individuals as historical actors, rather than mere objects of state and settler aggression."[5] James Hill's *Creek Internationalism in an Age of Revolution* responds to DeLay's call by centering Muscogee Creek, Seminole, and Miccosukee communities within the diplomatic history of the late eighteenth- and early nineteenth-century Atlantic world. Hill demonstrates that these peoples, despite divergent views on the future sociopolitical structure of their societies, understood the necessity of building extensive, international diplomatic networks and deploying European legal and political concepts to preserve their autonomy in the eyes of the United States. Indigenous leaders strategically advanced their diplomatic goals through travel and communication across vast maritime spaces and overland routes, while also building alliances and gaining access to important foreign trade goods.[6] But the United States' appetite for westward expansion rendered this period of internationalism short lived. As Deborah Rosen argues in *Border Law*, the first Seminole War (1818) ushered in a "sharp, racially inflected line between 'savage' and 'civilized' people" that allowed the United States to reject the universal application of international law to all populations within the nation, regardless of racial or cultural background. Moving forward, Rosen argues, US law would only fully protect peoples of European descent, removing Native Americans as equal participants in international diplomacy.[7]

Furthermore, only a few scholars have explored this period through a regional lens that equally considers the Indigenous, Spanish, and US perspectives. In *Bárbaros*, David J. Weber builds off earlier work on the Spanish borderlands, including his own *The Spanish Frontier in North America*, to investigate how the Bourbon reforms impacted a wide array of "independent Indians."[8] Weber highlights the Spanish perception of Native Americans as "*bárbaros*," or savages, and Spanish descriptions of independent Indians as unconquerable peoples living on the fringes of the Spanish empire. He argues that the failure of eighteenth-century Spanish Indian policy was its pragmatism, or its inability to consolidate a single approach. Sometimes the Spanish used the "modern" approach of trade and gift-giving to foster diplomacy; other times they relied on their traditional method of warfare and violence to try to subdue Native populations.[9] DuVal's aforementioned work also focuses on the Gulf South during this period, exploring the intersecting lives of Natives, Spaniards, Loyalists, and enslaved peoples. Her research illuminates the uniqueness of the Gulf, emphasizing that its inhabitants were largely disconnected from people in the northeast or mid-Atlantic regions, developing a distinct give-and-take culture that shaped interactions and behaviors. DuVal defines this culture as one of "advantageous interdependence" and uses it as a jumping-off point to consider the changing nature of dependence and independence throughout the course of the Revolution. She argues that independence was not always ideal, as it signified isolation, fostered a lack of diplomatic or exchange relationships, and potentially rendered populations vulnerable.[10]

This chapter builds upon this research by offering a nuanced historical analysis of the pivotal years following the independence of the British colonies in North America. It considers diverse cultural perspectives, local-level needs, ideological shifts, economic demands, and regional geopolitics, revealing that European and Native American groups adjusted to the changes that followed the American Revolution in ways that combined previous knowledge with new ideologies. Experience taught both Spanish officials and Native peoples that a cohesive, single economic or political policy was unrealistic, as the Gulf South's diverse Indigenous societies functioned through local-level diplomacy that ebbed and flowed according to individual needs and interests. In 1783 neither the prominence of the United States nor the independence of Spain's colonies was a forgone conclusion. In no way resigned to accept the dissolution of their cultural and political sovereignty, Creeks and many of their Indigenous neighbors looked to the nearby Spanish, as well as the burgeoning United States, to shape both the preservation of their own sovereignty and the contours of nation-building in the larger Atlantic world.[11] Following trends in recent scholarship, this chapter utilizes concepts of nation-building and territorial sovereignty to recenter Spain as a major political power in the region after the American Revolution and analyze the Creeks, and other Native groups, as prominent political and economic agents rather than passive observers.

* * *

The fear of rapid US expansion from the Appalachian region to the Mississippi River seemingly made the Spanish and the Creeks natural allies, but the latter were apprehensive of fickle European diplomacy. Scattered through modern-day Georgia, Alabama, and northern Florida, Creek towns functioned as autonomous political entities through the majority of the eighteenth century. This measure of political decentralization allowed Creek *micos*, or village headmen, to independently sign treaties with Euro-American outsiders without approval from the Confederacy as a whole. Consequently, many Creek headmen viewed the end of the American Revolution as the beginning of a new epoch requiring innovative methods of adaptation to protect their independence in ways that were reminiscent of the diplomatic strategies they employed in earlier decades. But the persistence of community-level authority and coalition-based politics within the Creek Confederacy threatened the efforts of Indigenous leaders who sought to delineate Creek nationhood in ways that Euro-Americans could understand. European and American outsiders commonly viewed these differing approaches to sovereignty as divisive and incorrectly assumed these activities spread chaos among Creek peoples. These views prompted some, like US Indian agent Richard Thomas, to lament that "the [Creek] Indians appear to me to be playing the old game of double, promising and never performing."[12]

From an Indigenous perspective, however, the Creeks *were* performing. They were successfully adapting and executing the local, decentralized, balance of power diplomacy that had served their interests earlier in the eighteenth century and had afforded the larger Confederacy flexibility during the American Revolution. Yet, Thomas was not the first to incorrectly interpret Creek sociopolitical strategies. His words—"playing double"—echo those of observers from earlier decades, who often lamented the Creeks' commitment to diplomatic neutrality, a formal, protective policy that forced Europeans to negotiate with individual town leaders and protected the Confederacy from committing to a strict alliance with any single empire. In 1766, for example, Governor Johnstone of British West Florida complained that the Mortar, an Upper Creek headman with French proclivities, provoked indignation among his people by entertaining "idle talks" about the British following their occupation of former French land claims. The Mortar, Johnstone complained, was only concerned with "his own power, which depends on confusion among his people from whence all kinds of disorders has spread through his country."[13] At the onset of the American Revolution, Britain's deputy superintendent of Indian affairs to the Creeks, David Taitt, complained of the Lower Creeks' refusal to commit to a Loyalist alliance. "The Creeks," he wrote in 1776, "would much rather wish to enjoy the advantages of a neutrality by being paid from both parties."[14]

In other words, competing national and coalition interests preserved the important political divisions that had safeguarded Creek autonomy in an earlier era, fortuitously mimicking the diplomatic neutrality of previous decades.[15] Many Creek headmen simply regarded the American victory as the beginning of a new era that would require the adaptation of existing sociopolitical and diplomatic strategies. Spanish officials shared this view, particularly in their diplomacy with Gulf South Native populations. The Spanish approach toward the Creeks was an adaptation of a strategy that had been evolving for some time as part of the larger reforms implemented by the Bourbons, beginning in 1714 with the conclusion of the War of Spanish Succession and confirmation of Bourbon rule in Spain. Early attempts at sociopolitical reform on the Peninsula had ended in mass riots, effectively shifting focus to the condition of the colonies by the 1760s.[16] There was little doubt at this time that Spain was wholly reliant on the colonies to generate revenue.[17] The general goals of the Crown's new policies in the Americas centered on increasing royal administrative control over the colonies while maximizing the revenue they generated for Spain.

Indigenous populations figured prominently in the Bourbon monarch's new imperial designs. Under Habsburg rule, the Spanish had focused on conquest, both physical and spiritual. But the influence of the Enlightenment and the growing preference for rationalism rendered reliance on conflict a relic of Habsburg inefficiency. Within the larger context of the Bourbon fiscal reforms, the Spanish considered Native peoples outside of their direct control an untapped resource. Rather than subjugating and exploiting them for labor, they sought to incorporate Indigenous populations into their economic system.[18] By exchanging hostility for friendship, the Spanish could "conquer" through commerce. Trade, not conquest or conversion, would be the bond that tied Native Americans to the Spanish. The influence of Enlightenment ideology on these new policies was, however, never substantial enough to be politically liberal. Charles III was a staunch absolutist, and he geared his reformist tendencies toward tightening control to strengthen peninsular Spain both politically and economically. Bourbon Indigenous policy, while perhaps more benevolent on the surface, was still exploitative. The Bourbons based their policies on a transitional socioeconomic philosophy that bridged Habsburg paternalism with more modern Bourbon ideologies in an attempt to align Spanish colonialism with the successful policies of Spain's European counterparts with shared economic interests in the Americas.

This transitional approach in Bourbon economic thinking in conjunction with Spanish–Indigenous diplomacy was best outlined in a treatise attributed to reformer and Spanish official José del Campillo y Cosío titled *Nuevo sistema de gobierno económico para la América* (New System of Economic Government for America).[19] Campillo felt that it was only natural that a "warrior spirit" prevailed throughout the sixteenth century, when Charles I was charged with "having to subdue millions of Indians and their caciques," who in turn "defended their liberty with ferocity." Now, through better policy, he argued, the Natives could be "encouraged" to work through better treatment and the introduction of trade, so they could become "useful Spanish vassals" rather than remaining oppressed and having "their character looked down upon, as it has been and is to date."[20]

The main issue for Campillo regarding Spain's Indigenous policy was the continual insistence on territorial hegemony requiring the subjugation of Native peoples. He advocated that Spanish officials should follow the lead of the French, who "do not intend to subdue the natives, but rather obtain their friendship and trade."[21] Indigenous groups could be encouraged through trade to provide a diverse range of products for the empire to supplement the wealth of the mines. Yet while the recommended strategy advocated better treatment of Native Americans, the exploitation and inherent paternalism of the Habsburgs remained; Campillo advised incorporating Indigenous groups into his New System "to civilize them," that they might become "powerful and indispensable auxiliaries . . . for themselves, for commerce, for [Spanish] policy, and for the augmentation of the Royal Treasury."[22]

* * *

Implementing a single economic or political policy for Native peoples in the Americas was unfeasible, as Spain's empire was vast and its Indigenous populations diverse.[23] Significant disparity existed between the territory Spain claimed and that which it controlled. The Spanish alleged dominion over all lands beyond 270 leagues from the coast of Africa dating back to the Treaty of Tordesillas in 1494. And unlike their European counterparts, the Spanish had declared that all Native inhabitants within these claims were subjects of the Crown, limiting the political and diplomatic options available to them in their interactions with Indigenous people.

Yet in practice, Spanish officials understood these claims were dubious, and other European powers had long ignored them. The scarcity of Spaniards inhabiting the Gulf South greatly weakened their claims. Participation in the Seven Years' War resulted in the forfeiture of Florida to the British in exchange for Havana, which was quickly captured after Spain entered the conflict.[24] British possession of land along the Gulf of Mexico was perilous to Spanish designs, as it posed a potential disruption of their supply lines to and from New Spain. Were it not for the French cession of the Louisiana territory in 1762, Spanish claims to any continental territory east of the Mississippi would have been nonexistent.

Recognizing the strategic value of the region, Charles III remained undaunted by the failures of the French, but the transfer of power in New Orleans moved slowly.[25] The newly appointed Spanish governor of Louisiana, Antonio de Ulloa, nevertheless wasted little time attempting to establish control of the Mississippi. He sent an expedition up the river to establish a fort in the Illinois Country, with explicit instructions on how to deal with the Natives.[26] The directives reflected the policies outlined in the New System, with certain adaptations indicative of the policies local Spanish officials later applied east of the Mississippi. Ulloa ensured that the expedition was heavily stocked with gifts for the Indigenous population. He also advised his officials to dispel them judiciously, so as to be sure to have more gifts to give when the Natives returned.[27] Gift-giving reflected the governor's desire to avoid conflict at all costs. Ulloa regularly reminded his subordinates of the professed emotional nature of the Natives, and he feared that Indian groups in the Mississippi region would turn to war at the slightest affront to their pride.[28] His apprehension over the potential conflicts that frequently arose between Natives and Europeans was so profound that Spaniards were ordered to wait and seek redress through the headmen and never attempt to resolve the problem themselves, even in disagreements between individuals where the Natives were deemed to be entirely in the wrong.[29]

The importance of gift-giving also demonstrates that establishing a Spanish presence in the Gulf South was only achievable with the consent of the local Indigenous communities. Colonists were to tread carefully with Natives in the Mississippi region, understanding that courting these ethnically, politically, and linguistically diverse peoples would require a different approach than those used with larger Indigenous polities. Forts were to be established, not as a means of aggression or even defense, but rather as a foundation to solidify claims and establish inroads for future endeavors. As Ulloa advised, "It must be inferred that at the same time one sets about the founding of a fort, that he is aiming also at the creation of a settlement; for a settlement will be of great aid and help in maintaining a fort."[30] In order to carry out Spanish designs along the Mississippi, these forts were to serve as fledgling communities and trading centers that encouraged future settlement while establishing and maintaining ties with resident Indigenous groups.

The governor sent secret instructions to establish two forts at the mouth of the Missouri River at the confluence of the Mississippi solely to deter British encroachment, which had begun in the area in the hopes of securing alliances with the Natives since the Spanish had "not had any settlement or defense hitherto."[31] The Spanish were invariably clear about their boundaries and their claims to all lands west of the Mississippi. They naively tried to force local populations to respect these boundaries in the hope of deterring them from seeking trade with the British to the east. As the governor pointed out, "each nation has its territory and its savage tribes with whom to trade."[32] In what was going to become a fundamental aspect of Spanish policy moving forward, the diplomatic and commercial relationships with Native groups served as a proxy for territorial hegemony, with the overall goal of rerouting Indian trade goods through New Orleans. Attempts to control the Mississippi

River far inland from the ocean show the strategic importance of the waterway both commercially and politically, which were becoming increasingly intertwined in Spanish Indigenous policy.

Unlike the populations west of the Mississippi, large Indian confederations east of the Mississippi were well versed in dealing with European administrators.[33] For the majority of the eighteenth century, these peoples, particularly the Creeks, had maintained authority in the region by remaining neutral in their dealings with neighboring Euro-Americans.[34] This strategy encouraged Creek headmen to act according to community-level interests, ensuring access to numerous sources of trade goods and guaranteeing that the larger society did not wholly align with one foreign power.[35] The end of the Seven Years' War, however, threatened Creek neutrality as the Peace of Paris effectively removed the French from the region and relegated the Spanish to New Orleans and Cuba. Upper Creeks believed these shifts might prove beneficial, as the British occupation of Pensacola could bring greater quantities of goods at better prices closer to their communities.[36] Lower Creeks viewed British expansion as a threat, fearing a rapid expansion of Anglo-American settlement. Thus, through the 1760s and 1770s, many Muscogee and Hitchiti-speaking peoples shifted their focus southward, maintaining ties to the Spanish through voyages to Cuba.[37] Travels to Cuba ensured that Creeks could mimic the play-off system of earlier decades, despite the European geopolitical reconfigurations that resulted from the Seven Years' War. This strategy directly benefited those who made the journey, while also protecting the sovereignty of the Confederacy as a whole.[38]

Spain also had something to gain from this relationship. Having been completely caught off guard by the rapid attack that led to the loss of Havana in the previous conflict, officials in Cuba were determined not to let it happen twice. Since the British had forced the exchange of Florida in return for the vital port city in Havana, the Spanish often lacked information from the continent. Cuban merchants and fishermen were always in close contact with Indigenous groups, engaging in frequent trading along the coast. This network became Spain's continental eyes and ears, offering knowledge of British naval movements, sightings of pirates and corsairs, and invaluable information on the changing sociopolitical contexts in North America.[39] Indigenous groups astutely used this to their advantage, often working their way onto Cuban ships with the pretext that they had "grave matters" to address with the Cuban governor.[40]

The outbreak of the American Revolution forced Spanish and Indigenous groups into difficult decisions. According to José de Gálvez, Spain's minister of the Indies, the British presence in the Mississippi region threatened Spanish commerce and "the security of our richest possessions." Gálvez also expressed a desire to quickly take back Florida, Mobile, and Pensacola, which were "the keys to the Gulf of Mexico," and to "cleanse the English from the shores of the Mississippi, which ought to be viewed as the wall which guards the vast empire of New Spain."[41] The American rebellion likewise threatened Indian territory, as Natives across the continent understood that a US victory would bring a flood of land-hungry settlers onto their lands. But the conflict also provided new opportunities for Native Americans to preserve their independence. In a sense, the Creeks and their Indigenous neighbors had been preparing for this moment since 1763, adjusting to the geopolitical shifts that followed the Seven Years' War in ways that allowed for the maintenance of their power and autonomy through neutrality.

* * *

After the French and Spanish entry into the conflict and the American victory at Yorktown, the British grew wary of fighting a global war and entered negotiations for peace. Unsurprisingly, the European powers disregarded the Native groups who had been integral to their designs. There were no Indigenous representatives present at the Peace of Paris, nor were any Native groups mentioned in the terms of the Treaty of Paris. Creek headmen were furious upon learning their allies had disseminated lands still firmly under Indigenous control. The French, on the verge of bankruptcy, had paid dearly for their involvement in the war and abandoned most of their promises to Spain in search of an easy exit. By sheer coincidence, Spanish designs most closely overlapped with those of the Creeks. The Spanish pushed hard, but unsuccessfully, to set the boundary of the United States in accordance with the Proclamation Line of 1763. The delegation from the United States, negotiating independently, proved more successful and was awarded with a westward boundary at the Mississippi River to the great trepidation of Spain and the Upper and Lower Creeks.

Despite setbacks in their Paris negotiations, the Spanish still found themselves well situated to continue their socioeconomic reforms and strengthen their grip on their colonies in the Americas. Spain's major objectives in the region after the American Revolution were tied to the same economic and political goal: the protection of New Spain (Mexico). By holding Cuba and New Orleans and regaining their territories in Florida while removing competing European interests, the Spanish controlled most of the land adjacent to the Gulf of Mexico and therefore were even better suited to accomplish their goals in 1783 than two decades prior. Yet despite ending the war on good terms with the United States, they also knew peace would be fragile. The hurried geopolitical shifts caused Spanish officials to view the region east of the Mississippi as vital to their political and economic objectives.

Spanish officials' attitudes about the Mississippi led to two significant regional goals within their postwar policies. The first was to impede, if not outright prevent, the westward expansion of the United States, as allowing the United States to assume control of the Mississippi River would pose an immediate threat to New Spain. The second was to control the region's other Gulf-flowing waterways to maintain access to area resources, protect lines of commerce, and improve economic relations with the Creeks. While New Spain was the chief income producer, Spanish commerce in the Americas had always run through Cuba. Lower Creeks, whose towns were positioned along the Chattahoochee, Flint, and Ocmulgee rivers, hoped the renewed Spanish presence in the region would result in their reoccupation of San Marcos de Apalache, a fort and trade outpost at the junction of the Wakulla and St. Marks Rivers. This fort provided the Spanish with direct access to Lower Creek communities via the Apalachicola river system. The Spanish occupation of British outposts at Mobile and Pensacola also allowed for renewed Spanish connections with the Upper Creeks and Choctaws through the Tombigbee, Alabama, Coosa, and Escambia Rivers and their tributaries. New emphasis on commodities and agriculture had increased the value of plantations and cattle ranches in the Spanish Caribbean.[42] This economic intensification was not lost on Spain's enemies, which is why the Caribbean had lured buccaneers and privateers in the previous decades and why Havana remained a prime British target.

While the Treaty of Paris's delineation of the Mississippi River as the western boundary of the United States tempered Spanish optimism, they were still dealt a good hand geographically to assist in accomplishing their goals. The Spanish could not assert authority over the entire river, but they were in possession of its mouth at New Orleans, offering them considerable control over its navigation. They also continued to benefit from their established relations with the Indigenous groups along the Mississippi River Valley. The Spanish postured as an empire in control of all continental territory in the Gulf of

Mexico and the vast majority of the mainland in the Circum-Caribbean, but regardless of territorial claims, their objectives relied almost entirely on the sociopolitical leanings of the Creeks and other Indigenous groups in the continental interior. Just as they had in the decades before the American Revolution, Native peoples "shaped the contours of empire" in the late eighteenth-century Gulf. Though Spanish and American bureaucrats held distinctive visions for empire and nationhood in this region, the fulfillment of their respective goals was expressly connected to the actions and political decisions of Creek leadership.[43]

Spain's policies in the region drew from a combination of colonial experience and an adaptation of their Indigenous policies within the Bourbon reforms. Now lacking direct competition from Britain and feeling confident that the United States was a weak and fragile state preoccupied with westward expansion, the Spanish quickly sought out Native alliances. They encountered willing partners eager to establish trade relations among the Upper Creeks. In March 1784, Alexander McGillivray, a biracial and bicultural Creek headman from the Upper town of Little Tallassee, informed Esteban Miró, Bernardo de Gálvez's replacement as governor of Louisiana, in his capable Spanish that he desired an alliance, as "the protection of a great monarch is preferrd [*sic*] to that of a distracted Republic."[44] While McGillivray is often regarded as a shrewd, albeit controversial, diplomat, this was the second time in two decades that his people found themselves navigating Euro-American geopolitical shifts. The American weaknesses McGillivray mentioned, namely the lack of clear governmental direction and astronomical debts, were problems on which McGillivray knew the Spanish depended.[45]

The treaty that emerged between the Creeks and Spaniards at Pensacola in 1784 was less indicative of the "greatest harmony, union and good friendship" that existed between them, and more reflective of coinciding economic and political objectives. The Spanish achieved the peace necessary to maintain access to local resources, and both could claim to maintain their respective sovereignty as the Creeks promised "to obey the laws of the Great King of Spain, in those points which are compatible with our character and circumstances."[46] The Creeks locked in "permanent and unalterable" access to trade goods through commerce with Spain, and Spanish officials made clear that unlike the Americans, they "do not exact from the nations of Indians any lands." In exchange, they secured Creek support against any encroachments in their claimed territories.[47] Equally, the Creeks received Spanish support for their own territorial claims.

The Spanish moved quickly to follow their success in Pensacola and held a similar congress in Mobile. Spain established treaties with the Chickasaws, Alibamons, and the Choctaws, though the latter insisted on special supplies on their visits to Mobile and New Orleans.[48] By supporting notions of Creek, Choctaw, and Chickasaw political and territorial autonomy and sovereignty, the Spanish undercut the claims of the United States and helped create the protective buffer zone they desired between the Appalachian Mountains and the Mississippi River. They also maintained free access to the region's waterways on the Mississippi and along the Gulf Coast. This allowed them to maintain commercial ties and trade with the Indigenous groups through the port towns of New Orleans, Mobile, and Pensacola while retaining access to the fisheries vital to sustaining Cuba.[49] To the Spanish, territorial hegemony in the continental interior, or even along the coastal regions they could legally claim, was unnecessary as long as the Creeks, Choctaws, and Chickasaws remained allied sovereigns.[50] The Spanish therefore were willing to engage in treaties, which by their nature recognized Indigenous sovereignty, something that would have been illogical under previous Habsburg models, which treated all Natives as either subjects or enemies of the Crown.[51]

The United States pursued treaties of their own with Native Americans in the region. The Hopewell Treaties of 1785 upheld the "perpetual friendship" of the Cherokees and Choctaws and the United States, and established US protection. The objectives of the United States differed from those of the Spanish, due mainly to their desire for land. When Spanish officials offered "protection," it was intended to grant authority to Indigenous groups over their lands and justify political and commercial intrusion, especially in disputed regions along the northern boundary of West Florida where Spanish claims were weak. The United States, however, obtained title to land that the Natives had never relinquished. By asserting the authority to offer protection and to specify hunting rights and territorial boundaries, the United States was attempting to solidify the legitimacy of their own claims to territorial sovereignty over the lands in dispute.

Though treaties may have been beneficial to Spain, the United States, and the Natives who signed them, they reflected larger diplomatic problems within Indigenous communities. Through the American Revolution, most Native peoples east of the Mississippi existed within societies that allowed local leaders to pursue their own interests in relations with Europeans. In the towns that comprised the Creek Confederacy, individual headmen lived in competition with one another for authority and used sustained economic and political relationships with the British, French, and Spanish to influence what happened within their villages. Creek micos did not rely on force to create social and political order or administer justice, a practice that often became a point of contention between Native headmen and British officials following the Seven Years' War. Rather, Indigenous leaders utilized persuasion and negotiation to gain the support of their people, ensuring that internal and external alliances remained both adaptable and flexible.[52]

Native responses to the official US victory often varied within larger tribal designations and are best understood on a coalition, clan, or town-by-town basis rather than a tribe-by-tribe one. While McGillivray's desire for a Creek-Spanish alliance in the years following American independence appeared to benefit the whole, his efforts were largely self-serving, as this relationship would not only bolster the prominence of his town, but also increase his personal wealth and standing within Creek society. As someone raised in both the Creek world of his mother and the colonial world of his father, McGillivray saw the United States' formative period as a chance to redefine his own society's political identity in ways that aligned with Euro-American philosophies, essentially creating a centralized "Creek Nation." To achieve this, McGillivray undermined the legitimacy of the Lower towns and their chiefs whenever possible. Writing to US treaty commissioner William Clark in April 1785, McGillivray asserted that Lower Creek headmen "will not be called to meetings by Traders or others." In August 1786 he advised William Telfair, governor of Georgia, not to negotiate with the Tame King or the Fat King, rival micos of the Lower Creek towns of Great Tallassee and Cusseta, respectively.[53] "Be not deceived by those men any longer," McGillivray stated, "they have no authority to give away."[54]

McGillivray's effort to create a Creek nation relied upon "the drawing of territorial boundaries, the creation of institutions of national leadership, and the inventions of ideologies that legitimize the existence thereof."[55] McGillivray highlighted Creek sovereignty in communications with Spanish and American representatives, describing them as a "free nation" and often referencing their "natural rights" in relation to territory and property, despite agreements outlined in European treaties.[56] In a 1785 letter to Miró, a frustrated Alexander McGillivray criticized Britain's willingness to agree to the Treaty of Paris. "If the British nation has been compelled to withdraw its protection from us," McGillivray stated, "She has no right to give up a Country she could never call her own."[57] McGillivray reiterated this point months later in an address to the Spanish, arguing that "it is a notorious fact known to the Americans . . . that his Brittanick Majesty was never possessed, either by session purchase or by right of conquest, of our Territorys and which the Said treaty gives away."[58] Writing to Andrew Pickens, a representative from South Carolina, McGillivray stressed that his people knew the exact bounds of their territory and "shall pay no regard to any limits that may prejudice our claims, that were drawn by an American and confirmed by a British negotiator. . . . They have been ours from the beginning of time and I trust that, with the assistance of our friends, we shall be able to maintain them against every attempt that may be made to take them from us."[59] Creeks recognized that US citizens unquestionably connected private property with freedom. By employing similar language in his correspondence with American officials, McGillivray demonstrated that Native peoples were also willing to equate land ownership with both individual and national independence.

The emergence of a nation-minded faction widened existing divisions within Creek society, threatened established coalition-based power structures, and adversely impacted sociopolitical relationships within the Confederacy as many micos resisted consolidation under the guise of a national identity. But while Creek elites debated the future political direction of their towns, life remained unchanged for nonelites who generally "carried on as they had for generations."[60] The fact that some Creeks would selectively adopt characteristics of Euro-American legal and political culture while others preserved the coalition-based sociopolitical structure of previous generations "demonstrated that while they were not entirely free of American influence, neither were they entirely at the mercy of it."[61]

While the conflicting and transitional nature of nationalism and coalition-based diplomacy protected Creek autonomy, these contrasting approaches to Creek sovereignty simultaneously exposed the limits in American and Spanish Indigenous policy. For the Americans, the goal was to establish trade relationships and acquire new territory, either through land-for-debt agreements or gradual settlement. But competing interests often undermined these objectives. On one hand, the states frequently entered into their own negotiations with Indigenous groups, usually with local headmen who held limited authority. It is no surprise, then, that boundary disputes continued to be the main source of conflict between Indigenous peoples and US citizens during the 1780s, while also negatively impacting relationships within Native communities. Three treaties in particular—Augusta in 1783, Galphinton in 1785, and Shoulderbone in 1786—revealed the extent of these rifts, as each ended with cessions of Creek land to the state of Georgia without the consent of the entire Creek Confederacy. By 1787, escalating border violence as a result of contested land cessions forced US commissioners to implore that Indian agents "use the utmost care to ascertain who are the leading men among the several tribes—the real head-men and warriors."[62] By identifying these "real" headmen and placating them with presents, American officials hoped they could attach most of the region's Natives to US interests and resolve the internal divisions that cast doubt on important territorial gains.[63]

Among the Creeks, internal divisions largely emerged over differing opinions of McGillivray. The Tame King and the Fat King, the same Lower Creek micos whose economic and diplomatic power McGillivray attempted to encumber in his correspondence, described him as duplicitous and "an Indian that has brought the nation to a grate [*sic*] disturbance."[64] They likewise took advantage of any opportunity to blame the Upper Creek towns for provoking their young men to act violently toward Georgians. In June 1787, Georgia's governor, George Matthews, chastised these headmen for being negatively influenced by the Upper towns who "continue to create differences between you and us. . . ." "You ought not to think of making us accountable for any measures of the Upper towns, our brothers," the Tame King and the Fat King replied, ". . . they can answer for themselves."[65]

While this discord highlights the difficulties that resulted from diverging nation-minded and coalition-based interests among Creek leaders and autonomous towns, Spain found it advantageous to maintain a united Indigenous front against the United States, who in turn sought to cultivate Native conflict to facilitate their agenda. But while Spain sought to support Indigenous territorial sovereignty because "nations" with a defined territory helped solidify land claims and served to check American expansion, they did not specifically need unified Creek, Choctaw, or Chickasaw autonomy, as long as they all maintained their opposition to the United States.[66] The problem for Spain, however, was that maintaining Indian opposition to the United States required a steady supply of goods. As had previously been the case in the Upper Mississippi region, Spanish officials knew that any possible alliance with Native nations was dependent on trade. The minister of the Indies, José de Gálvez, expressed to his governors that "the security and happiness of the said province was intimately linked to good harmony with the numerous savage Indian nations that circle it. And that harmony could only maintain itself through the opportunistic giving of gifts to the principal leaders and a trade of goods and effects suited to their needs and wants." For this reason, he appointed a new lieutenant governor in Louisiana and West Florida due to "his contacts and circumstances necessary to acquire stockpiles of goods in Europe with the appropriate economy, proportion, and arrangements." Gálvez also earmarked 80,000 *pesos fuertes* from New Orleans to provide gifts to the Natives and a sum of goods worth 200,000 pesos to be supplied to the traders. To ensure that Spanish supply lines could always maintain a steady stream of trade goods, the minister proposed to set aside an additional 100,000 pesos in reserve. The ultimate goal was for the Natives to understand "that His Majesty would give them the corresponding gifts and provide them with a Commerce so extensive and inalterable that it would fulfill all of their desires within eighteen months."[67]

Meeting the trade demands of their Indigenous allies in the region proved more difficult than Spain had anticipated. Spanish settlements frequently lamented their limited supply and the potential "embarrassment" to Spain if they could not supply the Natives with the demanded goods. Spanish officials also understood that if the Natives could not get these supplies from Spain, they would turn to the Americans, as friendship and loyalty could "only be achieved by way of gifts and a trade well-sustained."[68] Thus, when McGillivray advocated for the Scottish merchant and British Loyalist William Panton to maintain his trade with the Creeks and Seminoles, the Spanish relented, granting Panton's firm a monopoly on Indian trade in the region. They understood the influence Panton had among the Creeks and feared that rumors of American overtures to the merchant's firms were true. Yet the Scottish merchants were a necessary risk, since the Americans would "not fail to devote all their energies to capturing this trade," not only for economic purposes, but to "advance their political plans in the future."[69] Fortunately for the Spanish, Panton maintained a hatred for the United States.

While trade monopolies ensured a steady supply of goods at the broad level of society, they did not always meet the needs of local Indigenous communities. Additionally, seeking out trade and relationships of exchange held other sociopolitical ramifications for local headmen, as the monopoly system further limited the means by which they could secure goods. Creek leaders traditionally gained authority through access to foreign commodities, which they would redistribute within their communities in exchange for support. Presents "helped to promote peace and workable alliances between peoples who would otherwise remain strangers and potential enemies" and allowed Native leaders to gain authority and power through consensus.[70] Thus, Creek parties continued to make voyages to Cuba throughout the 1780s and 1790s. While many Spanish officials in Havana lamented the practice, suggesting the Natives were "attracted by greed for the presents that by virtue of royal orders are given to them there," they deemed it necessary to keep up appearances and achieve diplomatic goals.[71] The Spanish provided room and board for their Creek guests and supplied customary gifts, such as cloth, vermillion, rum, gunpowder, and shot. They became so commonplace that Spanish officials usually attempted to expedite the visits, sending them home with their gifts "as quickly as possible."[72]

Remarkably, Spain had been successful throughout the 1780s at using trade to maintain positive relations with the Creeks despite their own inability to uphold sufficient supply. Yet the growing threat of war with England and their amicable relations with the United States made Spain soften their stance to avoid outright conflict with the Americans, opening the door for both Upper and Lower Creeks to begin questioning whether Spain could continue to meet trade demands. McGillivray, potentially motivated by self-interest as much as his desire to create a Creek Nation, entered into a treaty with the United States in 1790. In the Treaty of New York, the Creeks gained greater recognition of their territorial sovereignty in exchange for ceding a large section of their lands to the United States.[73] Taken aback by the amount of land ceded by McGillivray and the Creeks, Spanish officials scrambled to form a response, negotiating a new treaty in 1792 to reinforce the agreements established in the prior decade. McGillivray's action tested the limits of a Euro-American-modeled "Creek Nation," as many headmen never supported the ceding of lands and had long challenged the notion of McGillivray's authority to speak collectively for the Creek Confederacy. For years McGillivray had steered local micos toward the Spanish. He was made a Spanish colonel and earned a yearly pension of six hundred dollars. Now, he stood to benefit financially from the new arrangement, earning a US military commission along with a substantial annual payment of twelve hundred dollars.[74] McGillivray maintained a show of loyalty to Spain, though he henceforth referred to his pension as a trader's pittance.

Both Spain and the United States supported the political sovereignty of the Creeks in part because it allowed both nations to continue to negotiate with them despite their respective alliances. Yet by the 1790s, Spain began to push even harder for Native unity to create a cohesive front in a last-ditch effort to curb American expansion. After securing the 1792 treaty with the Creeks, the Spanish negotiated with the Choctaws, Chickasaws, and Cherokees in an attempt to form a Pan-Indian alliance. On paper, the Spanish were successful, as all the major regional Indigenous groups agreed to such an alliance in 1793 through the Treaty of Nogales. The Creeks, Choctaws, Chickasaws, and Cherokees even agreed to cede some land for the Spanish to construct a fort in the hope of protecting their interests on the Mississippi.

The harmony was short lived, however, as past conflicts quickly reemerged and fomented discord among the Natives. Spain's position further weakened following Pinkney's Treaty in 1795, where it relinquished claims in West Florida to the lands north of the 31st parallel. Additionally, the United States was officially granted free navigation of the Mississippi River. While Spain still maintained much of the coastline and the port cities of Mobile and Pensacola, they could no longer claim possession of significant portions of this territory, which in reality had always remained under Indigenous control.[75] Spanish officials in the region continued to push for Native unity to maintain the buffer zone, though the empire's economic and political situation was proving untenable. As late as 1800, around the same time Spain was secretly drawing up a treaty to return Louisiana to France, the recently appointed governor of the territory, the Marquis de Casa Calvo, continued to advocate for the traditional Spanish policy of establishing diplomacy through trade. This was even more imperative now, as "only a few depraved Seminoles" lived in the areas Spain could still claim. In contrast, "the nations of the Choctaws and Chickasaws are entirely included in the American territory . . . likewise the part of the Tallapoosas (Creeks), best, most numerous, and most friendly to Spain, and all the Cherokees, have remained in the Territory of the United States."[76] Since Spain lacked any territorial control over the region, it was only through the maintenance of trade through Panton, Leslie and Company that they could "gain and maintain a preponderancy and friendship among those tribes, in order to constitute them a barrier between the American and Spanish possessions, forming, as they must do, an intermediate country, which it is our interests to keep in our dependence."[77] Within a few years, the territory would pass again from French to American hands, thus undoing decades of Spanish policy in the area and officially ending any Spanish designs on controlling the Gulf South.

* * *

Spain's conceptualization of national sovereignty in the Gulf South served two purposes. By claiming jurisdiction over the land along the coast and the Mississippi River, the Spanish were protecting their access to waterways and limiting American expansion into the Gulf. They also bolstered Indigenous land claims by asserting that Natives were under their protection. At the same time, as explicitly stated in the treaties, Spanish and Indigenous groups created a complementary relationship, with the Spanish exercising little hegemony in the interior while maintaining Native support through gifts and trade. Additionally, the Spanish recognized Indigenous territorial sovereignty, even though this acknowledgment could potentially weaken their own claims, in an attempt to undermine those of the United States.

Like Spain, the United States applied nationalism in distinct ways. Jeffersonian Republicans, who valued the individualism derived from private land ownership and agriculturalism, were perpetually interested in obtaining territory south and west of the Appalachians. The formation of both a national identity and recognition of the sovereignty over territories awarded in the Treaty of Paris, territory that the Native groups still firmly controlled, was vital to their objectives. The fledgling US government struggled to negotiate with Indian groups who were often involved in their own conflicts and compromises with individual states. Therefore, it served American interests to recognize Native national sovereignty, as this legitimized territorial acquisitions on a federal level. The United States also desired access to the Gulf through the Mississippi River and wanted to use trade to make diplomatic inroads to Indigenous groups. But where Spain had hoped to utilize trade to create Native buffer states and deter Indigenous peoples from seeking out exchange with Americans, the United States sought trade with the Natives to create a dependence on their goods. Unfortunately for Spain, the empire proved incapable of keeping up with Indigenous trade demands. They also proved unable to keep the United States from accessing the Mississippi and the Gulf. While US political elites were divided in their visions for the economic path that America should follow, both westward expansion and the rapid growth of trade were detrimental to Spanish and Indigenous designs.

Unsurprisingly, Indigenous groups again found themselves in the middle. Espousing nationhood worked well for leaders like McGillivray in that it afforded a level of power through unanimity that assisted in negotiations with other nation-states. It also allowed for such leaders to facilitate the supply of trade goods, which increased their authority and legitimacy. On the other hand, the creation of a Creek Nation implied a level of centralization that ran contrary to traditional Indigenous sociopolitical structures. The persistence of coalition-based diplomacy allowed for some headmen to improve their status, and the desires of opportunists like McGillivray did not always align with the needs of all the communities within their territory.

While the Creeks, Choctaws, Cherokees, and Chickasaws fiercely defended their borders for political reasons, diplomacy remained a largely economic endeavor. Ironically this served to both support and undermine Native nationhood. Treaties with Indigenous groups functioning as nation-states secured favorable conditions for trade and the acquisition of desired commodities. But they did little to curtail pragmatic policies at the local level. In the end it was Spain's failure to profit from their land claims and protect their exclusive access to vital waterways that undercut Spanish and Indigenous designs in the region. A lucrative agricultural trade never developed in Louisiana, and Spain was continually forced to loosen its restricted monopoly and outsource its trade to maintain supplies. Most of the non-Native interior populations consisted of British remnants, French transplants, or American settlers. Spain's inability to develop profits weakened its ability to meet Indigenous trade demands and opened the door for the British and the Americans to flourish in the region. Even last-ditch efforts at Pan-Indian alliances were weakened by Native disunity, fostered by decades of conflict. By 1800, Spain had all but given up its designs in the region, granting access of the Mississippi to the United States and returning Louisiana to the French. The coalition-based political structures and neutral diplomatic strategies long utilized by Indigenous groups collapsed, leaving the Creeks and their Indigenous neighbors to find new methods of adaptation and adjustment moving into the nineteenth century.

Notes

1. Kathleen DuVal, *Independence Lost: Lives on the Edge of the American Revolution* (Penguin Random House, 2016), 30–31.

2. The American Association of Geographers defines the Southeastern United States as the present-day states of Alabama, Florida, Georgia, Kentucky, Maryland, Mississippi, North Carolina, South Carolina, Tennessee, Virginia, and West Virginia. Geographically, it is bordered by the Atlantic Ocean to the east, the Mississippi River to the west, the Gulf of Mexico to the south, and the Ohio River to the north.

3. DuVal, *Independence Lost*; Alan Taylor, *American Revolutions: A Continental History, 1750–1804* (W. W. Norton, 2016); Colin Calloway, *The Indian World of George Washington: The First President, the First Americans, and the Birth of the Nation* (Oxford University Press, 2018). See also Francis Prucha, *The Great Father: The United States Government and American Indians* (University of Nebraska Press, 1984); Reginald Horsman, *Expansion and American Indian Policy, 1783–1812* (University of Oklahoma Press, 1992); David Andrew Nichols, *Red Gentlemen and White Savages: Indians, Federalists, and the Search for Order on the American*

Frontier (University of Virginia Press, 2008); Jeffrey Washburn, "Directing Their Own Change: Chickasaw Economic Transformation and the Civilization Plan, 1750s–1830s," *Native South* 13 (2020): 94–119.

4. Brian DeLay, "Indian Polities, Empire, and the History of American Foreign Relations," *Diplomatic History* 39, no. 5 (2015): 927–42, esp. 939.

5. DeLay, "Indian Polities," 940.

6. James L. Hill, *Creek Internationalism in an Age of Revolution, 1763–1818* (University of Nebraska Press, 2022). See also Eliga Gould, *Among the Powers of the Earth: The American Revolution and the Making of a New World Empire* (Harvard University Press, 2012); Leonard Sadosky, *Revolutionary Negotiations: Indians, Empires, and Diplomats in the Founding of America* (University of Virginia Press, 2009).

7. Deborah A. Rosen, *Border Law: The First Seminole War and American Nationhood* (Harvard University Press, 2015), 7–8.

8. Among the earliest advocates of a broad regional study over national histories was Herbert Bolton in his study of the Spanish borderlands. David J. Weber later built on this tradition while including the contributions of the Indigenous and ethnically mixed populations. See Herbert Eugene Bolton, *The Spanish Borderlands: A Chronicle of Old Florida and the Southwest* (Yale University Press, 1921); David J. Weber, *The Spanish Frontier in North America* (Yale University Press, 1992).

9. David J. Weber, *Bárbaros: Spaniards and Their Savages in the Age of Enlightenment* (Yale University Press, 2005).

10. DuVal, *Independence Lost.*

11. Steven Peach, *Rivers of Power: Creek Political Culture in the Native South 1750–1815* (University of Oklahoma Press, 2024).

12. "Richard Thomas to Benjamin Hawkins," 1 July 1798, in *Letters of Benjamin Hawkins, 1796–1806*, 485. https://archive.org/details/lettersofbenjami00hawk/.

13. Ibid., 523–26.

14. "David Taitt to John Stuart," 1776, in Kenneth Gordon Davies, ed., *Documents of the American Revolution, 1770–1783* (Irish University Press, 1972–1981), 12: 159.

15. Steven C. Hahn, *The Invention of the Creek Nation, 1670–1763* (University of Nebraska Press, 2004).

16. Stanley J. Stein and Barbara H. Stein, *Apogee of Empire: Spain and New Spain in the Age of Charles III, 1759–1789* (Johns Hopkins University Press, 2003), 81–115. Charles III entrusted many of his domestic reforms to Sicilian Leopoldo di Gregorio, whose policies created unrest. Subsequent riots led to his removal, but not before his appointment of José de Gálvez in the Indies, which allowed for a concentration of Bourbon economic reforms in the colonies.

17. Barbara H. Stein and Stanley H. Stein, *Edge of Crisis: War and Trade in the Spanish Atlantic, 1789–1808* (John Hopkins University Press, 2009), 4.

18. Weber, *Bárbaros*, 36.

19. There is debate regarding both the authorship of the manuscript and the extent of its influence on subsequent Bourbon policy. Scholars who support its influence suggest that it was likely circulated before its eventual publication in 1789. See Weber, *Bárbaros*, 181–82.

20. Joseph del Campillo y Cosío, *Nuevo sistema de gobierno económico para la América* (Imprenta de Benito Cano, 1789), 14–15.

21. Campillo y Cosío, *Nuevo sistema*, 16.

22. Campillo y Cosío, *Nuevo sistema*, 71.

23. Weber, *Bárbaros*, 8–11.

24. Elena A. Schneider, *The Occupation of Havana: War, Trade, and Slavery in the Atlantic World* (University of North Carolina Press, 2018), 210.

25. Ralph Lee Woodward Jr., "Spanish Commercial Policy in Louisiana, 1763–1803," *Louisiana History: The Journal of the Louisiana Historical Association* 44, no. 2 (Spring 2003): 143–45.

26. The Illinois Country was the general name for the territory encompassing the Mississippi River Valley. The Spanish continued to use the term after the French ceded Louisiana. It was also known at the time as Upper Louisiana. The area described here as part of Ulloa's directives would be in the US state of Missouri today.

27. "Ulloa sends an expedition to the Illinois Country to Establish a Fort and Settlement and his rules for government of the same," in Louis Houck, ed., *The Spanish Regime in Missouri* (R. R. Donnelley and Sons, 1909), I:6, 11.

28. "Ulloa sends an expedition to the Illinois Country," 10.

29. "Ulloa sends an expedition to the Illinois Country," 11.

30. "Ulloa sends an expedition to the Illinois Country," 13.

31. "Secret instructions of Ulloa to Captain Rui, dated January 7, 1767, relating to the construction of two forts at the mouth of the Missouri," in Houck, *The Spanish Regime in Missouri*, I:23.

32. "Ulloa sends an expedition to the Illinois Country," 13.

33. Elizabeth Ellis, "Petite Nation with Powerful Networks: The Tunicas in the Eighteenth Century," *Louisiana History* 58, no. 2 (Spring 2017): 133–78, esp. 136–37. See also Elizabeth Ellis, *The Great Power of Small Nations: Indigenous Diplomacy in the Gulf South* (University of Pennsylvania Press, 2022).

34. Comprised of approximately fifty-nine towns, Upper Creek villages were scattered along the Tallapoosa-Coosa-Alabama river system in what is today western Alabama. The twenty or so Lower Creek towns were positioned along the Chattahoochee, Flint, and Ocmulgee Rivers, east of the Upper Creeks on the modern-day Alabama-Georgia border. Creek influence extended across the flatlands of current-day Alabama and Georgia, east to the Atlantic Ocean, south to the Gulf coast of the present-day Florida panhandle, west to the Mississippi River, and north to the foothills of the Appalachian mountains near the Cumberland plateau. It is important to note that this general, bilateral division of "Upper" and "Lower" Creek can be problematic because it overlooks the nuances and complexities of Creek society. However, in keeping with the existing historiography, this chapter employs these terms when necessary.

35. Verner Crane, *The Southern Frontier, 1670–1732* (University of Alabama Press, 1928), 185.

36. The current path originated in Augusta, Georgia, passing through the Lower towns before reaching Upper Creek communities.

37. The Spanish did assume control over New Orleans after 1763, which was theoretically easier for Creeks to access than Cuba. However, it took a number of years for the transfer of power from the French to the Spanish to take hold, and the transition was difficult. Creeks additionally used travels to Havana as a way to try to access power, as recognized in both the spiritual and physical worlds.

38. John E. Worth, "A History of Southeastern Indians in Cuba" (working paper, University of West Florida, Pensacola, FL, October 2004), 1–16; John E. Worth, "Creolization in Southwest Florida: Cuban Fishermen and 'Spanish Indians,' ca. 1766–1841," *Historical Archaeology* 46, no. 1 (2012): 142–60; Hill, *Creek Internationalism*, chapter 2; Claudio Saunt, *West of the Revolution:*

An Uncommon History of 1776 (W. W. Norton, 2014); Bryan C. Rindfleisch, "The 'Owner of the Town Ground, Who Overrules All When on the Spot': Escotchaby of Coweta and the Politics of Personal Networking in Creek Country, 1740–1780," *Native South* 9 (2016): 54–88; Jennifer Monroe McCutchen, "'More Advantageous to Be with Spaniards': Gunpowder and Creek-Spanish Encounters in Cuba," *Terrae Incognitae* 52, no. 3 (2020): 245–60.

39. Light Townsend Cummings, *Spanish Observers and the American Revolution, 1775–1783* (Louisiana State University Press, 1991), 199–200.

40. AGI Santo Domingo 1221, transcribed by John E. Worth, Randell Research Center, Florida Museum of Natural History.

41. "José de Gálvez to the Governor of Havana; San Ildefonso, 29 August 1779," AGI Cuba 1290, translated and transcribed by James L. Hill. These sources are published online through the University of North Florida. Please see "Synopsis of Official Spanish Correspondence Pertaining to Relations with the Uchiz Indians, 1771–1783," Florida History Online, https://history.domains.unf.edu/floridahistoryonline/projects-proj-b-p-html/projects-uchize-index-html/; and "The Indian Frontier in British East Florida; Letters to Governor James Grant from British Soldiers and Indian Traders," Florida History Online, https://history.domains.unf.edu/floridahistoryonline/projects-proj-b-p-html/projects-grant-index-html/. For a description of the defensive and commercial significance of New Orleans and the Mississippi River region as a protective border and a conduit for trade, see also Kim M. Gruenwald in this volume, "When a River Is a Border: Rivalries and Commercial Networks in the Riverine West," 142, 143, 144, 146, 148, 151, 157, 159.

42. Stein and Stein, *Edge of Crisis*, 4.

43. Peach, *Rivers of Power*, 5.

44. Alexander McGillivray was part of a growing population of biracial Natives born to Indigenous mothers and European fathers in the mid-eighteenth century. McGillivray's father, the prominent Scottish trader Lachlan McGillivray, supported his European-style education in Charleston, as well as a business apprenticeship in Savannah. But as the son of a Creek woman, Alexander's Indigenous relatives viewed him as fully Creek and member of the powerful Wind Clan. Alexander's lived experiences in the Indigenous world of his mother and the Euro-American world of his father allowed him to learn both "the unwritten rules and expectations" of Creek life, while also gaining knowledge that ensured he would be comfortable navigating the colonial world of his father. This knowledge and experience expertly prepared Alexander for a leadership role in the Creek Confederacy in the years following the American Revolution. John Walton Caughey, ed., *McGillivray of the Creeks* (University of Oklahoma Press, 1938), 74 (quote). McGillivray was fluent in English and near fluent in Spanish. See James L. Roark et al., *The American Promise: A History of the United States, Ninth Edition* (Bedford/St. Martin's, 2023), 1:247.

45. Caughey, *McGillivray*, 65.

46. *Spanish Treaty with Talapuche Indians, Articles of Convention Signed at Fort of Pensacola, Florida*. -05-31, 1784. Manuscript/Mixed Material, *Library of Congress*, https://www.loc.gov/item/mtjbib000919/.

47. *Spanish Treaty with Talapuche Indians*; DuVal, *Independence Lost*, 259.

48. Jack D. L. Holmes, "Spanish Treaties with West Florida Indians, 1784–1802," *Florida Historical Society* 48, no. 2 (1969): 144.

49. AGI Santo Domingo 1210, transcribed by John E. Worth, Randell Research Center, Florida Museum of Natural History.

50. DuVal, *Independence Lost*, 259.

51. Weber, *Bárbaros*, 8.

52. Andrew K. Frank, "Taking the State Out: Seminoles and Creeks in Late Eighteenth-Century Florida," *Florida Historical Quarterly* 84, no. 1 (Summer 2005): 10–27, esp. 11–12.

53. Alexander McGillivray to William Clark, 24 April 1785, accessed August 22, 2021, https://dlg.usg.edu/record/dlg_zlna _tcc902/fulltext.text.

54. Alexander McGillivray to William Telfair, 3 August 1786, accessed August 22, 2021, https://dlg.usg.edu/record/dlg_zlna_tcc904/fulltext.text.

55. Hahn, *Invention*, 8.

56. Caughey, *McGillivray*, 73, 92.

57. Caughey, *McGillivray*, 73.

58. Caughey, *McGillivray*, 91.

59. "McGillivray to Pickens (September 1785)," in John T. Juricek, ed., *Georgia and Florida Treaties, 1763–1776*, vol. 12, *Early American Indian Documents: Treaties and Laws, 1607–1789* (University Publications of America, 2002), 387.

60. Kevin Kokomoor, *Of One Mind and of One Government: The Rise and Fall of the Creek Nation in the Early Republic* (University of Nebraska Press, 2019), 29.

61. Kokomoor, *Of One Mind*, 29.

62. George Washington and Henry Knox to Benjamin Lincoln, Cyrus Griffin, and David Humphreys, "Instructions to the Commissioners for Treating with the Southern Indians," 29 August 1789, accessed 22 August 2021, https://memory.loc.gov/cgi-bin/ampage.

63. Letter, "Instructions to the Commissioners for Negotiating a Treaty with the Tribes of Indians in the Southern Department, for the Purposes of Establishing Peace Between the United States and the Said Tribes," 26 October 1787, accessed 22 August 2021, https://memory.loc.gov /cgi-bin/ampage.

64. "Talk from the Lower Towns in the Creek Nation," 11 August 1786, accessed 22 August 2021, https://dlg.usg.edu/record/dlg_zlna_tcc026.

65. "Talk of the Fat King to his honor Governor Matthews and the Council," 27 July 1787, accessed 22 August 2021, https://memory.loc.gov/ammem/amlaw/lwsp.html.

66. DuVal, *Independence Lost*, 256–59.

67. "José de Gálvez to the Intendant of Louisiana," AGI Cuba 182A, f 639–644, *Florida History Online*, https://www.unf.edu/floridahistoryonline/.

68. "José de Gálvez to the Intendant of Louisiana."

69. "Zéspedes to Gálvez," AGI Santo Domingo 86-6-6 in Lawrence Kinnaird, ed., *Spain in the Mississippi Valley*, vol. III, *Annual Report of the American Historical Association* (United States Government Printing Office, 1945), pt. 2:111.

70. Steven J. Peach, "Creek Indian Globetrotter: Tomochichi's Trans-Atlantic Quest for Traditional Power in the Colonial Southeast," *Ethnohistory* 60, no. 4 (October 2013): 605–35, esp. 608.

71. "Las Casas to Alcudia," AGI Estado, Santo Domingo leg. 14, in Kinnaird, *Spain in the Mississippi*, IV pt. 3:135.

72. Multiple documents express a similar sentiment; the Natives were "provided with their daily necessities and returned to their country as quickly as possible." See the "Correspondence of the Captain General of Cuba, José de Ezpeleta," AGI Cuba 1387 for examples.

73. "Proclamation, 14 August 1790," *Founders Online*, National Archives, https://founders .archives.gov/documents /Washington/05-06-02-0122.

74. Caughey, *McGillivray*, 290.

75. This included the influential region around Natchez.

76. J. M. White, M. de Manuel y Rodríquez, I. Jordán de Assó y del Río, I., Texas., Mexico., Spain., France., Great Britain. (1839). *A new collection of laws, charters and local ordinances of the governments of Great Britain, France and Spain.* Philadelphia [Pa.]: T. & J. W. Johnson. 324.

77. White, *A new collection of laws*, 324.

CHAPTER 10

Comercio Libre

Revisiting a Concept on Trade and Borders in Creek Homelands

RACHEL B. HERRMANN

Comercio libre is a policy with unwieldy historiography. It has confounded scholars for several reasons: The name of the policy has been mistranslated; the Spanish introduced it incrementally throughout the Gulf South; unofficial relaxations to rules preceded the royal orders that formally wrote more liberal trade conditions into law; the main company responsible for trade between Spanish ports and Creek (or Muscogee) towns became exempt from *comercio libre*'s import and export duties; and historians have not sufficiently considered rivers in their discussion of the policy. This essay explains how *comercio libre* affected Creek townspeople during a time (c. 1763–1804) when Spaniards, Britons, and Americans disputed the riverine borders of Creek homelands in Florida and Louisiana. Creeks' knowledge of and control over rivers and swamps enabled them to take advantage of imperial competition and Spanish weakness. After 1784, this geopolitical power offered them better terms of trade in comparison to non-Indigenous people. The decentralized nature of Creek networks of kinship and politics forced the Spanish to liberalize their trade policy in ways that allowed Creeks to determine who and what crossed their borders.[1]

"Free trade" had such a different meaning in the eighteenth century that I choose not to translate *comercio libre* as "free trade" as other scholars have done; retaining the Spanish better reflects the policy's range of applications. In 1962 Jack Holmes proposed that historians use the term *libre cambio*, or liberal trade. When he identified the error of translating *comercio libre* as "free trade," both he and the translator he cited unfortunately consigned the important insight to a footnote. Nowadays, free trade and protectionism are oppositional ends of national trade policy. The eighteenth-century *reglamentos* (regulations or rules) for *comercio libre*, however, did not propose tariffs for revenue only, as free trade was defined in the century before the Second World War; nor did they propose to eliminate tariffs, as free trade is understood today. The title of the policy guidance for 1778—relevant because it increased confusion across a contested border by altering trade regulations in Louisiana but not Florida—is *Regulation and royal tariffs.*[2] *Comercio libre*, similar to other mercantilist policies of the era, had a protectionist core.

As Fernando Aguerre Core observes, *comercio libre* manifested in contradictory ways across Spain's empire.[3] Recent work explains the policy in places with Atlantic and Gulf South connections. David Narrett characterizes *comercio libre* in the Louisiana-Florida borderlands as a "commercial door half-open," meaning that in the 1780s it granted some non-Spanish merchants limited access to Spanish ports, the ability to import manufactured goods and export certain raw materials, and the permission to pay lower duties on an ad hoc basis. Paul Gilje casts 1795 as a watershed moment in which Spain granted duty-free trading rights—which he calls "free trade"—to American merchants in New Orleans. Elena Schneider and Adrian Finucane deal with Cuba. Finucane examines the period prior to *comercio libre*, exploring uneasy, interdependent relationships between Spanish and English merchants. Schneider shows how *comercio libre* increased the availability of enslaved laborers and expanded slavery; Havanans sought continued access to British trade and protected markets in Spain's American empire.[4]

Other scholarship has expanded historians' thinking about the merchants and filibusterers involved in the Creek deerskin trade during the time of *comercio libre*, but this essay departs from this work in its reinterpretation of tariffs. José Antonio Armillas Vicente opined in 1976 that opposition to Spanish tariffs stemmed from a combination of Creek "ignorance" and greedy American offers. William Coker and Thomas Watson dealt with the Scotsmen who obtained tariff exemptions from the Spanish through their work for the prominent trading firm Panton, Leslie and Company. Coker, Watson, and Narrett write of the opportunism that drove merchants' and adventurers' power grabs in the Louisiana-Florida borderlands. James Hill convincingly argues that when Creek leaders including Alexander McGillivray invoked the term "free trade," they were referring to the "notion that nations had the right to select their own trade partners," and this ability is what made them international. But Hill suggests that McGillivray did not mean for the term to refer "to an antitariff ideology," and here we disagree.[5] Sources in Spanish archives, which have been incompletely excerpted and published, document the half decade that McGillivray spent arguing against the levying of Spanish tariffs. This chapter's revisiting of these sources finds that Creek leaders and predominantly anonymous Indigenous women responded knowingly to *comercio libre* and campaigned in talks and letters to obtain lower barriers to trade. *Comercio libre*'s policymakers rarely granted full exemptions from tariffs, making it even more significant when they opened the door to do so in 1789.

This revisionist interpretation of tariffs first delineates the borders of Creek homelands and discusses Spanish, English, and American claims to land and water in those homelands where *comercio libre* operated. These were homelands where non-Indigenous people imagined river borders that for decades they failed to map and to control, and where they struggled to enforce the policies they enacted. On the coast between the Atlantic Ocean and the Mississippi River, Anglo-Europeans claimed four colonies: Georgia, East and West Florida, and Louisiana, which also encompassed parts of present-day Alabama and Tennessee. Semi-aqueous, "sustenant" places in these homelands were characterized by fertile, riverine land and towns and ocean-facing Spanish and English forts and trading houses. Although these claims were bounded by rivers and characterized by disputes over river boundaries, Robbie Ethridge explains that from river mouths on the Gulf Coast, European and American vessels were incapable of navigating above fall lines and further into Creek country. To travel long distances, Creeks and non-Indigenous traders used overland paths, which scholars interpret as physical roads and metaphorical relationships of kinship, alliance, and war. Using a path required permission and good relationships, but using a waterway required relationships, knowledge, and technology that Europeans lacked. Creeks, by contrast, possessed excellent riverine knowledge and power because of the positioning of their towns and agricultural fields above fall lines, the lightness of their water vessels, the seasonal floods that required learning about fords, and the wild water plants that women harvested to sustain communities.[6]

After considering these watery boundaries of Creek homelands, this essay draws on work about the Creek deerskin trade to explain continuities and changes during the period of *comercio libre*. In so doing, it pushes back against scholars' contentions about the trade in dressed deerskins. According to current scholarship, throughout the eighteenth century Creeks sought peaceful, plentiful, and steady trade across borders. Trade followed clan networks and served the interests of different towns. This historiography concludes that by the era of *comercio libre* deerskins had depreciated almost 50 percent compared to pre-Revolutionary levels, and merchants sought undressed deerskins instead of the dressed deerskins that women prepared. Hunters acquired more debts, and young men supplanted the elders who had strived to maintain a gendered balance in the trade.[7] This chapter discusses manuscript Spanish price schedules for dressed and undressed deerskins, and interprets them as evidence of Creek women seeking in 1784 to maintain trading power. Indigenous women's marriages to Anglo-American traders had allowed them to preserve relationships of kinship and obligation earlier in the century; by century's end those marriages enabled them to exploit Spanish dependence on foreign merchants and to maintain some trade in dressed deerskins. Finally, this essay explains why these relationships were ultimately incapable of protecting Creek homelands.

* * *

Before the era of *comercio libre* the Spanish were ambivalent about trading with Native Americans in the Gulf South. Then, during the American Revolutionary War they captured Mobile (in 1780) and Pensacola (in 1781), becoming claimants to British trading places. Some background on earlier eighteenth-century borders, geographical knowledge, trading places, and movement of towns and people is necessary to understand a shift in Spanish attitudes: from a policy of ambivalence to a willingness to liberalize trade for Creeks.

Indigenous negotiators limited Europeans' efforts to survey and mark the boundaries they claimed. Between 1763 and 1783 when the British claimed East and West Florida, the leader named Emesteseguo from the town of Little Tallassee on the Coosa River successfully confined British colonists between fall lines in Creek country and the sea coast. Representatives of the French and Spanish empires relied on the evidence of writers who had never traveled to the Americas and mapmakers whose knowledge was limited to coasts and some rivers as far as the tide flowed. Under the 1763 Treaty of Paris, the Spanish relinquished their claims to Florida and reacquired Louisiana from the French, who had earlier failed to clarify Louisiana's boundaries. The treaty confirmed the Mississippi River as the border between Spanish claims to the west and British claims to the east without identifying a Louisiana-Florida border. The British hoped to open "corridors of settlement" along West

Florida's rivers and new trade paths to the Creeks. These ambitions resulted in the acquisition of geographic information about forts, harbors, and rivers. After 1783, when the Spanish regained their Florida claims, Spanish and US officials disputed west-east boundaries between Louisiana and West Florida, and north-south boundaries between Georgia and both Floridas.[8]

From Indigenous perspectives this space had long possessed clear coastal and riverine borders that differed from European borders. Creek leaders repeatedly explained that they had rented but not sold Atlantic- and Gulf-facing coastal land to the British and Spanish. As late as 1792, two speakers named Okillissa Chopka and Tustunie Opoia sent a talk reminding the Spanish, "The Land your people live on at St. Augustine, St. Marks & Pansacola is ours & only lent by us to you."[9] The Tombigbee River was Creek country's westernmost border and its easternmost was the Savannah, which demarcated Creek lands from Cherokee lands. Between these two rivers were three south-flowing river networks that sustained Creek towns; one, the Oconee, Ocmulgee, and Altamaha, met the ocean on the Atlantic. The other two, the Chattahoochee, Flint, and Apalachicola; and the Coosa, Tallapoosa, and Alabama flowed into Gulf waters. Facing west from Creek country were the Choctaws, and facing north were Cherokees. Although European mapmakers were able to depict the locations of Creek towns, accessing them was another matter.[10]

As Elizabeth Ellis explains, these coalescent societies of the Gulf South were comprised of multinational settlements adjacent to foreigners living under different governments, which facilitated access to international markets. It was as common for inhabitants of Indigenous towns to move on rapidly as it was to relocate after staying more than a decade. Towns themselves moved, and consequently appear and disappear from colonial maps. Decentralized Indigenous power was exercised at the level of the town (*talwa*) and satellite town (*talofa*). Creeks, Cherokees, Chickasaws, Shawnees, Natchez, Seminoles, Yuchis, and Yamasees populated Creek *talwas*, which were more than just villages: the term referred to the larger group of people who shared council fires and square grounds, where political and cultural activities occurred. The Upper Creeks, whom the Spanish called the Talapuches or the Talapuz, had towns sited west of the Lower Creeks, along the Tallapoosa, Coosa, and Alabama rivers, with important towns at Little Tallassee, Muccolossus, Okfuskee, Okchai, and Tuckabatchee. Lower Creeks, whom the Spanish called Uchizes, had key towns on the Chattahoochee, Flint, and sometimes Ocmulgee rivers, including Apalachicola, Coweta, and Cussita. The Creeks who in the nineteenth century developed separate identities as Seminoles lived in the early eighteenth century around San Marcos de Apalache near the confluence of the Wakulla and St. Marks rivers, and along the St. Johns River (also known as the Alachua) in Florida. By the late eighteenth century Seminole towns are discernible on the Apalachicola, St. Marys, Suwanee, and Flint, and south from the Apalachicola River to Tampa Bay.[11]

Over the eighteenth century Creeks' interest in Gulf trading places increased while their towns moved and they became reacquainted with Spanish trading partners. In the first part of the century, Lower Creeks relocated several towns to the Ocmulgee and Savannah Rivers to privilege interaction with Atlantic-facing English colonies. The 1773 New Purchase, a "land-for-debt agreement" in which British deerskin traders called in debts and pressured Creeks and Cherokees to cede more than two million acres on the Savannah River, helps to explain voyages by these townspeople to Cuba. Their diplomats boarded Havana-bound Spanish fishing vessels to search for an anti-British alliance and additional markets. In 1775 Escuchape (or Escotchaby), the Lower Creek mico (headman, or chief) of the town of Coweta (which by then was located on the Chattahoochee River), visited Havana to explain his town's shifting trade priorities. Escuchape's interpreter said that the English had declared war, and his townspeople had killed them "along with their shops and for this reason planned to spread their Establishment" to "Tampa Bay." He and his war captains offered to negotiate on behalf of "the Indians of the Talapuz Nation." From Tampa, they proposed maintaining with Cuba "a reciprocal Trade of Pelts, Horses and other fruits." Cuba's governor declined Escuchape's export offer, saying that the Spanish and the English were at peace. He was unenthusiastic about the arrival of other visitors and offers to export to Havana "horses, Deer Pelts and other Goods which abound in their Lands." Meanwhile in Upper Creek homelands, after 1763 communities led by Emesteseguo built new connections to the British via the southern path to Mobile and Pensacola. They cultivated alliances with new claimants after the Spanish captured Pensacola, by which point Pensacola had replaced Charleston as the deerskin trade's most important port.[12]

Iberian policymakers' interest in the deerskin trade increased when it became clear that Americans sought the Creek homelands that constituted Spanish and US claims. Offers such as Escuchape's to export deerskins became appealing to the Spanish officials who learned about Anglo-American land hunger through reports from Upper Creek leaders. Strategists viewed Native Americans in Florida as subjects whose geopolitical power could be used to create "a barrier" that would preserve the zone that policymakers aspired to control: far west of Creek country, between New Orleans and the mines and markets of New Spain.[13]

The Bourbon Crown's new policy of *comercio libre* first focused on these more valuable trading places, which were accessible via Creek paths and waterways. The policy's implementation in Cuba (1765) and Louisiana (1768) resulted in increased migration to places where the import and export duties on some goods had been reduced, and Spanish officials granted exemptions on duties to certain people trading certain goods.[14] The Spanish imposed less favorable trading conditions on American merchants than those they negotiated with Native Americans. The US government attempted to claim more river space. Before 1795 Spaniards denied US ships the principle of navigational rights, preventing vessels from navigating around Florida and to New Orleans. The Spanish channeled exports of American goods into Louisiana via two rivers—the Ohio, and then down the Mississippi—and charged duties as part of the policy of *comercio libre* (15 percent before 1788, 6 percent thereafter). While he was secretary of state, Thomas Jefferson cited the Law of Nations to try to persuade the Spanish that the Mississippi River should be "considered as a streight of the sea." The land, he wrote, only rose above water level "here and there, in spots and slips," implying that this was ocean space that should be shared. The Spanish position was that the Mississippi was not the sea; access via the coast remained forbidden to US pilots.[15]

By the time the Spanish started paying attention to Creeks, many aspects of their town politics and trade practices resembled those earlier in the century. Politics still operated through the towns, which retained designations of red for war or white for peace and treaty making; and Beloved Men (or town councilors) still introduced themselves to Spanish, British, French, and American people by explaining what town they represented. Trade paths still ran between towns and colonial ports in Charleston, Savannah, Mobile, and Pensacola. Deerskins went to leather brokers in London, Bristol, and France for resale to leather manufacturers. The trade involved both men and women; men butchered hides from the white-tailed deer (*ítchu*), which they delivered to their wives' towns, where women processed them into dressed deerskins. For most of the century the trade operated on credit, which hunters needed to obtain ammunition and powder to hunt, and goods and manufactured cloth to supply their kin. British traders required credit and multiple years of work to counteract logistical shipping and spoilage problems from worms. Before the New Purchase, trade debts had operated as a form of relationality and interdependence, and people waited multiple years before requiring repayment.[16]

Creeks continued to use marriage to regulate trade; women collectively managed land, and always married outside of their clan. Metawney, the daughter of a Coweta headman and Escuchape's sister, married the Irishman George Galphin. From the plantation called Silver Bluff that Galphin built on Metawney's land, Galphin nurtured Lower Creek connections. When the American Revolution broke out, he threw his support behind the Americans, who misunderstood the nature of town politics and who sought his assistance to negotiate the neutrality of the entire Creek nation. After 1783, once the Spanish claimed both Floridas, their trade strategy with Creeks depended on a merchant whose traders were the men married to Creek women. In 1783 the Spanish recognized Creek women's exogamy (though they failed to understand it) by describing traders as "white men married to Indian women and domiciled in said Nation." On the Georgia-Florida border these men became known as Crackers. Through marriage, Creek townswomen opened their towns to their husbands, who then became honorary members of their wives' clans and eligible to claim hospitality from kin when traveling. A mico whose close female relatives married white traders could expel fraudsters, monitor licenses, and kill rulebreakers. Creek women politicked, forged political networks, solicited business from clan relations, and served as interpreters. They benefited from favorable trade rates and gifts for their families.[17]

Key aspects of gender and leadership roles had started to change. More trade in raw deerskins occurred beyond a town's boundaries and without the labor of women, or the mico's consent. Demand for raw deerskins increased in European markets around midcentury, which devalued women's prepared hides. Townspeople from Okfuskee preferred to sell undressed deerskins by the 1760s and 1770s. A shift in politics had also occurred, which deemphasized middle age as a measure of leadership ability. In the 1780s Spanish and American officials misconstrued the power of younger Beloved Men and their capacity to represent one nation constituted by unified towns. Officials' perceptions in turn granted some of those men more actual power to obtain trade gifts for redistribution to their kin, and to change Spanish trade rules. While some Creek men became wealthy merchants, slavers, and planters, Creek women sought to enhance older and develop newer forms of economic power. Some began to grow, spin, and trade cotton, while those considered in this chapter sought from the Spanish the opportunity to continue trading dressed deerskins.[18]

When the Spanish applied *comercio libre* to Creek homelands in East and West Florida that they claimed on paper but did not control in practice, they first granted several key foreigners authorization to trade. The terms of the 1783 Treaty of Paris required Britons in the Floridas to leave within eighteen months. William Panton and his partner, Frank Leslie—Scots merchants who had moved to East Florida as Loyalists during the war—avoided this obligation because the Creek women married to Panton's traders demanded that they be allowed to stay. Spanish colonial governors pointed to Spain's lack of manufactured goods and deerskin markets and argued for a savvy diplomatic policy change that permitted the continued residence of non-Spanish traders working for the British firm Panton, Leslie and Company. Thus British subjects secured permission—unusual at the time because officials granted it explicitly, rather than ignoring illicit trade—to take over Atlantic-facing storehouses in San Agustín (Saint Augustine) and on Río San Juan (the St. Johns River) to work in the Indigenous territory that the Spanish claimed. The Spanish were used to trading at a loss to maintain what remained of Spain's claims to land, and they increased their spending on Native American affairs from $4,000 in 1769 to $55,209 in 1794.[19] By virtue of Creek intermarriage, the relaxations to *comercio libre* that followed allowed more trade between Spanish ports and select foreign ports; the next allowed for the reduction in duties; and the most generous changes allowed some people exemptions from import and export duties entirely. Creek townspeople fell into the last camp of duty exemptions because the Spanish lacked knowledge about two related, important topics.

The Spanish did not know what they needed to know about rivers in Florida, and they were ignorant of Indigenous trading preferences. Riverine knowledge was necessary to resolve boundary disputes with the United States. The Spanish described the Apalachicola, St. Marys, and St. Johns Rivers in Creek and Seminole homelands as so labyrinthine that they created problems claiming land and conducting trade. The Apalachicola led to Lower Creek towns on the Flint and Chattahoochee Rivers and had been a recognized border between East and West Florida since 1763. The St. Marys was significant because since 1763 (but not for the first time) it had served as a north-south boundary between Georgia and East Florida, but it was also a Seminole and Cracker waterway settled by families and their free-ranging cattle. After 1782 when the Spanish claimed the Floridas, the Americans claimed a boundary further south, on the St. Johns, which Seminoles also controlled. A confidential 1788 report complained about the lack of any boundary "between East

Florida and the Americans" other than the St. Marys, and observed that neither this river nor the Apalachicola had "their courses" established with any "degree of precision."[20]

Spanish struggles to map and to monitor waterways allowed nonstate actors to attempt to claim territory and to offer better trading terms to Creeks and Seminoles. From towns on the Flint River, the Seminole leaders Okaiegigie (Thomas Perryman) and Kenhagee (William or Billy Perryman) forged relationships with the adventurer William Augustus Bowles that sought to address Lower Creek and Seminole townspeople's disillusionment with Spain and the United States. The Spanish interrogated Bowles's associates, whom they suspected of trading with Creeks, reminding them that foreign boats were prohibited from trading on the Gulf coast. In 1799 Bowles was said to be at the mouth of the Apalachicola River. In 1801 rumors placed him at the mouth of the Manatee River on Tampa Bay with Kinache, the mico of the town of Mikasuki (or Miccosukee), "with the intent" to establish "a house of commerce." The last governor of Spanish Louisiana asserted that Bowles had aspired to open trade on the entire coast from Panzacola to San Agustín. Bowles issued numerous proclamations declaring open ports at the mouths of the Apalachicola and Duck Rivers and Tampa Bay, and in a 1799 proclamation he introduced import duties for revenue only on liquor and all foreign goods. He is probably the sole non-Indigenous person in this chapter whose trade policies aligned with free trade as people understood it at the time. His proclamations responded to Creek and Seminole repulsion at export tariffs levied on their deerskins.[21]

In addition to seeking optimal trade conditions, Creeks and Seminoles also scuttled American and Spanish attempts to demarcate their homelands. During the joint Florida boundary survey of 1798 to 1800 to map boundaries agreed in the 1795 Spanish-American Treaty of San Lorenzo, astronomers serving both countries faltered in the face of imagined and real interference. Various advisors recommended that the surveyors avoid Native American paths and instead enter the Pearl, Mobile, and Apalachicola Rivers from the sea. In 1799 it took the surveyors three days to find "the real and navigable channel" of the Chattahoochee River with the help of a pilot who, it transpired, "knew no more of the coast" than the surveyors. In the end the surveyors fixed one boundary marker in the Okefenokee Swamp, abandoned the survey line further west, sailed around the Florida peninsula, and tried again from the east. These difficulties arose when Creeks from "the shore of taukelaukene"—possibly townspeople of Toccogulegau on the Flint River—stole the surveyors' domesticated animals and baggage, and when others set fire to trees in the swamp.[22]

As a result of these ongoing gaps in geographical knowledge, the Spanish liberalized their trade regulations in Creek territory. The Scots-Creek leader Alexander McGillivray—also known as Hoboi-Hili-Miko, or the Good Child King—sent letters educating Spanish governors and officials about Creek trading preferences and river travel. In 1801 Juan Ventura Morales misremembered the then-deceased McGillivray as the principal headman of the Upper Creeks, Lower Creeks, and Seminoles, but in reality he represented only some Abeika, Tallapoosa, and Alabama towns. McGillivray, like his mother, Sehoy Marchand, belonged to the influential Wind Clan—as his father, the Scots trader Lachlan McGillivray, would have belonged when traveling to trade. He was part of the new generation of Beloved Men who drew their authority from their matrilineal clan, but also from their youth and unprecedented property-holding.[23]

At least in 1784, McGillivray needed help communicating his ideas to Spanish officials. He had learned Greek and Latin at school in Charleston.[24] He admitted, however, that he spoke "no Spanish nor French." Instead, he relied on interpreters. In 1788 over "a matter of great Importance to us," he requested translations of Spanish correspondence. In 1789 he worried to Panton about the shortage of interpreters, and the following year he complained to Leslie that the Spanish considered him only "worth the miserable pay of an interpreter."[25] He needed translators, acknowledged that their pay was low, but also felt that he served a higher calling. One of his sisters, Sophia Durant, was known to speak for him during Creek council meetings, and there is at least one mention in McGillivray's correspondence of his sending a letter "by my Sister." Sophia Durant's work in this role suggests both the continuing importance of women interpreters in Creek diplomacy, as well as some of their male relatives' dismissal of their talents.[26]

Through Sophia and other linguists, beginning in 1784 Alexander McGillivray promoted Gulf trade locations, explaining to the Spanish that his townspeople preferred trading in West Florida over trade with the Americans. He proposed that the Spanish bring trade goods to Mobile, from whence he would "Supply my People by Water Carriage," which was "preferable to pack Horses." The fall lines on the Tallapoosa, Coosa, and Alabama Rivers posed no impediment to his plans to redistribute goods to his townspeople. In 1785 he expressed his hope on behalf of the Creeks, Chickasaws, and Cherokees that trade would be "established on the most permanent footing as the Spanish had promised them" at a treaty.[27]

During the lead-up to and negotiations during the 1784 Treaty of Pensacola, McGillivray advocated for better trading conditions. Mentions of dressed deerskins are evident in the treaty documents. He claimed in January 1784, "As for Raw & drest deer Skins, I can purchase any Quantity whatever, if they would turn to good account."[28] The treaty's first article established a defensive alliance in which Creek and other Indigenous signatories agreed to defend Spanish territory west of Creek homelands, and the Spanish pledged protection against Georgians' encroachment. The treaty also lifted restrictions on where British traders working for the Spanish could live, work, sail, and import, and provides scholars with evidence of Indigenous women's continued interest in selling *peletería pasada* (dressed deerskins), in addition to men's work selling raw deerskins in the hair, or *al pelo*. The treaty's eleventh article stated that traders were not permitted to trade secretly "in the woods and other private places" and had to trade in towns. Earlier scholarship averring that the Spanish did not distinguish between dressed and raw deerskins at the treaty does not consider the tariff documentation produced, which contains price schedules for goods exchangeable for both types of hides. *Peletería pasada* is listed in the price schedule (*tarifa*) as exchangeable for stroud and duffel blankets, ruffled shirts, binding, and trading guns, but not specified as exchangeable for plain or check shirts, calico, cotton, silk handkerchiefs, powder, balls, flints, hatchets, or looking glasses. John Walton Caughey, who published many of McGillivray's letters as well as a version of the price schedule, translated this latter document from Spanish to English by referring to "dress Skins" rather than "dressed," which has likely contributed to scholars' confusion.[29] Of course it is possible that McGillivray overstated his abilities to supply dressed deerskins, and it is important to acknowledge that treaties were notoriously difficult to fulfill and enforce. There is nevertheless more evidence of dressed deerskins in Creek and Spanish trading spaces than scholars have appreciated.

By 1785 Panton, Leslie and Company, which the Spanish came to call *La Casa de Panton* (Panton's house), had opened or taken over trading factories in San Agustín, on Río San Juan, in St. Marks (which became known as San Marcos de Apalache) at the confluence of the St. Marks and Wakulla rivers, and trading with "los Indios Talapuches" in Pensacola. The Spanish further liberalized trade by permitting the company's vessels to trade with foreign ports, including the Caribbean, the United States, and London. Then, the

Spanish began to grant the company exemptions from import and export duties, as *comercio libre* permitted in exceptional circumstances.[30]

Those duty exemptions resulted from McGillivray's antitariff arguments, which he contextualized with Indigenous trading preferences. In 1785 McGillivray asked in a letter what good it would do to "venerate the present duties imposed . . . especially those of the export of skins" and urged the adoption of "the most comfortable terms possible," including the removal of "dutys on the export of skins." In 1787 he reminded the Spanish that San Marcos had been established "upon the Condition" that the Spanish commandant at the fort did not interfere with Native Americans' trade—a provision that explicitly mentioned "dutys or Imposition or restrictions . . . laid on our Imports & exports here" as examples of interference. A year later McGillivray said that San Marcos was a Creek port that should be freed from Spanish duties because the British had restored the district to the Creeks before Spain regained the Floridas. In 1789 he acknowledged that the Spanish viewed duties as a "counterweight" to trade gifts and spending on diplomacy, but insisted, "This in reality is too much to load on a Poor Indian, whose constitution is such that he is always in need of favors rather than in a state to pay duties." This last letter, which Caughey excerpted in his collection of McGillivray's correspondence, excludes the previous sentence. On March 23 of that year, a response to McGillivray's advocacy arrived as a royal order that granted La Casa de Panton an exemption from import and export duties on articles "of the Indian trade." In September these exemptions were explicitly applied to the company's trading houses in Panzacola and Movila, and to San Marcos in 1790. Panton received retroactive exemptions from duties for the year 1788, because this was the year in which he submitted his request. Historians and some Spanish officials interpret the company's trading rights as a monopoly. However, the Spanish court never granted Panton this exclusive privilege, and dubbing the arrangement a monopoly fails to center Creeks' expectations that they would trade simultaneously across imperial borders.[31]

Additional Creek intervention in trade practices is evident in their use of price schedules. Anglo-Europeans published price schedules for trade goods throughout the colonial period in response to Native Americans' complaints about trade abuses. Issues with the 1784 *tarifa* indicate that Spanish price schedules similarly failed to eliminate corruption. After 1789 Panton refused to commit to Spanish price schedules agreed during the treaty, stating that he would seek to sell as cheap as or cheaper than American traders. Charleston merchants, meanwhile, sold blankets to Native Americans at a markup of

300 percent, and a 1790 observer estimated that Panton sold wares to Creeks at 500 percent of the cost. Creeks traded across non-Native borders by maintaining the ability to disregard price schedules, but so doing also resulted in higher prices on goods.[32]

US War Department officials knew that Creeks preferred Spanish trade, and the government accordingly adjusted its tariffs. Panton would claim that since the end of the American Revolution, his company had exported at least 124,000 deerskins per year, of which half to three-quarters probably came from Creek towns. At the 1790 Treaty of New York between the United States and the Creeks represented by McGillivray, the United States changed its tariffs. A secret article of the treaty—that required George Washington to seek Senate approval—designated people who could import up to $60,000 worth of duty-free goods for the Creek trade. It also offered McGillivray, Panton, and Leslie the opportunity to move their operations back to the United States. But though some Upper Creek micos and the US government agreed to the treaty on paper, Lower Creek and Seminole representatives were not invited to the treaty, and Bowles was still at large. The US government's Creek factory—once it was established on the St. Marys River at Colerain in 1796—exported just 25,000 deerskins annually.[33]

By 1794 Spain's First Secretary of State would explain that *comercio libre* was "the only way to preserve Louisiana, and to limit the advances and inordinate ambition of the United States." The Spanish Crown introduced the *cédula* of 1793 and extended *comercio libre* to Spanish Florida within this context, having failed to do so explicitly when it introduced *comercio libre* to New Spain in 1789.[34] The new *cédula* permitted Spanish residents in Louisiana and the Floridas to trade with people in Europe and America, but still required them to pay tariffs. For Creeks, very little about trade relationships changed with the *cédula*. In January 1794 Panton received reassurance that the new rules did not apply to his company. Because of their waterways and their exploitation of *comercio libre*'s exemptions, Creeks increased their capacity to trade on the Gulf, enjoyed the lowest duties and then duty-free trading, maintained a steady supply of goods, and retained their abilities to trade across fictional non-Native borders.[35]

* * *

This essay conveys the complexity of *comercio libre* by explaining with greater precision how it worked along Creek and Seminole trade networks on the rivers of their homelands. Creek townspeople not only received tariff exemptions made available through *comercio libre*: they also protected some women's production of dressed deerskins and negotiated flexible price schedules that maintained international trade. It is true that the demand for deerskins *al pelo* and a shift toward more centralized, patriarchal governance had pushed many women out of their role in the deerskin trade. It is also clear, however, that in the Gulf spaces that Spain claimed, intermarriage with Creek women still secured and informed trade. *Comercio libre*, which was not free trade, was instead a more flexible trade policy that allowed tariff exemptions in limited circumstances. The best and freest trade conditions for Creeks in riverine towns emerged because they could opt out of *comercio libre*'s tariffs and trade with multiple nations.

It should now be possible for readers to visualize in space the places where *comercio libre* applied. In July 1789 Vicente Folch said that according to his "way of thinking," McGillivray hoped "to obtain a seaport in order to escape the tariffs."[36] In reality, McGillivray needed no seaport because Creek townspeople escaped tariffs via rivers. Duty exemptions took effect above the fall line on waterways that flowed south into the Gulf and the Atlantic. These exemptions manifested at the storehouses that Panton and Leslie opened before and after changes to the policy: on the St. Johns River, the Wakulla River near San Marcos, and in San Agustín (1783); in Panzacola (1785); Movila (1790); New Orleans and Havana (1790); and on the Mississippi at Las Barrancas, or Chickasaw Bluffs (1795).[37] To see the deerskin trade's borders and the people who crossed them requires an act of imagination. Readers should picture deerskin factories in ports claimed by Spain, Britain, and the United States, traders with Muscogee wives circulating between towns, and Beloved Men and their translated letters on the move between towns and Gulf nodes working to obtain tariff exemptions. Such acts of imagination become more feasible after comparing archival sources with older collections of published letters.

Although Creeks won some of the best trading concessions at the time and were able to maximize the trade benefits of *comercio libre* from 1784 to the century's end, they still lost their land. The people with whom they traded started charging interest on debts around the same time that they won the tariff exemptions, and there was a disruption to how credit and debt worked to maintain relationships. From the 1780s to the 1790s, Creek leaders used trade to preserve their homelands, but as townspeople's debts to traders increased, so did the pressure to cede land. Ultimately, the pressure came from people who worked for Spain but were not Spanish. Officials had started to allow Panton's traders to extend lines of credit in 1784, the same year that the Treaty of Pensacola pledged protection against land-hungry Georgians. Panton, Leslie and Company also started charging 6 percent interest on Creeks' debts without their knowledge. The nature of debt changed to make long-term credit unviable.[38] Continuing the practice that the British promulgated with the New Purchase of 1773, Scots traders working for the Spanish south of the border and government officials working for the United States would demand land when Muscogees proved unable to settle the debts that had increased without their knowledge.

As Spanish and US negotiators finalized riverine boundaries, they claimed more Creek homelands. The 1795 Treaty of San Lorenzo fixed the US-Spanish northern border of Florida. It agreed that trade with Indigenous peoples should be mutually beneficial to Spain and the United States, which is when representatives from both nations began working together to survey Creek waterways. The merchant John Forbes, who took over when Panton died in 1801, formed good relationships with US officials. When Forbes claimed that Lower Creeks and Seminoles owed him money, Benjamin Hawkins, US agent to the Creeks, suggested land cessions as a solution. The 1804 Forbes Purchase extracted from Lower Creeks and Seminoles 1.4 million acres in West Florida between the Wakulla and Apalachicola Rivers. That year, the US Congress passed the Mobile Act, declaring that all navigable rivers that emptied into the Gulf of Mexico were rivers in US territory.[39] As readers know, a country's ability to claim riverine space did not confer the ability to control it. US Indian Department officials had opened a factory on the St. Marys River, but their knowledge of the waterway remained incomplete because its source was the swamp that surveyors had abandoned. Their hunger for knowledge about Indigenous waterways remained insatiable.

Notes

For comments the author thanks Alejandra Dubcovsky, Rebecca Earle, Kim Gruenwald, Jessica Choppin Roney, Marc-William Palen, David Silkenat, and attendees at the American History seminar at Edinburgh University and the British Group of Early American Historians' conference at Cardiff University. The Leverhulme Trust funded the research in Seville on which this chapter is based (Grant ref: RF-20202-150\).

1. Arthur Preston Whitaker, "Historical Introduction," in Arthur Preston Whitaker, trans. and ed., *Documents Relating to the Commercial Policy of Spain in the Floridas, With Incidental Reference to Louisiana* (DeLand, 1931), i–lix, esp. xix.

2. Jack D. L. Holmes, "Some Economic Problems of Spanish Governors of Louisiana," *Hispanic American Historical Review* 42, no. 4 (1962): 521–43, esp. 529n43; Whitaker, "Historical Introduction," xxiin4, xxiii; *Reglamento y aranceles reales para el comercio libre de España a Indias de 12. de octubre de 1778* (Madrid, 1778).

3. Fernando Aguerre Core, "Comercio libre y redes transatlánticas: la emigración española a Montevideo—de 1790 a 1810—y el tránsito de la autonomía a la independencia," *Colonial Latin American Review* 27, no. 2 (2018): 243–60, esp. 243; Ernesto Bassi, *An Aqueous Territory: Sailor Geographies and New Granada's Transimperial Greater Caribbean World* (Duke University Press, 2016), 29–30; Stanley J. Stein and Barbara H. Stein, *Apogee of Empire: Spain and New Spain in the Age of Charles III, 1759–1789* (Johns Hopkins University Press, 2003), 58; Janice Borton Miller, "The Struggle for Free Trade in East Florida and the Cédula of 1793," *Florida Historical Quarterly* 55, no. 1 (July 1976): 48–59. For a similar point about the amorphousness of free trade in the American context, see Paul A. Gilje, *Free Trade and Sailors' Rights in the War of 1812* (Cambridge University Press, 2013), 35, 337. For the Anglo-American context see Marc-William Palen, *The "Conspiracy" of Free Trade: The Anglo-American Struggle over Empire and Economic Globalisation, 1846–1896* (Cambridge University Press, 2016); Palen, *Pax Economica: Left-Wing Visions of a Free Trade World* (Princeton University Press, 2024), esp. 5.

4. David Narrett, "William Panton, British Merchant and Politico: Negotiating Allegiance in the Spanish and Southern Indian Borderlands, 1783–1801," *Florida Historical Quarterly* 96, no. 2 (Fall 2017): 135–73, esp. 152 ("commercial door"); David Narrett, *Adventurism and Empire: The Struggle for Mastery in the Louisiana-Florida Borderlands, 1762–1803* (University of North Carolina Press, 2015), 117; Gilje, *Free Trade and Sailors' Rights*, 131; Elena A. Schneider, *The Occupation of Havana: War, Trade, and Slavery in the Atlantic World* (University of North Carolina Press, 2018), 192, 252, 295; Adrian Finucane, *The Temptations of Trade: Britain, Spain, and the Struggle for Empire* (University of Pennsylvania Press, 2016).

5. José Antonio Armillas Vicente, "La Gran Confederación India: Interacción Hispano-Angloamericana con las naciones indias del Sudeste norteamericano a fines del S. XVIII," in *Estudios Sobre Politica Indigenista Española en América*, vol. II, ed. Seminario de Historia de America (Valladolid: Universidad de Valladolid, 1976), 249–66, esp. 251; William S. Coker and Thomas D. Watson, *Indian Traders of the Southeastern Spanish Borderlands: Panton, Leslie & Company and John Forbes & Company, 1783–1847* (University of West Florida Press, 1986); Narrett, "William Panton, British Merchant and Politico," 135–73; Narrett, *Adventurism and Empire*; James L. Hill, *Creek Internationalism in an Age of Revolution, 1763–1818* (University of Nebraska Press, 2022), 2, 14, 69 (quotes).

6. Thomas Wickman, "Our Best Places: Gender, Food Sovereignty, and Miantonomi's Kin on the Connecticut River," *Early American Studies* 19, no. 2 (Spring 2021): 215–63, esp. 218 ("sustenant"); Lissa K. Wadewitz, *The Nature of Borders: Salmon, Boundaries, and Bandits on the Salish Sea* (University of Washington Press, 2012); Andrew Lipman, *The Saltwater Frontier: Indians and the Contest for the American Coast* (Yale University Press, 2015); Renaud Morieux, "Anglo-French Fishing Disputes and Maritime Boundaries in the North Atlantic, 1700–1850," in *Governing the Sea in the Early Modern Era: Essays in Honor of Robert C. Ritchie*, ed. Peter C. Mancall and Carole Shammas (Huntington Library, Art Collections, and Botanical Gardens, 2015), 41–75; Joshua L. Reid, *The Sea Is My Country: The Maritime World of the Makahs, an Indigenous Borderlands People* (Yale University Press, 2015); Bassi, *An Aqueous Territory*; Susan Sleeper-Smith, *Indigenous Prosperity and American Conquest: Indian Women of the Ohio River Valley, 1690–1792* (Omohundro Institute of Early American History and Culture and the University

of North Carolina Press, 2018), esp. 6; Michelle Currie Navakas, *Liquid Landscape: Geography and Settlement at the Edge of Early America* (University of Pennsylvania Press, 2018); Steven C. Hahn, "The Cussita Migration Legend: History, Ideology, and the Politics of Mythmaking," in *Light on the Path: The Anthropology and History of the Southeastern Indians*, ed. Thomas J. Pluckhahn and Robbie Ethridge (University of Alabama Press, 2006), 57–93, esp. 62; Joshua A. Piker, "'White & Clean' & Contested: Creek Towns and Trading Paths in the Aftermath of the Seven Years' War," *Ethnohistory* 50, no. 2 (Spring 2003): 315–47; James Taylor Carson, *Making an Atlantic World: Circles, Paths, and Stories for the Colonial South* (University of Tennessee Press, 2007); Robbie Ethridge, *Creek Country: The Creek Indians and Their World* (University of North Carolina Press, 2003), 121–23; Charlotte Biggs, "Aspirational Designs, Interdependency, and Borders: An Analysis of Muscogee Border-Policy (1764–1790)" (Master's diss., Cardiff University, 2022); Alejandra Dubcovsky, *Informed Power: Communication in the Early American South* (Harvard University Press, 2016), 15; Steven Peach, *Rivers of Power: Creek Political Culture in the Native South, 1750–1815* (University of Oklahoma Press, 2024).

7. Kathryn E. Holland Braund, *Deerskins & Duffels: The Creek Indian Trade with Anglo-America, 1685–1815*, 2nd ed. (University of Nebraska Press, 2008 [1993]), xix, 11–12, 58, 71, 178; Joshua Aaron Piker, *Okfuskee: A Creek Indian Town in Colonial America* (Harvard University Press, 2004); Claudio Saunt, *A New Order of Things: Property, Power, and the Transformation of the Creek Indians, 1733–1816* (Cambridge University Press, 1999); Claudio Saunt, "'Domestick . . . Quiet being broke': Gender Conflict Among Creek Indians in the Eighteenth Century," in *Contact Points: American Frontiers from the Mohawk Valley to the Mississippi, 1750–1830*, ed. Andrew R. L. Cayton and Fredrika J. Teute (University of North Carolina Press, 1998), 151–74, esp. 161.

8. Rebecca Earle, *The Body of the Conquistador: Food, Race and the Colonial Experience in Spanish America, 1492–1700* (Cambridge University Press, 2012), 11; S. Max Edelson, *The New Map of Empire: How Britain Imagined America Before Independence* (Harvard University Press, 2017), 5 ("corridors"), 87, 96, 168; Finucane, *The Temptations of Trade*, 153; Narrett, *Adventurism and Empire*, 16, 24; Thomas D. Watson, "Strivings for Sovereignty: Alexander McGillivray, Creek Warfare and Diplomacy, 1783–1790," *Florida Historical Quarterly* 58, no. 4 (April 1980): 400–14, esp. 400–401.

9. "A Talk from the Kings, Chiefs & Warriors of the Lower Creek Nation to Captn Pedro Olivier Comisario Español, July 3, 1792," in John Walton Caughey, *McGillivray of the Creeks*, 2nd ed. (University of Oklahoma Press, 1959 [1938]), 314.

10. Hill, *Creek Internationalism*, 25; Braund, *Deerskins & Duffels*, 3; Edelson, *The New Map of Empire*, 87, 93, 96, 168.

11. Steven C. Hahn, *The Invention of the Creek Nation, 1670–1763* (University of Nebraska Press, 2004); Elizabeth N. Ellis, *The Great Power of Small Nations: Indigenous Diplomacy in the Gulf South* (University of Pennsylvania Press, 2023), 5, 12, 126, 236; Hill, *Creek Internationalism*, xv, 9, 216n9; Braund, *Deerskins & Duffels*, 6–7, 27; [Spanish interrogation of William Young, Juan Mincer, Josef Weaver, Fort San Marcos de Apalache, 4 March 1793], Papeles de Cuba, 1447, Archivo General de Indias (hereafter PC, AGI); Jennifer Monroe McCutchen, "'More Advantageous to Be with Spaniards': Gunpowder and Creek-Spanish Encounters in Cuba, 1763–1783," *Terrae Incognitae* 52, no. 3 (2020): 245–60, esp. 251; [Spanish translation and extract of a letter from William Panton, sent by Francisco Luis Héctor de Carondelet to the Conde de Aranda, New Orleans, 8 January 1793], PC, 1447, AGI; Brent R. Weisman, "The Plantation System of the Florida Seminole Indians and Black Seminoles During the Colonial Era,"

in *Colonial Plantations and Economy in Florida*, ed. Jane Landers (University Press of Florida, 2000), 136–49, esp. 138, 142.

12. Braund, *Deerskins & Duffels*, 38; James L. Hill, "'Bring them what they lack': Spanish-Creek Exchange and Alliance Making in a Maritime Borderland, 1763–1783," *Early American Studies* 12, no. 1 (Winter 2014): 36–67, esp. 39; McCutchen, "'More Advantageous to Be with Spaniards,'" 255 ("land-for-debt"); Chad McCutchen and Jennifer Monroe McCutchen in this volume, "'Playing the Old Game of Double': Navigating Creek and Spanish Geopolitics in the Post-Revolutionary Gulf South," 215–41; Edelson, *The New Map of Empire*, 283; Bryan Rindfleisch, "The 'Owner of the Town Ground, Who Overrules All When on the Spot': Escotchaby of Coweta and the Politics of Personal Networking in Creek Country, 1740–1780," *Native South* 9, no. 1 (2016): 54–88; "Don Rafael de la Luz, Interim Senior Assistant to the Plaza of Havana, 2 May 1775," and "Don Rafael de la Luz, rank of Infantry Captain of the Royal Armies and senior Assistant of the Plaza of Havana; Havana, 14 January 1776," in "Official Spanish Correspondence Pertaining to Relations with the Uchiz Indians: Section 1 Letters from 1771–1776," *Florida History Online*, https://history.domains.unf.edu/floridahistoryonline/projects-proj-b-p-html/projects-uchize-index-html/correspondence-pertaining-to-relations-with-the-uchiz-indians/#2-May-1775 and https://history.domains.unf.edu/floridahistoryonline/projects-proj-b-p-html/projects-uchize-index-html/correspondence-pertaining-to-relations-with-the-uchiz-indians/#14-January-1776 [accessed 16 December 2024]; Piker, "'White & Clean' & Contested," 327; J. Leitch Wright Jr., "Foreword," in Coker and Watson, *Indian Traders of the Southeastern Spanish Borderlands*, ix–xiii, esp. x.

13. "Confidential instructions given by the King to the Suprema Junta de Estado," the Palace, July 8, 1787, in Whitaker, *Documents*, 63 ("a barrier"); David J. Weber, "Bourbons and Bárbaros: Center and Periphery in the Reshaping of Spanish Indian Policy," in *Negotiated Empires: Centers and Peripheries in the Americas, 1500–1820*, ed. Christine Daniels and Michael V. Kennedy (Routledge, 2002), 79–103, esp. 81; Armillas Vicente, "La Gran Confederación India," 249; Whitaker, "Historical Introduction," in Whitaker, *Documents*, xix, xxii; Carmen de la Guardia Herrero, "The Experience of a Loss: Spain, Florida, and the United States (1783–1833)," in *La Florida: Five Hundred Years of Hispanic Presence*, ed. Viviana Díaz Balsera and Rachel A. May (University Press of Florida, 2014), 141–64, esp. 143; Kathleen DuVal, *The Native Ground: Indians and Colonists in the Heart of the Continent* (University of Pennsylvania Press, 2006), 119–20; Paul W. Mapp, *The Elusive West and the Contest for Empire, 1713–1763* (University of North Carolina Press, 2011), 337.

14. Barbara H. Stein and Stanley J. Stein, *Edge of Crisis: War and Trade in the Spanish Atlantic, 1789–1808* (Johns Hopkins University Press, 2009), 131; Schneider, *The Occupation of Havana*, 252, 271, 279; Holmes, "Some Economic Problems of Spanish Governors of Louisiana," 527, 532–33.

15. Narrett, "William Panton, British Merchant and Politico," 152; Holmes, "Some Economic Problems of Spanish Governors of Louisiana," 533; Thomas Jefferson, "Report on Negotiations with Spain, 18 March 1792," *Founders Online*, https://founders.archives.gov/documents/Jefferson/01-23-02-0259 [accessed October 26, 2022]; Gilje, *Free Trade and Sailors' Rights*, 131; Kim M. Gruenwald, "When a River Is a Border: Rivalries and Commercial Networks in the Riverine West," 151–52.

16. Braund, *Deerskins & Duffels*, 7, 53–54, 61, 171; Piker, *Okfuskee*, 156.

17. Bryan C. Rindfleisch, *George Galphin's Intimate Empire: The Creek Indians, Family, and Colonialism in Early America* (University of Alabama Press, 2019), 117; Rindfleisch, "The

'Owner of the Town Ground,'" 54–88; Braund, *Deerskins & Duffels*, 46; Rachel B. Herrmann, *No Useless Mouth: Waging War and Fighting Hunger in the American Revolution* (Cornell University Press, 2019), 100; [Arturo O'Neill to Bernardo de Gálvez], Pensacola, 24 March 1783, 326r, PC 36, ED_143_R_081, AGI ("Hombres blancos casados con Indias"; translation mine); "Report by Don Vicente Zéspedes, St. Augustine, Florida, June 20, 1790—duplicate," in James A. Lewis, "Cracker, Spanish Florida Style," *Florida Historical Quarterly* 63 (October 1984): 184–204; Megan Kate Nelson, *Trembling Earth: A Cultural History of the Okefenokee Swamp* (University of Georgia Press, 2005), 46–7, 119; Braund, *Deerskins & Duffels*, 75, 78, 83; Theda Perdue, "'A Sprightly Lover Is the Most Prevailing Missionary': Intermarriage Between Europeans and Indians in the Eighteenth-Century South," in *Light on the Path: The Anthropology and History of the Southeastern Indians*, ed. Thomas J. Pluckhahn and Robbie Ethridge (University of Alabama Press, 2006), 165–78, esp. 170; Andrew K. Frank, "Taking the State Out: Seminoles and Creeks in Late Eighteenth-Century Florida," *Florida Historical Quarterly* 84, no. 1 (Summer 2005): 10–27, esp. 23; Hill, *Creek Internationalism*, 56.

18. Braund, *Deerskins & Duffels*, 21; Saunt, "'Domestick . . . Quiet being broke,'" 153, 157, 163; Piker, *Okfuskee*, 154.

19. "Summary of a representation by Gilbert Antoine de St. Maxent relative to the commerce of West Florida and Louisiana," October 4, 1781, and "Summary of a representation of 1778 by Manuel de las Heras relative to the decay of commerce, El Pardo, January 9, 1779 [to Count Floridablanca]," in Whitaker, *Documents*, 3–10, 23–29; Holmes, "Spanish Treaties with West Florida Indians, 1784–1802," *Florida Historical Quarterly* 48, no. 2 (October 1969): 140–54, esp. 149; Frank, "Taking the State Out," 24; Narrett, "William Panton, British Merchant and Politico," 136; Braund, *Deerskins & Duffels*, 56.

20. Lawrence Shaw Mayo, *The St. Mary's River: A Boundary* (Privately printed, 1914), 24; "Vicente Manuel de Céspedes, Governor of East Florida, to the Marqués de Sonora enclosing a description of East Florida," St. Augustine, May 12, 1787, and "Discourse by José Salcedo relative to the cession of East Florida to the English," San Ildefonso, August 20, 1788, in Whitaker, *Documents*, 49–61, 75–99 (quotes, 59, 95–97).

21. Cameron B. Strang, *Frontiers of Science: Imperialism and Natural Knowledge in the Gulf South Borderlands, 1500–1850* (University of North Carolina Press, 2018), 2–3; Hill, *Creek Internationalism*, 78–9, 84–5, 231n3, 233n16; [Spanish Interrogation of William Young, Juan Mincer, Josef Weaver, Fort San Marcos de Apalache, 4 March 1793], PC, 1447, AGI; [Vicente Folch to the Marques de Casa Calvo], Pensacola, 21 November 1799, ff. 200v–201r, PC, 2355, AGI; Jacob DuBreuil to the Marqués de Casa Calvo, San Marcos de Apalache, 29 March 1801, f. 14r, PC, 2355, AGI ("con el yntento de Formar alli casa de Comercio"); William Augustus Bowles, "Proclamation," Apalachicola, 26 November 1799, f. 431r, PC, 2366, ED_145_R_048; [Juan Manuel de Salcedo to the Marques de Someruelo], [n.p.], 9 July 1803, f. 491rv, PC, 2355, AGI.

22. Strang, *Frontiers of Science*, 130; [Stephen Minor to Manuel Gayoso de Lemos], Chiscotosa Camp, Chattahoochee, 5 August 1799, f. 651v, PC, 2355, AGI; [Information from Manuel Gayoso de Lemos sent to Francisco Luis Héctor de Carondelet], to Luis de Las Casas, New Orleans, 13 September 1796, PC, 1447, AGI; [Stephen Minor to Manuel Gayoso de Lemos], Chiscotosa Camp, Chattahoochee, 5 August 1799, f. 651r ("knew no more"), PC, 2355, AGI; Estevan Minor, Andrew Ellicott, Thomas Power, and David Gillespie, Cumberland Island, 9 April 1800, ff. 676r, 689r, PC, 2355, AGI; Nelson, *Trembling Earth*, 72; Estevan Minor and Nicholas D[aunoy?] to Vicente Folch, Flint River, 24 September 1799, 639v, PC, 2355 ("the shore of"); Steven C. Hahn, "'They Look upon the Yuchis as Their Vassals': An Early History

of Yuchi-Creek Political Relations," in *Yuchi Indian Histories Before the Removal Era*, ed. Jason Baird Jackson (University of Nebraska Press, 2012), 123–53, esp. 116.

23. Caughey, *McGillivray of the Creeks*, 3; [Juan Ventura Morales to the Secretary of State, New Orleans, 31 December 1801], Santo Domingo, 2670 (hereafter SD), C_02634, f. 168r, AGI; "O'Neill to Ezpeleta, October 19, 1783," in Caughey, *McGillivray of the Creeks*, 62; Saunt, *A New Order of Things*, 67–89; Braund, *Deerskins & Duffels*, 21, 45–46.

24. Caughey, *McGillivray of the Creeks*, 15.

25. "McGillivray to O'Neill, February 5, 1784," in Caughey, *McGillivray of the Creeks*, 70 ("no Spanish); "McGillivray to Miró, June 12, 1788," in ibid., 83 ("a matter"); "McGillivray to Panton, August 10, 1789," in ibid., 249; "McGillivray to Leslie, May 20, 1790," in ibid., 263 ("worth the").

26. "McGillivray to O'Neill, January 3, 1784," in Caughey, *McGillivray of the Creeks*, 67 ("by my"); Albert James Pickett, *History of Alabama and Incidentally of Georgia and Mississippi, from the Earliest Period* (Robert C. Randolph, 1896), 419; Felicity Donohoe, "'Decoying Them Within': Creek Gender Identities and the Subversion of Civilization," in *Native Diasporas: Indigenous Identities and Settler Colonialism in the Americas*, ed. Gregory D. Smithers and Brooke N. Newman (University of Nebraska Press, 2014), 187–205, esp. 204n27.

27. "McGillivray to O'Neill, January 1, 1784," in Caughey, *McGillivray of the Creeks*, 65 ("Supply my"); [Spanish translation, Alexander McGillivray to Arturo O'Neill], Pensacola, 24 July 1785, f. 1rv, PC, 2352 ("sea establecido en el pie mas permanente como de les prometio").

28. "McGillivray to O'Neill, January 3, 1784," in Caughey, *McGillivray of the Creeks*, 67.

29. Narrett, "William Panton, British Merchant and Politico," 139; Braund, *Deerskins & Duffels*, 173. For earlier scholarship on the treaty and tariffs see Holmes, "Spanish Treaties with West Florida Indians," 142; Coker and Watson, *Indian Traders*, 59, 59n27; Saunt, "Domestick . . . Quiet being broke,'" 157. For the treaty documents see "Spanish Treaty with Talapuche Indians, May 31, 1784, Articles of Convention Signed at Fort of Pensacola, Florida," *Library of Congress*, http://hdl.loc.gov/loc.mss/mtj.mtjbib000919 [accessed February 16, 2023]; "Tarifa para el comercio de la nacion Crick, ó Talapouche en el Congreso general celebrado en Panzacola los días treinta, y uno de Mayo y primo. de Junio de este año de mil setecientos ochenta y quatro," [after 1 June 1784], ff. 590rv, PC 117C, ED_144_032, AGI; John Walton Caughey, "Alexander McGillivray and the Creek Crisis, 1783–1784," in *New Spain and the Anglo-American West: Historical Contributions Presented to Herbert Eugene Bolton*, vol. 2, ed. William C. Binkley, Cardinal Goodwin, and J. Fred Rippy (1932), 263–88, esp. 285–86.

30. Narrett, "William Panton, British Merchant and Politico," 139; [Martin Navarro], Pensacola, 16 September 1785, 290v, SD 2670, C_02635, AGI; [Juan Francisco Arnaud de Courville, index of royal orders and papers related to permissions granted to Panton and Leslie], Pensacola, 23 December 1809, ff. 276rv ("los Indios Talapuches"), 283v, SD, 2670, C_02635. All changes to Panton and Leslie's trading rights are indexed and summarized in SD, 2670, C_02635, ff. 276–89, AGI.

31. [Spanish translation, Alexander McGillivray to Arturo O'Neill], Pensacola, 24 July 1785, f. 1rv, PC, 2352, AGI ("venerar los derechos puestos . . . especialm[te]. los de la exportacion de pieles" and "los terminos mas comodas"); "McGillivray to O'Neill, July 24, 1785," in Caughey, *McGillivray of the Creeks*, 94; Coker and Watson, *Indian Traders*, 128–9; "McGillivray to O'Neill, November 20, 1787," in Caughey, *McGillivray of the Creeks*, 163–65 ("upon the" and "dutys or"); [Translation of a letter from Alexander McGillivray to Vicente Manuel de Zéspedes, Little Tallassee, 6 February 1789, sent by Vicente Manuel de Zéspedes to José Manuel de Ezpeleta, San

Agustín, 3 April 1789], PC, 1395, AGI ("para contrapesar los regalos, y demas gastos" and "Esto en realidad es demasiado cargar sobre un Pobre Indio, cuya constitucion es tal, que se halla siempre mas bien necesitado de favores, que no en estado de pagar derechos"); Caughey, *McGillivray of the Creeks*, 221; [Juan Francisco Arnaud de Courville, index of royal orders and papers related to permissions granted to Panton and Leslie], Pensacola, 23 December 1809, f. 280v ("articulos del Comercio de Indios"); McCutchen and McCutchen, "'Playing the Old Game of Double,'" 232; Whitaker, "Historical Introduction," xxxii; Miller, "The Struggle for Free Trade in East Florida," 50–51, 129; Narrett, "William Panton, British Merchant and Politico," 166.

32. Braund, *Deerskins & Duffels*, 99, 112–116, 163, 190; Miller, "The Struggle for Free Trade in East Florida," 49; Narrett, "William Panton, British Merchant and Politico," 154–55.

33. [James Seagrove] to [Henry Knox], Rock Landing on the Oconee, 24 May 1792 and [Information delivered by James Leonard to James Seagrove, Rock Landing on the Oconee, 24 July 1792], *American State Papers*, Class II, Indian Affairs, ed. Walter Lowrie and Matthew St. Clair Clarke (Washington, DC: Gales and Seaton, 1832), 1:296, 308; Braund, *Deerskins & Duffels*, 72, 98, 175–77; Watson, "Strivings for Sovereignty," 412–13.

34. Miller, "The Struggle for Free Trade in East Florida," 55–56; John Fisher, "Imperial 'Free Trade' and the Hispanic Economy, 1778–1796," *Journal of Latin American Studies* 13, no. 1 (May 1981): 21–56, esp. 23; John B. Stetson, J. Franklin Jameson, Charles B. Reynolds, and George Parker Winship, "Foreword," in Whitaker, *Documents*, vii.

35. [Duque of Alcudia to Francisco Luis Héctor de Carondelet, New Orleans, 18 August 1794], PC, 1447, AGI ("el único medio de conservar la Luisiana"); Miller, "The Struggle for Free Trade in East Florida," 55–56; [Juan Francisco Arnaud de Courville, index of royal orders and papers related to permissions granted to Panton and Leslie], Pensacola, 23 December 1809, f. 283v, SD, 2670, C_02635 (Panton and the *cédula*). On fictional borders see Edward G. Gray, "Toward a Pre-History of Territory: Thomas Hobbes, the Maryland Palatinate, and the Colonial Boundary Problem," and Christian J. Koot, "Things to Think With: The Use of Borders on a Seventeenth-Century Map of New England," both in this volume.

36. "Folch to Miró, July 2, 1789," in Caughey, *McGillivray of the Creeks*, 242.

37. [Juan Francisco Arnaud de Courville, index of royal orders and papers related to permissions granted to Panton and Leslie], Pensacola, 23 December 1809, f. 283v, SD, 2670, C_02635; Whitaker, "Historical Introduction," xxxiv; Narrett, "William Panton, British Merchant and Politico," 136.

38. Coker and Watson, *Indian Traders*, 186–87; Narrett, "William Panton, British Merchant and Politico," 155; Hill, *Creek Internationalism*, 142, 243–44n21.

39. Coker and Watson, *Indian Traders*, 201, 235; Narrett, "William Panton, British Merchant and Politico," 137; Narrett, *Adventurism and Empire*, 159; Braund, *Deerskins & Duffels*, 177; Whitaker, "Historical Introduction," xxxvii; Gilje, *Free Trade and Sailors' Rights*, 131, 133.

CHAPTER 11

Possibilities and Peril

Exploring the Transnational Experiences of Black People in the Maritimes, 1783–1792

HARVEY AMANI WHITFIELD AND SARAH CHUTE

The history of Black people crossing the early American-Canadian border is usually framed as a crossing from American enslavement to Canadian freedom.[1] This threshold model of the border does not do justice to the diversity of outcomes inherent in transnational migration in the eighteenth-century Atlantic world. After the American Revolution, the Maritime colonies of Nova Scotia, New Brunswick, and Prince Edward Island were sites where indentured servants, enslaved, and re-enslaved people lived next to free Black Loyalists who sometimes gained meaningful freedom, formed families, and owned small farms. Watery borders shaped life in the Maritime colonies. The ocean readily enabled connections to other Atlantic locales, from Boston, to Kingston, to Liverpool. This connectivity meant the water borders held both possibilities and perils for unfree Black people in the Maritimes.

Historians Isaac Land and Michael Pearson have argued that "maritime influence" shaped individuals beyond the littoral; a "proximity of causation" could tie people to the ocean whether they were physically close to shore or not.[2] Such an expansive proximity challenged the security of free and enslaved Black people in coastal towns and on inland farms alike. Water could take people to enslavement in the West Indies, United States, or other parts of the Maritimes through sales and kidnappings, but it could enable escapes and migrations to freedom. John Morton argues in this volume that borderland residents and itinerant missionaries in New Brunswick and Maine were responsible for the on-the-ground understandings of international boundaries, largely dictated by their use of waterways that could connect or separate people.[3] Similarly, free and enslaved Black people in the Maritimes discussed and demonstrated their understandings of international and imperial politics—and thereby helped legitimize British colonial claims—whenever they referenced the Crown's obligations to Black Loyalists, many of whom had fled Patriot enslavers to fight and seek refuge behind British lines during the Revolutionary War.[4] At the same time, Black people still faced exploitation in the Maritimes, and thousands remained enslaved to white Loyalists as they migrated north from the United States. As our study shows, free and enslaved Black people in the Maritimes were quickly threatened by the Atlantic's fluid connectivity to other slave societies, but also sometimes chose to traverse those waters in search of greater security and autonomy. How meaningful and tangible were these colonies' borders, then? While archival fragments seldom include the direct words of Black people in the Maritimes, their diverse actions can tell us something about how they might answer that question. How and why Black people interacted with the aquatic highway of the Atlantic was deeply informed by the politics, laws, racism, and labor that took place within the jurisdictions of Prince Edward Island, New Brunswick, and Nova Scotia.

The connectivity to Atlantic trade and migration afforded by the coastal nature of Maritime port towns—as well as the constant influence of the land as a place of local displacement and settlement—reflect the significance of these histories' locations, where landed imperial powers could cast people forcibly across the Atlantic. These relationships thereby strengthened capitalist colonial power on the ground, whether in the form of carrying out criminal charges to uphold a social system, bolstering the wealth of merchants and enslavers who forged and exploited the ties between the regions, or denying self-sufficiency and the means for power to Black individuals locally. The environmental, spatial, and geographic dimensions to this human history reflect David Austin's identification of the historical and political experiences of Canada, or in this case the Maritimes, as "a laboratory through which we can understand Black diasporic experiences, perhaps in ways that studies of the United States or Britain, for example, have generally not afforded."[5] The perilous experiences in the Maritimes—the exclusions, exploitations, and denials of Black humanity—were deeply shaped by those colonies' surrounding Atlantic boundaries and by jurisdictional, economic, and sociopolitical claims steeped in notions of dominance on and over the land.

Peninsular and insular in geography, the Maritime coasts and islands were hardly isolated from other parts of the Atlantic world. Shaped by a nexus of ideologies that spanned the Atlantic, the region can be thought of as an extension of the plantation complex. The plantation complex usually centers upon Brazil or the greater Caribbean, but what we are showing is the significance of slavery and waterways to what Philip D. Curtin calls the "important trading partners" in the northern region of the plantation complex encompassing the Maritimes, New England, and part of northern New York, which were connected to the tropics by water.[6] Paul Musselwhite's chapter reminds us how plantations in the seventeenth-century English Caribbean cartographically stood as "nodes of communal authority" and later "a coherent vision of imperial jurisdictional space"—concepts that persisted rhetorically throughout the eighteenth-century British Atlantic.[7] For example, among Maritime enslavers, Caribbean plantations functioned as a rhetorical device and actual destination of private and court-sanctioned punishment and violence against Black people.[8] By looking at the northern reaches of the plantation complex, we gain a broader understanding of the many dimensions of enslavement in the Atlantic world. Connected by overseas routes of trade and migration, northern and tropical systems of slavery were mutually reinforcing. Our study approaches enslavement through the specific cases of three individuals—Jupiter Wise, Hector, and Lydia Jackson—to demonstrate the complex interplay between political jurisdiction, water borders, possibilities, and perils at this northern outpost of the British Atlantic. These short biographies are not widely known and deserve to be told in detail. Ultimately, a combination of personal contacts, the circumstances of control, and chance factored into whether they found possibility or peril in the waters that met the rocky Maritime shore.

Jupiter Wise

The mid-1780s case of Jupiter Wise demonstrates how local conflicts had Atlantic consequences for Black people in the Maritimes. This intricate case enriches our understanding of relationships between enslaved, indentured, and free people.[9] It also showcases how enslaved people in Prince Edward Island (PEI) thought about their bondage in relation to other places like the United States, specifically, Massachusetts, where slavery had been tested against the new state constitution and recently outlawed through a series of Supreme Judicial Court cases in the early 1780s.[10] Jupiter Wise's experience with the justice system in colonial Charlottetown demonstrates the region's close imperial connections to the West Indies.[11] Historians have used the example of Jupiter Wise as an illustration of slavery and resistance on PEI. Examining this case anew tightens our grasp on Black attitudes about enslavement on the island and about the function and flexibility of the colony's water border.

According to the Book of Negroes, in 1783, Jupiter Wise left New York for Shelburne, Nova Scotia, as a twenty-five-year-old "[stout] fellow" who was "Certified to be free."[12] The Book of Negroes is not simply a record of free and enslaved Black people; it also contains hints about how Black people were easily re-enslaved. The Book of Negroes listed free Blacks as being in the "possession" of a white Loyalist even if they had a GBC (a General Birch Certificate, or freedom pass issued in New York to Black people who could provide evidence that they had "resorted to British Lines" during the Revolutionary War).[13] Throughout the Maritimes, the difference between enslavement and servitude could be nebulous. Even those who were indentured servants were often treated as slaves and consistently faced the fear of being sold to the West Indies or elsewhere. Unlike in the other Maritime colonies, though, slavery was legally protected in Prince Edward Island. An act that dealt with baptism gave statute recognition to slavery, which allowed it to persist in the colony until legislation overturned it in 1825.[14] This recognition made it difficult for enslaved islanders to successfully challenge their bondage in court. For example, in 1802, the PEI Supreme Court returned an escaped slave to his enslaver, noting that the owner's proof of ownership was "perfect and compleat."[15]

A muster book reveals Jupiter lived in Birchtown, Nova Scotia, with a woman, Hannah Wise, who was probably his wife.[16] Coming to the Maritimes as a Black Loyalist, Jupiter Wise should have been able to enjoy his "Certified" freedom.[17] By 1785, however, he had left Birchtown and was living on Prince Edward Island as a "Black Servant" to Captain George Burns in Charlottetown.[18] Another record described him as a "Labourer."[19] This transition from freedom to servitude deserves scrutiny. First, the local markets were largely controlled by middling and elite white merchants who profited from the slave trade, some of whom were enslavers themselves. As Kim M. Gruenwald's chapter in this volume shows, merchants exploited their advantages in their most lucrative and efficient domains: watery spaces (in this case, the Atlantic) that served as "arteries of trade."[20] In both the colonial "Riverine West" and postwar coastal Maritimes, enslaved people of African descent were victims of merchants' capitalist and imperial competitions over water.[21] But the businesses of individual merchants were not the only ways in which Caribbean slavery shaped life in the Maritimes. Shirley Tillotson's research about the tropical origins of Nova Scotia's provincial revenue further attests to how the Maritimes were culturally and economically "part of an Atlantic-Caribbean region—a region in which the culture of the plantation naturalized white hostility towards independence for Black people."[22] Jupiter Wise may have suffered from the scarcity of resources allotted to Black Loyalist refugees or the hostility toward Black people that was endemic in Shelburne and eventually culminated in the 1784 Shelburne race riot. While the initial reason for Wise's move to Prince Edward Island is unknown, the process of going from freedom in Nova Scotia (promised by the British, though largely nominal) to a dependent relationship of servitude in a neighboring colony exemplifies how the Maritime political economy was ultimately inhospitable to Black Loyalist efforts for self-determination.

The court records from Prince Edward Island testify to this reality. The earliest events relayed in the court documents from Wise's trial come from James Stevens, who described himself as "A Free Black Man" and accused Jupiter Wise of burglary. Stevens explained that before he met Wise, he had joined a group of unfree islanders in making plans to escape the colony.[23] In the summer of 1785, a Black man named Mingo, enslaved to Captain Callbeck, made a tentative proposal to James Stevens to run away, having recently suffered beatings from Callbeck. Mingo and Stevens were cautious because they needed to recruit loyal co-conspirators who would be willing to help Mingo gain freedom. According to court records, the two hoped to "propose the scheme to Ben (Gov. Pattersons black Slave) but the aforesaid Mingo said he would first mention it to him as a joke, as he had no great dependence on his faith And also that he would speak to the Chief Justices Black Slave Peter." Stevens ended up seeing Ben before Mingo could make the sarcastic suggestion, and upon hearing the plans, "Ben instantly consented, and said that By God he had been a long time waiting to join some others in running away and further said he would have a Dance before he went off."[24]

As Mingo's intended joke to Ben suggests, trust was a powerful currency among enslaved people, who carefully spent or reserved it to protect themselves, their interests, and aspirations. Ben was likewise a judicious participant, having "observed that he would not take Peg (a female Slave of Governor Pattersons) as he could not trust her." Ben was eager to escape, ready to "sell all his Cloathes but two or three Shirts" to fund the escape and "proposed taking the Governors Green Barge and going to Boston where he could work for more Cloathes." Resulting from a series of freedom suits in the early 1780s, the abolition of slavery in Massachusetts may have made Boston an especially attractive destination to skilled, enslaved men like Ben.[25] When Stevens and Mingo proposed the scheme to Peter a week later, they warned him that "the first who flinched would be put to Death." Despite the dangers, Peter agreed to the conspiracy under the condition that they take the governor's sloop instead of his "Boat." During their conversations, the men were careful to avoid being overheard. They even made plans to defend themselves by making sure they had a musket, musket balls, powder—enough ammunition that "if any Boat pursued them . . . they would stand Sixteen rounds." However, when Ben later reported to Stevens that he could not find "a Boat" on Governor Patterson's property and Mingo made no more mention of it to Stevens, "there the matter dropped."[26]

According to the court record, "shortly after Captain Burns' arrival in Town," James Stevens had been in Burns's kitchen when Jupiter Wise, one of the captain's "Black Servants," called him over and requested he get a "Gimblet" so Wise could get "Rum which was not his Masters." Stevens consented to the plan, borrowed the drill-like tool, and returned with it. Wise then asked him for a candle, which Stevens obtained "from his Masters House." Taking with them an empty "Stone Jugg," Stevens followed Wise to a house "lately occupied by *Mr. Michael McDonald* in CharlotteTown" where Wise broke in and came out with the vessel filled with nearly two gallons of rum. The two hid the jug in a lot where it would be safe.[27]

The next Sunday, James Stevens, Jupiter Wise, Guy ("a black Servant of Capt. Callbecks"), and Joe ("another of Capt. Burns' black Servants"), met at "Raineys Hutt" in Charlottetown. Jupiter Wise and Guy suggested having a dance there the following night, offering "to bear each half of the expense of the Supper." The others pitched in, too: Stevens stated he could provide "half the rum" and Joe charitably "said he would offer a Dollar if it was wanted." Finishing up the rum they had at present, Wise mentioned to Stevens that "he knew where to get more." Carrying a mug to the yard where they had stored the stolen liquor, Wise filled the cup and returned to share it with the group at Rainey's hut, where shortly afterward James Stevens agreed to "invite the Company for the next evening in his own name." To prepare for the affair, Jupiter Wise promptly arranged for Stukely Burns—his master's son—to write "Cards of invitation" which were "sent round to their black acquaintances that evening and one to Mrs. Pollard Capt. Gores Servant Maid." The fact that the men extended a party invitation to a presumably white servant demonstrates that trust over shared, pilfered food and drink sometimes extended outside the circle of enslaved Blacks to unfree white associates. Over the course of that evening and the next morning, the company returned to the Rainey house to deposit their goods for the party. Four freshly killed fowls, "a piece of Beef" and "some Potatoes" were courtesy of Jupiter Wise. Guy brought some flour and butter wrapped in a handkerchief, tucked in a basket. Stevens stated that Wise directed him on the evening of the party to fetch the stolen rum where they had left it. The celebration then took place that night at the Raineys'. The sole details known about the party itself relate to how it was dissolved: Captain Callbeck "took away His black Slave the aforesaid Guy sometime after which they broke up and went severally home."[28]

Stevens's motives for making this statement are not certain, but it seems possible he was trying to protect himself from the consequences of having associated with Wise. A tavern-keeper and legislator named John Clark claimed that a man named James "in the Service of Mr. Walter Berry" had joined Wise and Clark's slave Thomas Williams in an attack against him and a white servant and soldier named Sylvester Petty. Four days later, Walter Berry witnessed James Stevens's statement about Wise's burglary. It seems probable, as Hornby suggests, that Stevens "may have been buying freedom from prosecution or his master's punishment by testifying against Wise."[29] Berry's influence over Stevens and among his own peers should not be understated; he had served as Speaker of the Legislative Assembly several years before.[30]

The accounts of John Clark and Sylvester Petty outlined their recollection of the events that took place on November 26, 1785.[31] On that Saturday evening, Jupiter Wise and a man enslaved to John Clark named Thomas Williams sat in Captain Burns's kitchen with some stolen items, playing cards. Williams had been missing and was believed to have thieved from the house of Dr. Walker, a garrison surgeon for whom Sylvester Petty labored. When Petty saw Williams and Wise, he explained to John Clark that their goods were the ones that had been stolen from the doctor. Petty feared being blamed—lashed—for the disappearance of these items, so exposing the Black men allowed Petty to protect himself.[32] The mutual vulnerability of Black slaves and white servants could pit these groups of lower-class people against one another for fear of retribution by their more powerful white masters. Wise, reacting to the accusation, cursed and assaulted Petty. Clark, in turn, used a large stick he was carrying to strike down Wise.[33]

Subsequently, Captain Burns's wife, whose husband held Wise as an indentured servant, called Clark into the parlor where slaveowner Thomas Haszard was paying her a visit. There, Mrs. Burns expressed regret that Clark "was going home alone, and that She had no White Man about the House to send with him, as she was apprehensive the black People had premeditated some design against him."[34] Sylvester Petty, meanwhile, waited outdoors. As John Clark left the house, Wise and Williams attacked Petty. Joined by "another Black who he Supposed is named James and is in the Service of Mr. Walter Berry," they then turned upon Clark.[35] A stick fight broke out. Wise hit Clark over the eye "and Clark also exchanged stick blows with James."[36] Jupiter Wise sprinted inside the Burns's house and returned to the street, brandishing a cutlass. He used the short sword against Clark, "who fended him off with his stick." Thomas Haszard then opened the door and illuminated the scene with a candle, which quelled the brawl. Wise bid Clark to join him in the darkness of the street, but Clark declined to fight further.[37] Soon thereafter, the local authorities arrested Wise and charged him with assault.[38]

On February 24, 1786, the court convicted Wise of felonious assault and sentenced him to death.[39] Wise "Pleaded ye Benefit of Clergy," which the court accepted and duly reduced his sentence.[40] Hornby explains that this plea "allowed first-time convicts of some felonies to escape the usual hanging," and that Wise was the first person on Prince Edward Island to request it.[41] Jupiter Wise was "to be Transport'd for seven Years to some one of his Majestys Islands in the Wst. Indias."[42] Of course, this sentence probably meant a lifetime sentence of enslavement as it is hard to believe that in the Caribbean he would have been freed and allowed to go where he pleased.

The shared grievances of the resisters in this intricate story—from Mingo's Boston-bound hopefuls to Jupiter Wise and his party invitees—about their experiences of subjugation in Prince Edward Island show us how these people challenged the wrongs against them and how they carved out landed, coastal, and maritime spaces of freedom amid their exploitation. Despite slavery's statutory recognition on the island, this jurisdiction was supposed to represent freedom to formerly enslaved African Americans who had escaped their Patriot enslavers and sided with the British during the Revolutionary War. As documented in the Book of Negroes, the voluntary migrations of free Black Loyalists suggest many Black colonists resettled with the hope of a better life, a life of meaningful freedom. Yet the imperial and economic connections between the Maritimes and West Indies ensured the ongoing insecurity of

Black Loyalists who, like Wise, were vulnerable to the loss of their autonomy or re-enslavement, often carried out through transatlantic voyages.[43]

This imperial relationship is one of the reasons why Wise was sentenced to the Caribbean, where he would have entered a competitive market that not only exploited Black laborers but was also driven by the lucrative demands of sugar production. Wise received seven years, the minimum sentence for transportation.[44] This type of punishment was "shaped by Atlantic labour markets."[45] As historian Hamish Maxwell-Stewart explains, transportation sentences for convicts like Jupiter Wise were often fixed at seven years "not for legal reasons but in order to competitively position convicts within the transatlantic market in unfree labour."[46] Whether in the Maritimes as an unfree Black Loyalist or in the West Indies as a convict and potentially re-enslaved, Jupiter Wise faced continual exploitation in the British Atlantic.

The water border facilitated the relationship between the Maritimes and the Caribbean. Although largely a colony based on subsistence farming, Prince Edward Island had ties to other parts of the empire, including through the fishing industry: to northern Atlantic places such as Halifax or London, but also tropical islands such as Barbados and St. Vincent.[47] Throughout the region, access to the sugar islands enabled Loyalists to obtain rum distilled in the British West Indies.[48] This connection, then, was an important precondition for Wise and his friends to even consider stealing the rum. Their actions make us wonder, what did the stolen products of enslaved labor from the Caribbean mean to unfree people in the Maritimes? Wise was ultimately sentenced to the region where that drink was produced, so, once in the West Indies, did the liquor or the place take on new meaning to him?

Like other parts of the Maritimes, the way in which the ocean surrounded the island meant its imperial connections to other parts of the British Atlantic were actualized by way of water. For Jupiter Wise, though, the water border ultimately offered more obstructions than avenues to freedom. Unfree Black islanders' proximity to the Atlantic meant the ocean functioned as a steady threat (for the reasons outlined above) as well as a source of opportunity for escape. James Stevens and his counterparts, who faced the vulnerability of residing on Prince Edward Island, attempted to take advantage of the fluidity of the water border.[49] The initial plans to steal and escape to Boston aboard Governor Patterson's sloop demonstrates how the group of enslaved and indentured people perceived their watery surroundings as an important channel to freedom. Bound on an island, it is logical that they would turn to the sea; a terrestrial escape on tiny Prince Edward Island would offer limited protection. Enslaved Black people had sought escape through these means before. For example, when Mingo proposed to James Stevens the idea of running away as a group, he warned him about saying too much to Ben because Ben had "before deceived some men who had intended to run away with a Schooner from Governor Patterson's farm."[50] The unfree community clearly recognized and attempted, at least a couple times, to profit from the water border's ability to facilitate escapes. In the end, the aquatic connectivity between Prince Edward Island and the Caribbean made Wise's ultimate sentencing there possible, linking him all too easily to a dangerous environment with arduous tropical labor conditions.

We know that the land border and imperial networks influenced Wise after his sentencing and before his arrival in the Caribbean. The summer after his sentencing, he and another convict named John O'Neil escaped from the jail where they had been awaiting transportation. They did not get very far. A sheriff caught them in Pictou, Nova Scotia, and brought them back to PEI.[51] It seems as though Wise and O'Neil hoped crossing the Northumberland Strait would offer better opportunities to take control of their futures. Not only was Nova Scotia larger than PEI, but it also had larger Black communities as well as more robust shipping networks where they could potentially connect to other places. The border, these imperial connections, and the brutality of plantation labor in the Caribbean were possibly the reason Wise tried to flee from his imprisonment: to escape such a fate. More simply, perhaps he was seeking to reunite with Hannah Wise, who presumably remained in Birchtown. Jupiter Wise, like other Black islanders, recognized that the Atlantic waters surrounding the colony offered a slim, perhaps final chance at freedom: the opportunity to circumvent punishments that colonial society could enforce ashore. His recapture shows the extent of the danger he was willing to risk in order to have some control and a sense of security in his life. It is unknown what happened to Jupiter Wise after this.[52]

These factors exemplify how the jurisdictional border—the legal, political, and ideological distinctions between Prince Edward Island and elsewhere—distinguished the colony as a space where Black people were vulnerable to enslavement locally and in other parts of the British Atlantic. Although we do not know their exact thoughts about PEI as a jurisdiction where slavery and servitude were legal, we do know that several of them planned to leave the island on a vessel for Boston "where [they] could work."[53] The motives underlying this proposed trip are unclear, but it is possible that Boston seemed to offer better opportunities, especially in light of Massachusetts's recent abolishment of slavery. This conception of the jurisdiction goes against the way we typically think of the significance of the Canadian border: a bulwark against American slavery, a distinct threshold to security and freedom. The border and the lack of protection it provided to Black Loyalists, despite the promises made to them by the British government about land and freedom, mattered. The border encouraged Mingo and his companions to contemplate sailing to Boston. It led Wise to orchestrate a nighttime celebration for joy, movement, and resistance. It pushed him to defy an imprisonment that endangered him within a British Atlantic system of enslavement and servitude. Ultimately, Wise's story allows us to reevaluate the jurisdiction's relevance.

Within the jurisdiction of Prince Edward Island, subject to the laws of the colony, the watery border ultimately endangered Wise because it linked the plantations of the West Indies to the island. Some enslaved islanders saw possibilities with water escape, but for Wise, embroiled in a system of legal punishment and constantly threatened by the legality of slavery on the island, his biography demonstrates that his life was increasingly in peril from the time he landed in the Maritimes.

Hector

Not every enslaved person operated in close connection to others. Many found opportunities to resist by taking solitary action and running away by themselves. A man named Hector chose this path. He was enslaved in New Brunswick to Frederick William Hecht, a Loyalist from New York. In the *Saint John Gazette*, Hecht placed the following runaway advertisement:

> Five Guineas Reward.
>
> RUNAWAY from the subscriber, on Saturday evening the 26th ult. A negro man Slave, named Hector, by trade a cooper, a tall slender fellow, speaks English like the West India negroes, and is very talkative; he came from St. Augustine to this place, via New-York, on December last, had his feet frost bitten on the passage, and has a very lazy gait. Whoever secures and delivers him up to me shall have the above reward.
>
> And as I have reason to suspect that he has been carried off by some vessel or other, I hereby offer the like rewards to any person

who can give information, so as the offender or offenders shall be convicted thereof.

Fred. Wm. Hecht.
Fort Howe, 13th July 1784[54]

This advertisement is an important piece of evidence for the vast Atlantic connections that enslaved Black people had in the Maritimes. From the Caribbean or possibly Africa, to East Florida, to New York, to finally, New Brunswick, Atlantic slavery dictated Hector's movements across various bodies of water.[55] Although succinct, the details in this advertisement provide a glimpse into his personality, skills, background, experiences, and resistance. Runaway slave advertisements like this one are a valuable way for historians to better understand the lives of enslaved people in the region.

In addition to offering fascinating biographical details of one enslaved person in the Maritimes, the advertisement for Hector illuminates the many different places a single individual experienced Atlantic slavery. The imperial connections between New Brunswick and other jurisdictions played a role in shaping Hector's life. For Hector, crossing into the Maritimes from New York represented a change of location but also the continuity of living under the British Crown. East Florida had been a British colony since 1763. Frederick William Hecht probably acquired Hector after he had arrived in St. Augustine in December 1782 and before he left the following November, so it is clear Hector more than likely had lived in St. Augustine when British control of the region was contested.[56] What colony, if any, in the West Indies he lived in prior to this is uncertain. Regardless, Hector had spent at least a portion of his life enslaved within various British colonies.

The restriction of willful movement inherent to enslavement had an effect in determining the spaces and places people like Hector occupied. As enslavers like Hecht decided upon the migrations of enslaved people and dictated where they worked, ate, socialized, slept, and breathed, enslaved people challenged their landed restrictions and proximity to their enslavers in numerous ways. These forced migrations to different parts of the British Empire caused the frostbite Hector had suffered from the prior winter's sailing. By the time Hector ran away several months later, this injury may have still been painful and possibly was what his "very lazy gait" indicated.[57] Fleeing from Hecht despite injury required courage and strength. Hector's actions reveal he did not accept the continuum of control that characterized his crossing into New Brunswick.

Hector's migrations suggest the durability of ties between different colonial regions.[58] These connections were predicated on economy. If Hector's skills as "a cooper" were applied in the Maritimes, the barrels he built were probably used to transport local salted cod to the British West Indies, which fed enslaved Africans there.[59] The Maritimes' "advantages derived from fish and lumber" regularly embroiled enslaved people in these daily operations.[60] For example, a 1786 runaway advertisement described a man named Tom, who ran away from Captain James McDonald of Shelburne, as a sailor.[61] Though enslaved people's labor could be maritime in nature, the reasons for these connections had everything to do with empire.

Hector's life reveals how living on one side of a British colonial border fostered deep connections to other lands connected to the Crown. Since Hecht enslaved Hector, the immediate explanation for these forced migrations is Hecht's own discretion. At the same time, however, it is crucial to also remember that Hecht was appointed as assistant commissary in 1783 to go to New Brunswick to "take charge of the Provisions, Fuel, &c. and be accountable for the same."[62] Even Hecht's duties and movements, then, were in deference to powerful imperial ties and demands. Hecht, who had lived in New York but also owned property in St. Augustine, had an "unaccommodating temper and ill humor," which undoubtedly made Hector's life in the interconnected British Atlantic miserable.[63] Clues from the advertisement suggest Hector attempted to liberate himself from this networked world by means of water.

Living so close to the Atlantic, enslaved people regularly took advantage of the numerous ships in the harbors or escaped via water routes. For example, in 1786, a man named Joe Charleton allegedly stole several items from a vessel in Saint John and took a "small skiff."[64] The next year, two enslaved teenage siblings named Sam and Beller escaped with a free Black man named Tony Smith "IN a BIRCH CANOE" from their master, Thomas Lester of Waterborough, New Brunswick.[65] It was common practice for advertisements to include a warning to "masters of vessels" who might try "to harbor or Carry [off]" the fugitives, either in cooperation with an escaped enslaved person or through force or deceit.[66] Rather than just giving the usual warning, however, Hecht gave a little more credence to the idea that Hector possibly escaped on board a ship, as he stated, "I have reason to suspect that he has been carried off by some vessel or other."[67] As Kevin Dawson has shown, enslaved people throughout the Atlantic world used their aquatic skills such as swimming, boating, and diving.[68] The aquatic culture of enslaved Black people on the Atlantic coast is captured in a 1751 advertisement in the *Boston Post-Boy*, which says "JUST arriv'd from Hallifax, and to be sold, Ten hearty Strong Negro Men, mostly Tradesmen, such as Caulkers, Carpenters, Sailmakers, Ropemakers: Any Person inclining to purchase may enquire of Benjamin Hallowell of Boston."[69] Over twenty years later, Lieutenant Governor John Wentworth described his enslaved men as "expert in boats."[70] It is possible Hector used his own aquatic skills on the water to facilitate his escape.

The advertisement reveals nothing about what that specific reason was. Perhaps he heard about enslaved people being captured or choosing to escape aboard ships. Perhaps Hector had connections to people who could give him access on board a boat, and Hecht knew this. Perhaps working on a boat would have been more comfortable than a job that required him to be on his feet because of his disability.[71] Whatever the cause for this reason, it shows that enslaved people and enslavers were familiar with the ease with which the watery borders of the Maritime colonies connected places like New Brunswick to other corners of the Atlantic world.

This fluidity helped facilitate the transatlantic culture of Maritime slavery. Atlantic connections shaped enslavement at the personal and systematic levels. The British Atlantic, specifically, the West Indies, had an immeasurable impact on Maritime slavery. On a personal level, for instance, the advertisement tells us that Hector had a West Indian accent, but it does not tell us about any Caribbean customs, knowledge, and background he would have eventually carried with him to New Brunswick. We know also that Hector's painful, frigid voyage caused frostbite, but historians will likely never find evidence for what other traumas and memories it may have inflicted upon him.

Systematically, Caribbean slavery shaped the role of slavery in the Maritimes, a colonial region built to support West Indian sugar production. In Shelburne, for example, nearly 60 percent of enslavers were either farmers, merchants, or traders.[72] With Nova Scotia and New Brunswick's "highly developed trade networks with the West Indies," these individuals could use these Atlantic connections "as an outlet for their crops or the goods they manufactured in their shops."[73] When enslaved people were involved in industries associated with this trade (as Hector possibly was, as a cooper), they deepened the interregional links. Empire and land borders formed the reasoning behind these Atlantic connections, but they were only possible through the connectivity inherent to the water that surrounded the colonies. Appreciating the role of Atlantic slavery in terms of how it shaped Maritime slavery at the personal and systematic scales gives us a better understanding of the porousness and interconnectivity of the Maritimes' watery borders.

Although brief, the advertisement for Hector gives us a glimpse into how the Maritimes' aquatic border connected different parts of the British Empire, shaping the life of one man over the course of at least three dramatic migrations across the British Atlantic. The imperial and economic rationale for those connections explains the continuity of control Hector experienced within the jurisdiction of the colony of New Brunswick and as he remained in continued bondage to Hecht, his Loyalist enslaver. It also explains why Hector likely took to the seas to escape this bondage. Though the nautical connections to places such as the West Indies endangered enslaved people in the Maritimes and could perpetuate their enslavement, it also afforded people like Hector the means for escape aboard ships out of that system. The fragmentary evidence we have of Hector's life testifies to this fact that land and water borders severed ties just as easily as they created them. While it is possible Hector was recaptured or re-enslaved, the evidence is certain that he escaped, demonstrating the possibilities the watery border afforded to Black people in the Maritimes.

Lydia Jackson

While it was enslaved men who usually attempted to escape aboard ships because they could become or pose as sailors, enslaved women like Lydia Jackson encountered more obstacles. Her story demonstrates the overlapping hardships and exploitation that Black Loyalists faced, the tenacious efforts of a re-enslaved woman to regain her freedom, as well as the ways in which the land and water borders shaped her experience in Nova Scotia. The tenuousness of slavery is highlighted by the numerous attempts of owners to gain statute protection for their property (in 1787, 1789, 1801, and 1808 in Nova Scotia, and in 1801 in New Brunswick). The bills were defeated each time. Yet, slavery in New Brunswick and Nova Scotia lasted until the late 1810s and possibly even the early 1820s primarily because owners constantly fought in the courts and legislatures to protect their alleged right to own slaves.[74] The other way slavery lasted was through re-enslavement of free Black people like Lydia Jackson, who was told she was a servant, but became virtually enslaved.

The story of Lydia Jackson comes from the diary of John Clarkson, a Royal Navy officer and one of the organizers of the Black Loyalist exodus to Sierra Leone in 1792.[75] In his journal, Clarkson explained that as a result of poverty and the desertion of her husband, Jackson indentured herself to a Loyalist in

Manchester named Henry Hedley. Most free Black Loyalists faced arduous conditions as the postwar recession restricted employment opportunities, and some sought to use an indenture to provide the short-term necessities of life. Jackson's reluctant concession to what she thought would be a one-year term of servitude demonstrates the way in which formerly enslaved people strove ardently for self-sufficiency, negotiating terms of labor to the extent their desperate situations could tolerate. It turned out that Hedley took "advantage of her ignorance" (her illiteracy) and tricked Jackson into signing an indenture for thirty-nine years, essentially making her a servant for life. The next day, he told her "that she was to serve out the year with Dr. Bulman of Lunenburg, Nova Scotia and sent her round for this purpose in a Schooner." Once in Lunenburg, Jackson learned "to her great astonishment that she had been articled for the term of thirty nine years, and that she had been made over to [Bulman] for the consideration of £20 which he had paid to Henry Hedley."[76] In effect, a woman who gained her freedom during the American Revolution had been re-enslaved in the Loyalist Maritimes. In this example, up to this point, it did not matter that she had crossed an international border.

Enslaved to Bulman, Jackson suffered from the most brutal mistreatment. He regularly beat "her with the tongs, sticks, pieces of rope &c. about the head and face" with the assistance of his wife.[77] Clarkson's diary is clear that Jackson regularly endured beatings for little or no cause whatsoever. In one instance, Jackson spoke to Bulman carefully and "with the least intention of giving offence," but "Bulman took occasion to knock her down, and though she was then in the last month of pregnancy, in the most inhuman manner, stamped upon her whilst she lay upon the ground."[78] Sadly, it seems as though she lost the baby she was carrying since Clarkson made no mention of a child.

With the assistance of an attorney in Lunenberg, Jackson "lodged a complaint" against her enslaver, but the case "was soon silenced by the overbearing manners & influence of Bulman, who then or soon after expressed his intention of selling her to some Planter in the West Indies to work as a slave." As the town was deeply enmeshed with maritime trade, the threat to sell Jackson into enslavement was not an idle or empty threat. Instead, Bulman sent Jackson to his farm, which was three miles outside of town, "giving authority and sanction to his servants to beat & punish her as they thought fit." After three years of this cruelty, she escaped to Halifax, where she presented a memorial to the governor to no avail.[79] She also petitioned for assistance from the chief justice, who apparently only "promised to enquire into the business." Finally, she came to Clarkson, who "immediately wrote to Dr. Bulman respecting her and consulted a Lawyer on the business who gave it as his opinion that her wages could be recovered for the time she lived under Bulman but the forms of law would most likely prevent its being finally settled, so as to enable her to go with me." Clarkson found "there was no chance of the business being settled" and "advised her to give it up and leave Bulman to his own reflections." Lydia Jackson's story must be considered in the context of Clarkson's following comment: "I do not know what induced me to mention the above case as I have many others of a similar nature; for example, Scott's case, Mr. Lee, Senr. case, Smith's child, Motley Roads child, Mr. Farish's negro servant, &c."[80] The point here is that Black people throughout the Maritimes were regularly re-enslaved and sometimes sold down to the West Indies. Abolitionist Thomas Clarkson explained the phenomenon of re-enslavement in Nova Scotia: "It was not long till these loyalists, many of whom had been educated with all the ideas of the justice of slavery, the inferiority of negroes, and the superiority of white men, that are universal in the southern provinces of America, began to harass and oppress the industrious black settlers, and even wantonly to deprive them of the fruits of their labour, expelling them from the lands they had cleared." Clarkson continued by noting that whites reduced "again to slavery those negroes who had so honourably obtained their freedom. They hired them as servants, and, at the end of the stipulated time, refused payment of their wages, insisting that they were slaves: in some instances they destroyed their tickets of freedom, and then enslaved the negroes for want of them; in several instances, the unfortunate Africans were taken onboard vessels, carried to the West Indies, and there sold for the benefit of their plunderers."[81] Clearly, enslavers in the Maritimes used the watery Atlantic highway to send re-enslaved Black people to the Greater Caribbean. This phenomenon connected slavery and enslaved people in the northern reaches of the plantation complex to its heart in the Caribbean. Arriving first over water to the Maritimes as free people, Black Loyalists also experienced the Atlantic as a watery conduit to enslavement in the staple-producing tropical regions of the British Empire.

Evaluating the life of Lydia Jackson enables historians to get a glimpse of the possibilities and perils associated with the borders in the Maritimes. As it was for Jupiter Wise and other Black Loyalists, the land border for Lydia Jackson signified the insecurity of living in Nova Scotia, where local figures of authority regularly disregarded the promises of freedom that British generals had extended to Black Loyalists. Having arrived in Nova Scotia three years after her escape from her enslaver in New York, Jackson's crossing into the colony demonstrates her hope, if not her trust, that she would enjoy such liberty. The natural and social environment attached to this border, however, did not yield an easy life for Black Loyalists. As they received paltry land grants (if they received any land at all), white Loyalists took advantage of and exacerbated formerly enslaved people's vulnerability in order to cushion their own lives, constructing a colonial hierarchy built on racism and zealous exploitation.[82] For Lydia Jackson, these preexisting and developing hardships of settlement in the Maritimes, compounded by the fact that her husband left her, meant Nova Scotia's frontier was a disenchanting landscape of inescapable challenges.

Doing as much as she could to secure her autonomy, Jackson recognized that there were spaces within the colony's landed boundaries where her chances of obtaining freedom were greater. She positioned herself to contact white people who could help her—first, the attorney in Lunenburg—an act of bravery. Her escape to then seek the aid of the governor, the chief justice, and eventually John Clarkson required her to turn a dangerous woody landscape separating her from Halifax's colonial officials into her pathway to potential freedom. Jackson's refashioning of the Nova Scotian land from a barrier to a byway demonstrates her hope in ultimately obtaining the justice she knew she was owed upon British soil, having entered the colony as a Black Loyalist. In a sad twist of irony, this hope renders Jackson as a figure who pursued to uphold and possibly earnestly believed in the purported meaning of the colony's border. The Nova Scotian authorities, on the other hand, disgraced themselves when they "paid little or no attention" to instances like Jackson's where Britain's promises were unfulfilled.[83] (Of course, however, such promises had never been made solely for the sake of the formerly enslaved, having been leveraged more as military strategy against the American rebels.[84]) Beliefs and actions of what it meant to live on Nova Scotian or Maritime soil—or in other cases, colonial Canada or the Bahamas[85]—reveal how Black Loyalists thought about themselves in relation to the rest of the British Empire and how they assigned meaning to the border.

Like the land border's signification as a domain of British influence, the water border signified Nova Scotia's ties to other parts of the British Atlantic world. More easily than any other natural form, water connected enslaved people to other parts of the British Atlantic, for better or for worse. For example, Jackson experienced her sale to Bulman by way of a "Schooner" that transported her from Manchester to Lunenburg.[86] Later, Bulman's dangerous threat to send her to the West Indies intensified the power of the aquatic highways that could have resulted in Jackson's enslavement on a sugar plantation.[87]

Lydia Jackson's story is the ultimate example of how one person could experience both perils and possibilities as a Black person in the Maritimes. She experienced the perils of being re-enslaved, being beaten so much she lost her child, and running away to get her freedom. But water brought disappointment and danger as it did hope. For Lydia Jackson, her story ends with the ultimate possibilities of the water border. That Jackson cast her faith in Clarkson, a man who was arranging a transatlantic crossing to Sierra Leone, is significant. Possibly, Jackson believed the chance to join the exodus to Africa would enable her to finally have the security she sought. If that is the case, then water—the Atlantic Ocean and the distance from Nova Scotia's unfulfilled promises that it could afford—likely factored into her effort to garner Clarkson's support. Though Clarkson's writings are not clear about Jackson's future, it is hard to believe she would not have continued to pursue freedom in other, ongoing ways. The assurances of freedom associated with having crossed the border into the British colony were ones Jackson fought to have realized for her own life, whether by means of land or water.

Conclusion

The biographies of Jupiter Wise, Hector, and Lydia Jackson demonstrate how Black people—enslaved or minimally free—faced both possibility and peril. The balance could be tipped by luck, friendships, as well as political, economic, and social power. Charged with a felony, peril overwhelmed Jupiter Wise, and he probably ended up re-enslaved in the West Indies. Hector, on the other hand, experienced the possibilities of freedom when he got to the Maritimes and ran away from Frederick William Hecht. As a Black Loyalist, Lydia Jackson experienced possibility upon arrival in the Maritimes, but she soon suffered the peril of re-enslavement before she escaped and met John Clarkson. It is possible that she went to Sierra Leone with Clarkson or remained in Nova Scotia.[88]

These Black lives demonstrate how people negotiated and moved, voluntarily and involuntarily, over borders, along coasts, and across oceans. Although policymakers in Halifax and London could draw boundaries, people transcended these borders and helped map out realities on the ground. The areas encompassed within the lines of the Maritime colonies on a map were not merely spaces that were strictly "the Maritimes." The history explored in this chapter reveals that these were spaces that were connected to Boston, New York, St. Augustine, the greater Caribbean, and Africa. In other words, we do not want to think of the Maritimes as separate and cut off from the rest of the plantation complex when they were an integral part of it. The migratory lives of Jupiter Wise, Hector, and Lydia Jackson demonstrate the embodied reality of these transatlantic and transnational connections. Their movements across borders are significant because they reveal the obstacles and opportunities associated with borders and jurisdictions, particularly the far-reaching impact of slavery in the Atlantic world.

Black people who arrived in the Maritimes after the American Revolution attempted to inscribe meanings upon the jurisdictional boundaries of Prince Edward Island, New Brunswick, and Nova Scotia. The meaning that they hoped to embed in these colonies was freedom: freedom of movement, freedom to work and receive payment, and freedom to have control over their own lives. The three people in this chapter are examples of how possibility and peril were part of these attempts. In addition to their understanding of jurisdiction, water—whether as a means of escape or a conduit to enslavement—connected them to other legal jurisdictions that inhibited or facilitated their ability to have that freedom.

Notes

1. For work that has pointed to the migration of enslaved Black people from Canada to the United States and the selling of free Black people into colonial Canada, see Gregory Wigmore, "Before the Railroad: From Slavery to Freedom in the Canadian-American Borderland," *Journal of American History* 98 (September 2011): 437–54; Harvey Amani Whitfield, *The Problem of Slavery in Early Vermont, 1777–1810* (Vermont Historical Society, 2014). For a bibliography on Black Canadian history and slavery in Canada, see Suzanne Morton and Donald Wright, "Black History in Atlantic Canada: A Bibliography," *Acadiensis* 50, no. 1 (Spring 2021): 223–75. See also Harvey Amani Whitfield, *North to Bondage: Loyalist Slavery in the Maritimes* (University of British Columbia Press, 2016); Sarah Elizabeth Chute, "Bound to Slavery: Economic and Biographical Connections to Atlantic Slavery Between the Maritimes and West Indies After 1783" (MA thesis, University of Vermont, 2021); Franco Paz and Harvey Amani Whitfield, "On the Edge of Freedom: The Re-Enslavement of Elizabeth Watson in Nova Scotia," in *In Search of Liberty: African American Internationalism in the Nineteenth-Century Atlantic World*, ed. Ronald Angelo Johnson and Ousmane K. Power Greene (University of Georgia Press [Race in the Atlantic World, 1700–1900], 2021), 17–39; Harvey Amani Whitfield, "White Archives, Black Fragments: Problems and Possibilities in Telling the Lives of Enslaved Black People in the Maritimes," *Canadian Historical Review* 101, no. 3 (September 2020): 323–45; Catherine M. A. Cottreau-Robins,

"Exploring the Landscape of Slavery in Loyalist Era Nova Scotia," in *The Consequences of Loyalism: Essays in Honor of Robert M. Calhoon*, ed. Rebecca Brannon and Joseph S. Moore (University of South Carolina Press, 2019), 122–34; Charmaine Nelson, "A Tone of Voice Peculiar to New England," *Current Anthropology* 61, supplement 22 (October 2020): S303–S316; Charmaine Nelson, "Special Issue: Expanding and Complicating the Concept of Creolization," *African and Black Diaspora: An International Journal* 12, no. 3 (2019): 267–70; Harvey Amani Whitfield, "Runaway Advertisements and Social Disorder in the Maritimes: A Preliminary Study," in *Violence, Order, and Unrest: A History of British North America, 1749–1876*, ed. Elizabeth Mancke et al. (University of Toronto Press, 2019), 214–35; Afua Cooper, "'Deluded and Ruined': Diana Bastian—Enslaved African Canadian Teenager and White Male Privilege," *Brock Education Journal* 27 (2017): 26–34; Harvey Amani Whitfield, "The African Diaspora in Atlantic Canada: History, Historians, and Historiography," *Acadiensis* 46, no. 1 (Winter/Spring 2017): 213–32; Ken Donovan, "Female Slaves as Sexual Victims in Île Royale," *Acadiensis* 43, no. 1 (Winter/Spring, 2014): 147–56; Harvey Amani Whitfield and Barry Cahill, "Slave Life and Slave Law in Colonial Prince Edward Island, 1769–1825," *Acadiensis* 38, no. 2 (Summer/Autumn, 2009): 29–51; Harvey Amani Whitfield, "Black Loyalists and Black Slaves in Maritime Canada," *History Compass* 5, no. 6 (November 2007): 1980–97; Kenneth Donovan, "Slaves and Their Owners in Île Royale, 1713–1760," *Acadiensis* 25, no. 1 (Autumn 1995): 3–32; D. G. Bell, "Slavery and the Judges of Loyalist New Brunswick," *University of New Brunswick Law Journal*, 31 (1982): 9–42; Barry Cahill, "Habeas Corpus and Slavery in Nova Scotia: *R v. Hecht, ex parte Rachel*, 1798," *University of New Brunswick Law Journal*, 44 (1995): 179–209; Barry Cahill, "Slavery and the Judges of Loyalist Nova Scotia," *University of New Brunswick Law Journal*, 43 (1994): 73–135; Robin W. Winks, *The Blacks in Canada: A History* (Yale University Press, 1971); W. A. Spray, *The Blacks in New Brunswick* (Brunswick Press, 1972); T. W. Smith, "The Slave in Canada," *Collections of the Nova Scotia Historical Society*, 10 (1899): 1–161.

2. Isaac Land, "Port Towns and the 'Paramaratime,'" in *The Routledge Companion to Marine and Maritime Worlds, 1400–1800*, ed. Claire Jowitt et al. (Routledge/Taylor & Francis, 2020), 193.

3. John Morton in this volume, "Data Maps of Downeast Maine: Missionary Records from the Early Republic Borderlands," 89–111.

4. James W. St. G. Walker, *The Black Loyalists: The Search for a Promised Land in Nova Scotia and Sierra Leone 1783–1870* (University of Toronto Press, 1992), 2–4.

5. David Austin, *Fear of a Black Nation: Race, Sex, and Security in Sixties Montreal*, 2nd ed. (Between the Lines, 2023), 222.

6. Phillip D. Curtin, *The Rise and Fall of the Plantation Complex: Essays in Atlantic History*, 2nd ed. (Cambridge University Press, 1998), xi.

7. Paul Musselwhite in this volume, "Depicting and Defining the Plantation in the Early English Caribbean, 1625–1675," 121, 129.

8. Sarah Elizabeth Chute, "'To ship her to the West Indies, and there dispose of her as a Slave': Connections of Enslaved People to the Loyalist Maritimes and the West Indies," *Acadiensis* 51, no. 2 (Autumn 2022): 34–59.

9. Sarah Elizabeth Chute, "WISE, JUPITER," in *Dictionary of Canadian Biography*, vol. 4 University of Toronto/Université Laval, 2003–, accessed 16 January 2025, https://www.biographi.ca/en/bio/wise_jupiter_4E.html.

10. Emily Blanck, "Seventeen Eighty-Three: The Turning Point in the Law of Slavery and Freedom in Massachusetts," *New England Quarterly* 75, no. 1 (March 2002): 22–51.

11. There are other examples that demonstrate these regional connections. For example, an 1828 slave register indicates that four slaves belonging to Catherine Ormsby of Montserrat resided with Mr. Ormsby in Prince Edward Island. See Montserrat Slave Register, 13 December 1828, p. 285, T/71/450, TNA.

12. Book of Negroes, Sir Guy Carleton Papers, RG 1, Nova Scotia Archives, Halifax (hereafter NSA); for an online version, see https://archives.novascotia.ca/africanns/book-of-negroes/.

13. Walker, *The Black Loyalists*, 11; Passport for Cato Ramsay to emigrate to Nova Scotia, 21 April 1783, Gideon White family fonds, MG 1 vol. 948 no. 196 (microfilm 14960); Book of Negroes, Sir Guy Carleton Papers, RG 1, NSA. There is a column in the Book of Negroes called "Names of the Persons in Whose Possession They Now Are"; see, for example, https://archives.novascotia.ca/africanns/book-of-negroes/page/?ID=19. It is the column right before the page break. Most of the time this column is filled out but occasionally it is not.

14. An Act, *declaring that* Baptism *of* SLAVES *shall not exempt them from* BONDAGE, 1781, CO 228, Colonial Office Acts No. 1, Prince Edward Island, 1770–1781, MS. 1–66, Public Archives and Records Office, Prince Edward Island (hereafter PAROPEI).

15. Supreme Court, Prince Edward Island, RG 6.1, ser. I, sub-ser. I, vol. 10, p. 49, PAROPEI.

16. Muster book of free Black settlement of Birchtown, 1784, 7, MG9-B9-14, Library and Archives Canada (Ottawa). Copy available online at *heritage.canadiana.ca/view/oocihm.lac_reel_h984*, image 83.

17. Book of Negroes.

18. James Stevens, Information against Jupiter Wise for a Burglary, 30 November 1785, Supreme Court, M. McDonald vs. Jupiter Wise-1786, RG 1, PAROPEI.

19. Verdict, Supreme Court Fonds, Case Files, 1770–1899, 1785, The King vs. Jupiter Wise, RG 6.1, Series 5, Subseries 1, PAROPEI.

20. Kim M. Gruenwald in this volume, "When the River Is a Border: Rivalries and Commercial Networks in the Riverine West," 143.

21. Gruenwald, "When the River Is a Border," 160.

22. Shirley Tillotson, "Importing the Plantation: The Greater Caribbean Trade and Loyalist Nova Scotia's Public Revenue, 1789–1835," *Journal of the Royal Nova Scotia Historical Society* 25 (2022): 1–26, esp. 18.

23. James Stevens, Information against Jupiter Wise for a Burglary, 30 November 1785, Supreme Court, M. McDonald vs. Jupiter Wise, 1786, RG 6.1, Series 5, Subseries 1, PAROPEI.

24. James Stevens, Information against Jupiter Wise for a Burglary.

25. Blanck, "Seventeen Eighty-Three: The Turning Point in the Law of Slavery and Freedom in Massachusetts."

26. James Stevens, Information against Jupiter Wise for a Burglary.

27. James Stevens, Information against Jupiter Wise for a Burglary.

28. James Stevens, Information against Jupiter Wise for a Burglary; Jim Hornby, *Black Islanders: Prince Edward Island's Historical Black Community* (Charlottetown: Institute of Island Studies, 1991), 17.

29. Hornby, *Black Islanders*, 19.

30. James Stevens, Information against Jupiter Wise for a Burglary; Hornby, *Black Islanders*, 19; Chronological List of Past Speakers of the Legislative Assembly: 1773–Present," Government of Prince Edward Island Web Archive, accessed 30 November 2021, http://www.gov.pe.ca/photos/original/leg_speakers.pdf.

31. Information of John Clark, Supreme Court Fonds, Case Files, 1770–1899, 1785, The King vs. Jupiter Wise, RG 6.1, Series 5, Subseries 1, PAROPEI.

32. Hornby, *Black Islanders*, 18.

33. Information of John Clark.

34. Information of John Clark.

35. Information of John Clark.

36. Hornby, *Black Islanders*, 18.

37. Hornby, *Black Islanders*, 18.

38. Supreme Court Fonds, Minutes, 24 February 1786, RG 6.1, Series 1, Subseries 1, Vol. 2, 2–3, PAROPEI.

39. Supreme Court Fonds, Minutes, 24 February 1786, RG 6.1, Series 1, Subseries 1, Vol. 2, 2–3.

40. Supreme Court Fonds, Minutes, 24 February 1786, RG 6.1, Series 1, Subseries 1, Vol. 2, 2–3.

41. Hornby, *Black Islanders*, 16.

42. Supreme Court Fonds, Minutes, 24 February 1786, RG 6.1, Series 1, Subseries 1, Vol. 2, 2–3.

43. Carole Watterson Troxler, "Uses of the Bahamas by Southern Loyalist Exiles," in *The Loyal Atlantic: Remaking the British Atlantic in the Revolutionary Era*, ed. Liam Riordan and Jerry Bannister (University of Toronto Press, 2012).

44. Hamish Maxwell-Stewart, "Transportation from Britain and Ireland, 1615–1875," 183–210, especially 187, in *A Global History of Convicts and Penal Colonies*, ed. Clare Anderson (Bloomsbury Academic, 2018).

45. Maxwell-Stewart, "Transportation from Britain and Ireland," 187.

46. Maxwell-Stewart, "Transportation from Britain and Ireland," 187.

47. *Royal Gazette and Miscellany of the Island of Saint John*, 29 April 1793; Graeme Wynn, "1800–1810: Turning the Century," in *The Atlantic Region to Confederation: A History*, ed. Phillip A. Buckner and John G. Reid (University of Toronto Press, 1994), 211. For more on PEI during the eighteenth century, see J. M. Bumsted, *Land Settlement and Politics on Eighteenth-Century Prince Edward Island* (McGill-Queen's University Press, 1987).

48. S. Hollingsworth, *The present state of Nova Scotia: with a brief account of Canada, and the British Islands on the Coast of North America* (Printed for William Creech, 1787), 169–171; Chute, "Bound to Slavery," 57.

49. James Stevens, Information against Jupiter Wise for a Burglary.

50. James Stevens, Information against Jupiter Wise for a Burglary.

51. Executive Council Fonds, Minutes and Orders-in-Council, RG 5, Series 1, Vol. 1, 95; Hornby, *Black Islanders*, 18.

52. H. T. Holman, "Slaves and Servants on Prince Edward Island: The Case of Jupiter Wise," *Acadiensis* 12, no. 1 (Autumn 1982): 100–104.

53. James Stevens, Information against Jupiter Wise for a Burglary.

54. Runaway slave advertisement, *Saint John Gazette*, 15 July 1784.

55. J. Leitch Wright Jr., "Blacks in British East Florida," *Florida Historical Quarterly* (April 1976): 425–42; Linda K. Williams, "East Florida as a Loyalist Haven," *Florida Historical Quarterly* (April 1976): 465–68, 469–78.

56. Peter Wilson Coldham, *American Migrations, 1765–1799* (Genealogical Publishing Society, 2000), 156.

57. Runaway slave advertisement, *Saint John Gazette*, 15 July 1784.

58. Wright, "Blacks in British East Florida," 425–42; Williams, "East Florida as a Loyalist Haven," 465–68, 469–78.

59. Runaway slave advertisement, *Saint John Gazette*, 15 July 1784.

60. Samuel Goldsbury to Edward Winslow, 1 March 1785, in *Winslow Papers, A.D. 1776–1826*, ed. W. O. Raymond (The Sun Printing Company, 1901), 271.

61. Runaway slave advertisement, *Nova Scotia Packet and General Advertiser*, 6 July 1786.

62. Esther Clarke Wright, *The Loyalists of New Brunswick* (E. C. Wright, 1955), 97.

63. Wright, *The Loyalists of New Brunswick*, 97; Coldham, *American Migrations*, 156.

64. Runaway advertisement, *Royal Gazette and New Brunswick Advertiser*, 15 August 1786.

65. Runaway slave advertisement, *Royal Gazette*, 10 July 1787.

66. Runaway slave advertisement, *Nova Scotia Gazette and Weekly Chronicle*, 20 May 1783.

67. Runaway slave advertisement, *Saint John Gazette*, 15 July 1784.

68. Kevin Dawson, *Undercurrents of Power: Aquatic Culture in the African Diaspora* (University of Pennsylvania Press, 2019).

69. Advertisement, 23 September 1751, *Boston Post-Boy*.

70. John Wentworth to Paul Wentworth or his attorney, 24 February 1784, Wentworth Letters, vol. 49, NSA.

71. For more on disability, see Stefanie Hunt-Kennedy, *Between Fitness and Death: Disability and Slavery in the Caribbean* (University of Illinois Press, 2020).

72. Whitfield, *North to Bondage*, 65.

73. Whitfield, *North to Bondage*, 65.

74. *Journal and Proceedings of the House of Assembly* [hereafter JHOA], (Halifax: King's Printer 1787), 17 and 22, NSA; An Act for the Regulation and Relief of the Free Negroes Within the Province of Nova Scotia, In Council, 2 April 1789, RG 5, Series U, Un-passed Bills, 1762–1792, NSA; *JHOA*, 1801, 72, NSA; A Bill relating to Negroes, 6 February 1801, Legislative Assembly Records, RS 24, S 14-B 9, Provincial Archives of New Brunswick; An Act for regulating Negro Servants within and throughout this Province, 1808, RG 5, Series U, Un-passed Bills, NSA.

75. C. B. Fergusson, ed., *Clarkson's Mission to America, 1791–1792* (Public Archives of Nova Scotia, 1971).

76. Fergusson, *Mission to America*, 89.

77. Fergusson, *Mission to America*, 90.

78. Fergusson, *Mission to America*, 90.

79. We have been unable to locate this document and secondary sources about Lydia Jackson's life have not mentioned it, but that does not mean it does not exist.

80. Fergusson, *Mission to America*, 90.

81. Harvey Amani Whitfield, *Biographical Dictionary of Enslaved People in the Maritimest* (University of Toronto Press, 2022), 139.

82. Marion Gilroy and D. C. Harvey, *Loyalists and Land Settlement in Nova Scotia* (Public Archives of Nova Scotia, 1937; Clearfield, 2002).

83. Fergusson, *Mission to America*, 90.

84. Walker, *The Black Loyalists*, 2.

85. For works on slavery in Canada outside of the Maritimes, see Brett Rushforth, *Bonds of Alliance: Indigenous and Atlantic Slaveries in New France* (University of North Carolina Press, 2012); Frank Mackey, *Done with Slavery: The Black Fact in Montreal, 1760–1840* (McGill-Queen's University Press, 2010); Afua Cooper, *The Hanging of Angélique: The Untold Story of*

Canadian Slavery and the Burning of Montréal (University of Georgia Press, 2007); Marcel Trudel, *L'esclavage au Canada français: histoire et conditions de l'esclavage* (Presses universitaires Laval, 1960); Nelson, "A 'tone of Voice Peculiar to New-England.'" For the Bahamas, see Christopher Curry, *Freedom and Resistance: A Social History of Black Loyalists in the Bahamas* (University Press of Florida, 2017).

86. Fergusson, *Mission to America*, 89.

87. Fergusson, *Mission to America*, 90.

88. It is unclear from Clarkson's diary whether Lydia Jackson joined the voyage to Sierra Leone.

CHAPTER 12

Amphibious Tales

Villagers and Strangers in a Border-Crossing World

SAMUEL TRUETT

Historians of early North America have been caught in a tug-of-war between continental and oceanic tales, one centered increasingly on interior spaces of Indigenous America and the other on the waves of an expanding European world. Despite calls for a "vast Early American" history that embraces what Karen Wulf calls the "geographically expansive and interconnected realms" of land and sea, scholarly practice has reinforced older metanarratives. Continental approaches focus chiefly on lands north of Mexico, with oceanic strangers entering largely from the east, from a more inclusive yet still familiar Atlantic. Whether we look east from Indian Country or west across a multivocal sea of settlers, slaves, and others, the figurative beach (as Greg Dening called it), where strangers met, entangled, struggled, and changed shape in tandem, has moved very little. Oceanic East meets continental West in an Early American epic that culminates, more often than not, in the modern United States.[1]

As a borderlands scholar who studies North American entanglements with Latin America and the Pacific, I see these relationships differently. In this chapter, I begin with the Atlantic and sail east rather than west, taking as my traveling companion an English orphan named John Denton Hall. As a youth, Hall drifted through the China Seas and the Pacific basin to the California gold fields, and from there to northwestern Mexico, where he lived the rest of his life in a small peasant village. Taking his path and those of others in his cohort as my points of departure, I situate the aqueous and terrestrial horizons of North America in a frame that disrupts state-centered, modernizing histories. In a volume devoted primarily to earlier-modern American and Atlantic histories, I conclude with a series of tales that circumnavigates those spaces, leading to alternative ends.

I first met John Denton Hall years ago as a reader in the the Huntington Library, in San Marino, California, and was drawn in by his memoirs, *Travels and Adventures in Sonora*. "In the summer of the year 1850, being then a gold digger in Angels Creek," he began, "an opportunity offered to satisfy my wandering disposition by accompanying a Mexican gentleman, who was returning to his native state, Sonora." His new companion, whom he simply referred to as Don Chico, hailed from the foothills of the Sierra Madres. He and two sons, a brother, and three Indigenous Ópata servants invited Hall to return south to their home village of Cucurpe—in the borderlands between Mexican Sonora and the Apache-controlled lands to the north. "I could not speak a word of Spanish, nor my companions English," Hall would later confess. "But youth and the love of adventure . . . decided me."[2]

Hall imagined that he'd spend a few years in Sonora, recoup his California losses in a Mexican bonanza, and sail on. But as months turned into years, he began raising a family with a local woman, and soon found himself entangled in the intimate affairs of his adopted village. When mining ventures failed, he became a jack-of-all-trades: a smuggler, a surveyor, a healer, a teacher, a village scribe. For a time, he kept a diary, recording adventures with the enthusiasm of a sojourner. By the early 1860s, older and perhaps more resigned, he began to convert this into a backward-looking account—taking fuller measure of the world around him. He lingered repeatedly on the state's mineral wealth and his inability to turn it to his own advantage. He was hemmed in by his distance from outside markets, by revolts and chronic warfare, by unforeseen poverty. By January 1867, when he snuffed his candle for the last time, he had fit sixteen years into hundreds of manuscript pages. He affixed a preface, as if to prepare his tale for publication. And then he filed it away.[3]

In retrospect, it is surprising that Hall's memoirs saw the light of day. They gathered dust for years. But then in 1879, he took them down, jotted a few notes in the margins, and appears to have handed them with a hand-drawn map of Sonora to a US Army quartermaster named William Myers. Myers was passing through Arizona from San Francisco to a new post as depot quartermaster in Chicago, having paused briefly in Tucson to scout local mining properties. Crews were building the Southern Pacific Railroad east from California, opening new connections to regional mines. Hall was also in Tucson, but for different reasons. He and other villagers were waiting out a revolt in Sonora. If Myers and Hall discussed these troubles, they certainly also discussed the mines that Hall described in the pages he carried with him for safekeeping. Perhaps they agreed that others ought to hear about these details. In Chicago, Myers had the manuscript typeset, the map traced, and a small run printed in 1881. Years later, a surviving copy made its way west to the Huntington Library's rare book vaults.[4]

I initially took Hall's *Travels and Adventures in Sonora* for a garden-variety nineteenth-century booster tract. In his preface, still dated 1867, he led with a discussion of Mexican mines. This seemed appropriate for the 1860s, as expertise from the Comstock Lode and new mining machinery from San Francisco sailed south along the coast to Sonora. Among those visiting the state was a young French American mining engineer, Louis Janin Jr. Trained at the Mining Academy of Freiberg, Janin had been hired by California investors to assess the mining claims around Cucurpe. Janin stayed with Hall, and even mentioned "Hall's Book," which was then probably a modest account of Hall's adventures. Janin may have urged him to do more with it. Soon after his visit, Hall began weaving descriptions of mines into his tale, envisioning a book for a generation that was already projecting, if not yet building, cross-border railroads. The Mexican investment bubble soon burst, as civil wars broke out in both nations, but when railroads actually were built after 1879, Hall's memoirs—half travel narrative, half booster account—were ready to dust off and print.[5]

And yet Hall's memoirs were unlike any other booster or travel accounts that I had seen. Each of the six surviving copies appears to have been a presentation copy, owned privately by a prominent railroad or mining investor. They appeared briefly on the rare book market, years later, making them largely unknown. They were also unusual for their perspective. Mexican villagers rarely commit their tales to paper. Accounts like Hall's were typically written by outsiders—men passing through, making superficial observations to be carted away in short order to distant readers. Hall's portrait of Cucurpe was, by contrast, sensitive, precise, and immediate, reflecting his position as an adopted son. Through his intimate renderings of local life, he evoked George Simmel's figure of the *stranger*—the newcomer who "comes today and stays tomorrow." He was someone whose liminal and yet enduring (if not also useful) place in the village allowed him to bridge, to broker, to go between.[6]

His memoirs were also striking for what they left out. He began his story *in medias res*, at the point of leaving California. In the first few pages, he bids farewell to five nameless partners at Angel's Camp and rides south. He never says how he got there, and in the 300 pages of his extensive narrative, he says nothing about his Mexican family. He never fully names Don Chico—the intrepid villager who brought him south—and "Mr. C" is the name he gives to his closest companion, a carpenter from Texas who joined Hall and Don Chico in their mining ventures. Hall's decision to keep these personal details to himself makes my work hard. It has taken years to track down his wife and children; longer still to identify which of Cucurpe's many "Chicos" (a diminutive for Francisco) was Hall's Don Chico. And I've yet to fully unmask the enigmatic Mr. C.

Equally elusive, if also more intriguing, is Hall's globetrotting past. He drops tantalizing hints throughout his memoirs, a trail of breadcrumbs to other worlds. A brief sojourn among the Quechan (Yuma) Indians on the road to Mexico reminds him of his youth spent among the Indigenous peoples of coastal Borneo. He finds himself fumbling, instinctively, for words in Malay and Bengali. The ways that Cucurpeños use tortillas as utensils reminds him of the use of chopsticks in opium-trading "chow chows" on the coast of China. Sugar mills in the countryside below Cucurpe remind him of sugar mills on the Ganges River in the foothills of the Himalayas. Then there is the English sea captain he meets on the coast of Sonora, Sir Henry Keppel, with whom he had fought pirates in the China Seas. "I shall never forget his surprise," Hall says. "I must confess, I was not rigged in a sailor-like manner, my dress being a short Mexican jacket, long buckskin moccasins, and a tall Mexican hat."[7]

These glimpses of Hall's past life pulled me in. I began tracking clues, eventually setting off on a journey of my own, following ships, sailors, and drifters of all kinds around the world. With Hall as my traveling companion, I began cutting unfamiliar paths through the nineteenth century—between earlier and later modern worlds. My journey led me to see borderlands differently—as spaces bound to empires and nations, but also to autonomous itineraries, crossings, and histories that refuse state-centered frameworks. Seen at smaller scales, borderlands took shape less as spaces of differentiation or demarcation and more as spatial synapses where global vectors touched down and got tangled up in the contingencies of the local. Rethinking borderlands as portals to other realms, transfers in larger circuits, or webs of overlapping sovereignty has allowed me to ground my journey while also keeping an eye to paths that resist state-centered bifurcations, moving in unexpected directions.

My journey in search of Hall has also led me to a more *amphibious* approach, one that crosses land and sea while asking how terrestrial and aqueous spaces entangle. Such spaces of entanglement, whether straits, coasts, estuaries, watersheds, or interior networks connecting rivers, often structure mobility and flows but also serve as choke points, areas of friction, or places of shape shifting. They evoke in this last sense the Greek roots of amphibious, from *amphibia*, plural of *amphibious*, "living a double life." Such spaces, Lauren Benton, James Scott, Eric Tagliacozzo, and others argue, can manifest as regions of refuge and escape; zones of piracy, captivity, and apostasy; places last seen. Whether in the form of beaches, swamps, shoals, concealed coves, or riverine capillaries, they have the potential to destabilize core and periphery, inside and out, here and there.[8]

In what follows, I cross nineteenth-century seas, continents, states, and nonstate realms with Hall at my side, but I do so with an eye to his ever-shifting cohorts. Their own stories allow me to pull my own across a wider canvas—and they let me do so in ways that take nothing for granted. At times, his cohorts move through the world in familiar ways, connecting distant shores in ways that we might expect. At others, they veer off our mental maps, compelling us to rethink the building blocks of world history. Sailing at humanized scales into the nineteenth-century world, we find Atlantic and American realms entangled, but not always in ways we expect.

* * *

John Denton Hall was born in 1827, a child of the Atlantic world. He spent his earliest years in a well-heeled household of pottery manufacturers in Burslem, Staffordshire, village kin to Wedgwoods and Darwins. Their world and its wares floated west on the Trent & Mersey Canal to the port of Liverpool and points beyond. John Hall & Sons (headed by his grandfather and namesake) were known best for their blue-on-white "Oriental" scenes, inspired by exotic lands and destined for North America. Their pottery sailed mostly to Boston's North End, where Hall's uncle John and aunt Caroline (daughter of a New England glass and earthenware merchant) were fighting a losing battle for New England dinner tables. In 1832, shortly after Hall turned five, John Hall & Sons went bankrupt. When Hall's father ran off to America, presumably to avoid debtor's prison, his mother, Hannah, gathered her five children and joined her younger brother, Ashlin Bagster, in Liverpool.[9]

Most considered it a family affair, meriting only two handwritten pages to the Court of Review in Bankruptcy. But to understand the trauma that set Hall in motion, I must open a wider lens. I must sail with the tall ships and their cargos of earthenware and china, and trace the webs of transnational capital. From Liverpool, I see a family dashed on the shoals of an Atlantic world and then reunited by the industrial infrastructures of a new railroad age. Ashlin Bagster, Hall's then-twenty-two-year-old uncle, was a protégé of the English railroad pioneer, "father of the railways," Robert Stephenson. Ashlin had cast his lot with Stephenson in 1829 as he completed the Liverpool & Manchester Railway, England's first intercity railway. Not long after taking in Hannah and her children, Bagster and his household began to drift through the English countryside with Stephenson and his fellow engineers—first to Leicester (home of the coal-carrying Leicester & Swannington) and then to London, where Bagster became the manager of the new London & Birmingham Railway in 1837.[10]

Traveling east and then south across England, Hall's family never drifted far from the Atlantic currents that set them in motion. By 1837, railroads had collapsed space and time between Liverpool, Newcastle, Bristol, and London, giving England a new, modern coherence. But it was still an outward-looking coherence, tied as much to wind-powered vessels as to steam-powered engines. In Liverpool, railroad workers connected the mills of Manchester to the cotton plantations of the US South, and in Leicester, Swannington coal fired the machines that turned cotton into hose. In Birmingham, the "city of a thousand trades," trains pulled traffic southeast instead of west, but built manufacturing capacity that fed increasingly to the Americas. Stephenson's own career was tied to Atlantic crossings—having cut his teeth in railroad and mining ventures in Colombia in the 1820s. Here he joined other engineer-adventurers who envisioned new connections in contexts of Central and South American movements for independence. American and British infrastructures were, for Stephenson, part of a larger venture to make the Atlantic modern.[11]

It is hard not to think here of Hall's future path. When Stephenson looked to American mining and railroad ventures, he initially looked to Mexico, not Colombia. Those who went to Mexico in the 1820s inaugurated a legacy of foreign investment that culminated with the railroad and mining dreams that later pulled Hall's memoirs north. Perhaps stories of Stephenson's America circulated in the Bagster-Hall household. What is more certain is that Hall absorbed something from the engineer's technical worlds at a young age—perhaps attentive to his uncle's work, maybe focused by his mother, who supplemented their itinerant household's income as a teacher. This aptitude, more than anything, shaped his path ahead. Once settled in London in 1837, Hannah baptized John in the Church of England, a metropolitan rite of crossing from the Nonconformist Potteries. She then took him down to Newgate Street, near the Thames, and enrolled him in Christ's Hospital.[12]

Christ's Hospital was founded during the sixteenth century for poor boys cast into the streets by the dissolution of the monasteries. In 1837, it was still a charity school (a school for poor children), but it also took in children of better-off families that had fallen on hard times. Christ's Hospital housed the Royal Mathematical School, established in 1673 as a "nursery" for future navigators. Its forty most talented, the so-called "King's Boys," were trained under royal subsidy. Their reputation was boosted in the 1770s under William Wales, who sailed the Pacific with James Cook and brought to the school his knowledge of hydrographic surveying. With a knack for math, maps, and technical drawing, Hall advanced quickly, becoming a King's Boy at the age of twelve. In so doing, he joined the ranks of such students as Thomas Cass, John Septimus Roe, and John Bushnan, renowned for their surveys of New Zealand, Western Australia, and the Arctic. Following Wales, they trained for an amphibious empire that was moving out from the Atlantic to new frontiers.[13]

In entering Christ's Hospital, Hall secured the patronage of John Pirie, a prominent shipping agent in the East India trade, and soon to be Lord Mayor of London. How Hall came to Pirie's attention is not exactly clear, but it is likely that Stephenson or someone of his London railroad cohort made the introduction. He may have been introduced by Captain Constantine Richard Moorsom (the son of Admiral Sir Robert Moorsom), a director of the London & Birmingham Railway to whom Ashlin Bagster reported. With Pirie as a patron, Hall perhaps saw a future in the East Indies where, generations earlier, the East India Company had forged a lucrative trade between India and China. It was a path completed by family tragedy. The news came first of his father's death, somewhere in America, and then Ashlin's demise on railroad business near Derby. Then a few days shy of his thirteenth birthday, his mother, Hannah, passed in Birmingham. As aunts and uncles sorted younger siblings on a thread of households between London and Hampstead, Hall threw himself at his studies and the prospects of a life at sea.[14]

In less than a decade, a series of displacements had reoriented Hall from the Atlantic, by way of the English countryside, toward the trades of India and China. England had been pivoting east for generations, but Hall's path was shaped at intimate scales by such men as Stephenson, Moorsom, and Pirie—members of an enterprising circle that coalesced locally in the railyards, docks, and taverns of a world-facing London. Unbound from the webs of family, forged anew in the imperial crucible of the Royal Mathematical School, and soon launched from the East India Docks, Hall would drift from prior Atlantic worlds into the Indian Ocean and the China Seas. But he did so in ways that London's brokers and sea captains could not yet foresee.

It began with a thirty-six-year-old adventurer named James Brooke and a voyage from Singapore to Borneo. Brooke, the son of a prominent East India Company judge, had been raised in a mixed Anglo-Indian household to the north of Calcutta. The East India Company was in decline, having severed its maritime branch in 1834. In 1839, as Hall puzzled over sextants, almanacs, and *Bonnycastle's Algebra*, Brooke joined other free traders in the newly opened China Seas. He pointed his schooner east along the northwest Borneo coast to the mouth of the Sarawak River. He expected to trade with Indigenous Bidayuh and Chinese colonists upriver, but he soon found himself swept up in a fight between coastal Malay gentry and agents of the Sultan of Brunei, whose seat of power lay to the east. In return for his assistance, Brooke was given a position within the sultanate as a vassal ruler of Sarawak—a province that would later swell to the size of England.[15]

In 1842, as the self-styled "White Rajah of Sarawak" settled into the Malay village of Kuching, agents in London outfitted a 250-ton bark, *Ariel*, to prime the trade with Singapore. In January 1843, they visited Christ's Hospital. In March, two apprentices, Henry Steele and John Denton Hall, boarded the *Ariel* with their new master, a former East India mariner named George Steward. As the mouth of the Thames faded behind them, Hall may have recalled the eastern romances of his family's blue-on-white ceramics—but the romance of his own voyage was short-lived. After reaching Sarawak, Hall's master began feuding with Brooke over the future of the adventurer enclave (roughly twenty strong), and in 1844, Iban and Samal-Iranun raiders from nearby rivers and the Sulu Sea—Indigenous peoples the British called the "pirates" of Borneo—began challenging their power. Steward joined Brooke and Captain Henry Keppel, a British naval officer on the East Indies and China Station, in brutal campaigns against their new foes. His companions last saw him as an Iban combatant dragged his body from his gig into the jungles of the nearby Skrang River.[16]

Apprenticeships dissolved, Henry Steele stayed on the island with Brooke, and Hall plunged headlong into the China Seas. After landing in Singapore, he volunteered for duty on HMS *Samarang*, a surveying vessel tasked to map saltwater passages through the East Indian Archipelago in the wake of the

Opium War of 1839–42 (when the British sought new trade routes to the treaty ports of China). Parlaying his mathematical training into a position as assistant navigator, Hall drifted for two and a half years across a range of borderlands. Eastward the hydrographic crew sailed into the multiethnic fringes of the Brunei and Sulu sultanates; north to the Loo-Choo (Ryukyu) Islands that bridged China and Japan; south to the Sulu Archipelago and nearby Mindanao, a contested borderland (as it is today) between the Christian Philippines and Islamic Indonesia; then west across Dutch and English lines of influence into the heart of the Indian Ocean and points beyond.[17]

After the *Samarang* reached London in early 1847, Hall roomed at the city's fringe with other sailors near the docks. By April, he was back at sea as second mate on the *Royal Albert*, a frigate bound for India and China. Captained by Irishman Anthony Scanlan, known on the docks as the first to sail a cargo of Chinese tea direct to Dublin, the *Royal Albert* carried Hall into a trader's world based in the ports and estuaries of South and East Asia. For historians, this world is harder to see. It was preserved not in official accounts (trading knowledge was a guarded commodity) but by notes jotted at ports of call on crew agreements. They tell me the crew waited out monsoons in Bombay (where they boarded cotton and opium), then rounded India, threaded the Malacca Straits, and landed in Hong Kong. What they don't say is that the crew then moved cargo along the coast, through a shadowy littoral of islands, coves, and anchorages between Canton and Shanghai. They then returned south to winter with other sailors at Whampoa, in the Pearl River. As spring monsoons let up, they went to Calcutta and then up the Hooghly and Ganges rivers in search of homeward-bound freight.[18]

From when the *Ariel* sailed for Borneo in early 1843, to when the *Royal Albert* returned to the East India docks in late 1848, Hall is mostly absent from the written record. His movements on HMS *Samarang* are the easiest to sort, in part because this was a state-sponsored expedition. The primary goal was to gather hydrographic data for imperial expansion, and its commander, Sir Edward Belcher, submitted not only the requisite log but also countless maps to the imperial archive. Belcher and one midshipman published accounts of the voyage, with another account circulating in manuscript form. None name Hall, but he appears on the ship's muster and in Belcher's letter book. In Sarawak, Hall is invisible—mentioned only obliquely by Brooke, upon his arrival ("I have had the pleasure of making the acquaintance of H— and S—, both excellent and gentlemanly fellows"). He appears in the opium trade only as a name (misspelled) on the *Royal Albert* crew list.[19]

Such adventurer realms avoid scrutiny by state actors—and in the opium trade, in particular, Hall's path ventured off the London-centered grid into realms bound more tightly to Canton, Bombay, Calcutta, Singapore, Manila, Boston, and Edinburgh. Scanlan had ties to the London agency house that launched Brooke's first voyage to Borneo, but these were on-again, off-again connections. By the time Hall returned to London at the end of 1848, political crises in Europe had further unsettled the trade in India and China, and he soon slipped the city's orbit altogether. A few months later, he boarded a Liverpool bark for Lima, Peru. The 350-ton *Dickey Sam* sailed for Antony Gibbs & Sons, a leader in the Atlantic guano trade between Peru, Bolivia, and England. This trade, led in 1849 by the Spanish-born Guillermo Gibbs and organized around new Latin American monopolies, was just as decentralized as the China Seas trade—further underscoring the power of local markets and politics far from England in the global rewiring of British empire.[20]

Hall was more than an ordinary sailor on this voyage, and he may have been seeking to invest profits from China in a return to Atlantic worlds—indeed, a space not too distant from that which had first set Robert Stephenson in motion. He spent a month or so, after reaching Lima, in assessing trade opportunities along the coast. But by then, all eyes were on California. When the *Dickey Sam* anchored in Callao (Lima's port town) in August 1849, the crew could not have missed the Peruvian and Chilean fleets headed north to the gold fields. How Hall charted his own passage north remains a mystery. He may have boarded a local vessel, or hopped an English ship passing through the port. I have searched crew lists in the Board of Trade's archive with this latter scenario in mind, in vain. Many of these lists never made it back to England—and many rotted with their ships in San Francisco's harbor as sailors deserted in droves for the gold diggings. With them, Hall entered new orbits, far beyond the familiar littorals of London's world-facing empire.[21]

* * *

Hall may have seen California as place to get rich, have adventures, and sell the tale before sailing on. This was how Frank Marryat saw it when he landed in 1850, seeking gold and a setting for *Mountains and Molehills; or Recollections of a Burnt Journal* (1855). Frank was a son of the sea novelist Frederick Marryat, known best for his 1836 novel, *Mr. Midshipman Easy*, a pioneering work in the field. His own father, Joseph, had been a leading merchant and ship owner in the British West Indies, but the 1807 abolition of the slave trade reversed the family fortunes. Young Frank, like Hall, turned east. The two met as teenaged messmates on the *Samarang*, and Marryat reached San Francisco shortly before Hall left Angel's Camp for Cucurpe. Hall had surely read Marryat's 1848 *Borneo and the Indian Archipelago*, a portrait of his youthful escapades on the expedition. Marryat emulated his father, and Hall emulated both. He likely adapted his title, *Travels and Adventures in Sonora*, from the elder Marryat's *Narrative of the Travels and Adventures of Monsieur Violet in California, Sonora, and Western Texas* (1843), a tale of a French nobleman who fled France to seek a new life in America.[22]

If Hall charted a similar path to Frank Marryat, he lacked his pedigree and social connections (which had extended to the *Samarang*, commanded by Frederick Marryat's first cousin, Edward Belcher). Hall found himself more exposed than his messmate to "dame Fortune's 'smiles and frowns.'" If love of adventure guided his journey south, he was also pushed on the road by a great fire in San Francisco that June, "whereby I, as well as many others, found ourselves ruined." Riding south (and penniless) with his new friends, Hall wondered why he had traded the upper decks of a British vessel for the saddle of a Mexican mule. "I have taken the weather casing off the Horn and the Cape of Good Hope, but the pitching was nothing in comparison to what I experienced on the back of that confounded jade," he wrote. "I was not adept at that kind of navigation."[23]

Facing Apache raids and smallpox in Sonora, Hall returned briefly to California. He returned in 1852 with the Texas-raised Englishman Henry Clarke (Hall's Mr. C). If Hall brought training in math and surveying to Mexican mines, Clarke applied carpentry skills to the building of headframes, shafts, and makeshift pumps. Waiting for their ventures with Don Chico and others to pay, Hall and Clarke began building families with two local women, Francisca Palomino and María Concepción Bonillas. These were not typical marriages between European American outsiders and higher-ranking Mexican villagers, as often seen on the coast or inland trade centers, which pulled strangers into positions of influence. Palomino and Bonillas are surnames associated with lower-ranking servant families, the Palominos most likely an Ópata family from a neighboring village. The women may have been camp or village servants before becoming partners and mothers, bound by patron-client relations to families of influence, but limited in their access to village networks. For Hall and Clarke, social and economic mobility was tied more to their relationships to such high-ranking men as Don Chico and Eduviges Sinohui (a prominent Ópata with close ties to Hall), as well as the labor they gave to the village. Hall and Palomino were born into different social worlds, but they found their journeys to the village center similarly mediated and constrained; and it would ultimately be up to future generations to finish the task. In their household, Tomás arrived first, followed by Carlos, named after Hall's father and brother. Then came Guadalupe, María, and Enrique, the latter likely named after their fellow border crosser, Henry Clarke. This new generation, more than anything else, would build locally contingent relationships that anchored border-crossing fathers, mothers, and families in place.[24]

Hall's transformation from British sailor to Mexican villager was far from typical—and it was also far from inevitable. It was far from inevitable because the United States, a nation with imperial inclinations, was on the move. When Hall arrived in 1850, the nearest US settlement to Cucurpe was Doña Ana, New Mexico, some four hundred miles northeast through Apache territory and the rugged Sierra Madres. This changed in 1854, when the United States annexed 30,000 square miles of land from Mexico in the Gadsden Purchase (which became southern New Mexico and Arizona), moving its border closer to Cucurpe. Within months, newcomers from the eastern United States and California began trickling into the annexed territory to prospect for silver and gold. Hall might have moved north—as working families from Sonora did in the mid-nineteenth century—to live alongside white English speakers. The border raises questions. Why Mexico? Why not something easier? Given the isolation of Cucurpe from global markets, why did Hall not choose something more lucrative?[25]

The answer was bound up in legacies of violence not unlike those that he had witnessed in the China Seas. During the 1850s, Sonora was plagued by filibusters (originally from the Dutch *vrijbuiter*, or pirate) from the north. Among the most notorious was the adventurer William Walker, who briefly invaded Sonora and Baja California with fellow California-based filibusters in 1853 and 1854, before moving in 1855 to take Nicaragua. Walker's venture in Nicaragua resonated with Brooke's own in Sarawak, placing both in a larger "Anglo-world" context of nineteenth-century adventuring. Two years later, a Californian named Henry Crabb, who had previously been associated with Walker, followed his footsteps into Sonora. Hoping to create an independent "Republic of Sonora," Crabb and his freebooting cohort seized the borderland town of Caborca in 1857, only to lose their lives in a pitched battle with Mexican volunteers (many of them recruited from nearby Cucurpe).[26]

These events further unsettled a land wracked by generations of violence. Crabb's filibuster, in particular, inflamed many in Sonora against

Anglo-Americans, and for a while Hall and Mr. C sought refuge in Arizona. But Arizona was no better. In the former Mexican presidio of Tubac—now the local headquarters for US mining entrepreneurs in southern Arizona—they met Henry Titus, a soldier of fortune with Walker in Nicaragua with his eyes now set on former Spanish silver mines in the nearby Santa Rita and Patagonia Mountains. The three men initially hit it off, and in 1859, Titus hired Hall to bring Cucurpeños north to work his claims. That May, a mob of former Crabb supporters rode through the nearby Sonoita Valley, killing several Mexicans and terrorizing countless others. When Hall joined a posse to pursue the mob, he was thwarted by white Arizonans who labeled him a "greaser." The so-called Sonoita massacre marked a line in the sand—compelling Hall to decide where his affinities and loyalties lay. Packing his saddlebags, he headed south, bidding adios to his short-lived venture in the barbaric American West.[27]

And so, the seasons passed as the former British sailor became a Mexican villager, perched at the edges of his adopted community. The uncertainties of prospecting kept him and his companions-in-mining in motion from one local mountain range to the next. When mining schemes failed, as they almost always did, he returned to the village. By the 1860s, now finding Hall's Spanish up to the task, Cucurpe's *president municipal* (mayor) enlisted him as his secretary and scribe. "All the writing in the city is done by Mr. Hall, who is also school teacher," noted the mining engineer Louis Janin Jr. in 1864. Hall also worked as a healer (he preferred "country doctor") in Cucurpe and its vicinity, augmenting training from the China Seas with Indigenous botanical and medical knowledge, possibly gleaned from his wife, Ópata workers on his mining ventures, and Indigenous patrons in the village. Whether prospecting or healing, his paths crossed multiple borderlands—between mestizo and Ópata enclaves in the river valleys, and between these villages and highland Apache camps. Cucurpe rooted him in place at some seasons, while local conflicts pushed him and his family on the road at others, to seek refuge in better-protected towns such as Magdalena or Tucson.[28]

The structures of power, violence, and mobility in Hall's new home had much in common with worlds he left behind. In Sarawak and the China Seas, newcomers like Hall entered what the historian James Warren calls the "Sulu zone," extending out from coastal Borneo into the Sulu Sea. At its center was the Sulu Archipelago, dividing the Sulu Sea from the Celebes Sea to the south. These islands were a contested choke point between the Philippines and Indonesia, dividing Christian and Islamic realms and entangling webs of Spanish, Dutch, British, Chinese, Bugis, and Malay trade influence. At its heart was the island of Jolo. Arab travelers had imported Islam to the Tausug of Jolo by the fifteenth century, laying a foundation for the Sulu Sultanate. By the nineteenth century, the sultan ruled the archipelago and the northeastern coast of Borneo from the island's capital (also called Jolo)—an urban borderland divided into a "Malay" town, a ghetto of "birdcage-looking buildings" perched on a shoal for Chinese traders, and ships of all nations at the edges. It was a "rendezvous of pirates," wrote Frank Marryat—with those not practicing piracy "always ready to aid, assist, and protect those who do."[29]

What Marryat had in mind was a regime of saltwater nomads centered on the nearby islands of Balangingi and Basilan, and the southwest coast of Mindanao, to the east. This was a kinetic regime, to paraphrase Pekka Hämäläinen, in which people held power over maritime space by "keeping things—violence, markets, attachments, possessions, themselves—fluid and in motion." This regime swirled around a demand for *trepang* (sea cucumber), pearls, pearl shells, tortoise shells, and birds' nests, all of which could be traded in coastal China for tea; a demand for slaves to gather these goods; and a demand for arms for saltwater slaving. Slaves were procured largely by Samal and Iranun raiders, whose Basilan, Balangingi, and Mindanao refuges were connected with Malay-Tausug trading networks on Jolo—and Samal and Iranun raiders were augmented by Filipino *renegados*, captives who apostatized to Islam and then became raiders themselves, parlaying a local knowledge of reefs, coves, and coastal villages throughout the Philippine archipelago.[30]

If the Sulu Sea basin served as a maritime anchor for these denizens of the Sulu Zone, Iranun slaver-raiders also drew on other empire-repelling spaces and regions of refuge, all linked to a larger world of water. Several years before Hall and Marryat sailed through the Sulu Sea, British sea captain J. J. Blake complained that Iranun *prahus* (sailing boats) shook British and Spanish ships by vanishing into "the mangrove shores with which the islands thereabouts abound." At the center of this maritime geography of concealment was a "continued line of mangroves and swamp" on the coast of Mindanao, fed by the marshy drainage of "an immensely extensive inland lake [which] . . . they consider as their stronghold and their home." In these amphibious realms, the Iranun built and repaired *prahus*, raised families, and "carry on all intercourse with each other as an insulated and distinct community." Other regions of refuge included the long and narrow island of Banguingui at the far eastern side of the Sulu Archipelago—a crescent twelve miles long and two miles wide, "principally mangrove growing upon coral banks," Blake observed. It is "well calculated for protection and secure concealment," he wrote, sheltering as many as 200 Iranun *prahus* at a time.[31]

Christian (or Christianized) captives-turned-*renegados* were only one indicator of the deeper crossings of this shape-shifting realm. Villagers and sailors incorporated strangers of all sorts in this space in their efforts to build entourages and maintain power. Shortly after arriving in Sarawak, Hall's master George Steward and fellow apprentice Henry Steele forged sexual liaisons with Malay women, both men fathering local children. For Steele, who remained behind, this was a likely gateway for his fluency in Sarawak Malay—which secured him a place on the island as an interpreter and go-between. Malay men likely brokered these companionate unions to incorporate the men into their entourages, and while the women may have hailed from coastal gentry families, they may just as likely have been captives brought by Iban, Iranun, Samal, or Malay raiders from elsewhere. As Southeast Asian scholar John Walker observes, Sarawak Malay men often used captive women to pull outside men into mixed-community Malay entourages—and since Sarawak Malays were matrilocal, such women helped keep their male entourages anchored in place.[32]

These amphibious, shape-shifting dynamics took shape on multiple fronts. The British cohort of adventurers at the Sarawak River and the crew of hydrographic surveyors on the *Samarang* relied heavily on a Sulu Sea go-between named William Wyndham. Born in Ireland (or Scotland or England, the reports vary), Wyndham was said to have cut his mariner's teeth as a mate or midshipman under Lord Thomas Cochrane in the Chilean Wars of Independence (1818–22). He then crossed the Pacific, married a Filipina *mestiza* from Iloilo, purchased a schooner, and drifted south to Jolo. His personal ties to the Sultan of Sulu, an apparent mastery of the Spanish, Visayan, and Tausug languages, and his success in the pearl and tortoise shell trade of the Sulu Sea catapulted him to prominence. The sultan eventually granted him the position of *datu*, or aristocratic chief. "Dressed in Malay costume and from long residence among them," wrote Spenser St. John, Brooke's private secretary, Wyndham had taken on "the appearance and manner of a native."[33]

Wyndham seemed to shift and drift across the greater China Seas as comfortably as his Samal and Iranun neighbors did. As a merchant, he linked imperial and Indigenous markets within the China trade; as a *datu*, he controlled labor and mediated disputes; and as a trade and political representative of Manila, Singapore, or the Sultan of Sulu (as it suited him), he served as a diplomat, aided smugglers, purchased, sold, or redeemed captives, and harbored fugitives of all kinds. Not unlike Joseph Conrad's fictional character Tom Lingard—whose border-crossing ventures into the Malay world were drawn from stories Conrad later heard in the China Seas about William Wyndham—the sea "took him young, fashioned him body and soul." Wyndham seemed to sail into the right place at the right time, but the waters all about were teeming with his kind. Whether they sailed Chinese junks, Spanish brigs, Malay *prahus*, Bugis *pinisi*, US whaling ships, or British barks, they were pragmatic shape-shifters. Men of the sea, they were also denizens of Greg Dening's figurative "beach," men who prospered in the seams between cultures and regimes.[34]

In Sonora, Hall moved within a similar kinetic realm, a space shaped in part by the unsettling forces of Mexican, US, and Indigenous territorial expansion, and in part by local dynamics of mobility. Ever since Spanish newcomers first lay claim over Cucurpe in the early seventeenth century, this had been a contested terrain in the borderlands of O'odham, Eudeve, Teguïma, and Jova (the latter three glossed by the Spanish as Ópata) villages, with occasional intrusions of the more mobile, desert-living Comcaac (Seris) to the west and highland Jocomes to the northeast. At a divide between watersheds to the north and south, and a confluence of desert, upland steppe, and montane ecosystems, this was a multiethnic, polyglot space. The northward expansion of the Spanish reorganized these borderlands in open-ended ways, as reflected in Hall's networks with both Ópata and non-Ópata families. The southward expansion of various Apache groups added another layer of entanglement. When Hall came on the scene, Chokonen, Chihene, and Nedni bands (which US Americans glossed as Chiricahua Apaches) had already claimed prior Jocome homelands in the Sierra Madres as their own and were expanding their territorial claims in the wake of Mexican independence and renewed cycles of violence with both Mexican and US settlers.[35]

The highlands near Cucurpe where Hall, Don Chico, and Mr. C prospected for gold and silver—the Sierra Azul, Sierra San Antonio, and ranges farther afield—were within these expanding Apache territories. To the extent that Chiricahua raiders seasonally tapped Cucurpe and its sister villages for livestock and captives, these riparian communities were in a sense part of Apache territory, if only as villages held in a condition that the Mexicans would call vassalage. Regardless of where villagers and raiders marked the border between their worlds, this was a realm in which Apaches held power over others by staying in motion and obliging others to follow suit. Like the Samal and Iranun of the China Seas, Apaches absorbed goods and captives with great skill—focusing, like their marine counterparts, on women and children, captives who were more useful and could be more easily incorporated. They built regional reputations for their conversion of mestizos, Ópatas, and others into an Apache way of life (much as Sulu Sea regimes were famous for their *renegado* fleets). For Hall, it was as if grasslands had become seas of a different sort, with horses replacing Malay *proas* and grass replacing saltwater—embodying for imperial agents all that sedentary "civilization" was not.[36]

As with the Malay villages and Iranun fleets of the China Seas, these spaces and relationships looked different from different sides. What empires saw as realms of fugitivity, illegibility, and spaces beyond the pale—in the Sulu Sea, coral reefs, mangrove thickets, coastal swamps—were, for the other side, regions of refuge, passages to home, inlands. Spaniards saw the Sierra Madres east of Cucurpe as "craggy mountain strongholds" for "apostate" mission Indians and "savage Apaches." Well into the nineteenth century, Mexicans characterized these spaces as impregnable barriers to non-Indigenous settlers. Apaches saw it differently. With "trees, grass, game, one's friends and relatives, safety, and a happy life," the Sierra Madres was as close as it came to the "Happy Place," or Apache heaven, recalled Asa Daklugie, born into the Nedni band of northern Mexico in the 1860s. He recalled traveling south along the sierras with his father, Juh (a Nedni chief), to a cave "sacred to Ussen," the Apache god. Here was not only a strategic stronghold but a center of ceremonial space, a spiritual crossroads. And when the US-Mexico border was fixed in 1854, setting legal limits to how far US and Mexican armies could pursue them in times of war, political boundaries seemed to reinforce divides between those peoples in motion and those who sought to remove them or fix them in place.[37]

And yet distinctions between Apache and non-Apache spaces were never so clear-cut. In the uplands of Sonora, as in the archipelagos and seaways of Island Southeast Asia, denizens of sedentary and settler regimes rarely stood still. Ópatas and mestizos moved seasonally into Apache lands when Apaches followed the seasons to more distant mountain ranges on the US side, to hunt feral livestock that had wandered into Apachería and plunder old mines abandoned by prior Spanish or Mexican owners in the face of Apache expansion. The seasonal anchors of this kinetic world surely resonated for Hall, who had come of age in a world shaped by "pirate winds," as Dutch and English sailors called the China Seas monsoons that signaled seasons of maritime raiding. Mestizo and Indigenous captives also drifted in open-ended ways across landscapes of cultural interiority—including a former captive from the nearby village of Sinoquipe who was adopted into an Apache family, escaped, then returned to Apache lands as a guide for Cucurpe prospectors. These were some of the moving parts that turned Cucurpeños into consummate shape shifters—an aptitude that Hall took to heart as he tried to master on land what his Sulu Sea counterpart, William Wyndham, mastered at sea.[38]

* * *

Hall's amphibious, shape-shifting journey—his transformation from British sailor to Mexican villager—had consequences. He rode south from California on the back of a mule, hoping to strike it rich and sail on. But he never did. Years later, his children buried him in desert soil, a poor stranger in a foreign land. "I shall never forget his surprise," Hall had written of his chance encounter with Keppel in 1851, on the coast of Sonora. Hall stood before him, dressed in local leather head-to-toe, like a character in one of James Fenimore Cooper's terrestrial novels.[39] Like Wyndham, Hall had taken "the appearance and manner of a native." The resonances between Sonora, Sarawak, and the Sulu Sea, though worlds apart, were, no doubt, apparent to both men. They gave Hall's path powerful coordinates: the familiar drift of unfixed identities, the power of local protocols over distant dictates, the challenge of drawing distinctions, between here and there, inside and out, us and them.

And yet by the end of Hall's life, these things were far more fixed in place—and as states began to deploy new space-fixing technologies (railroads, steamships), so too did they have more say in the mapping of local geographies and identities. In Cucurpe, this shift was most clearly articulated in the conquest of the Chiricahua Apaches in the 1880s—a conquest that coincided with the suppression of Samal and Iranun raiding-and-slaving regimes on the far side of the Pacific. Whether through British merchant fleets, Spanish naval campaigns, or US and Mexican armies, empires and nations built their power by the century's end through brutal acts of incorporation and sorting. They banked on keeping others in motion, to be sure. But modern grids choked and reorganized these crossings. The shifts of the 1880s would tip the global balance of power toward larger-scale, state-centered networks based on new technologies and practices of territorialization and containment.[40]

In Hall's Mexican borderlands, these shifts were bound up with the railroad, that space-fixing technology that had first set him in motion as a youth in England—and which, by his own story's end, had caught up with him on the far side of the world. In 1879, when Quartermaster Myers took

Hall's manuscript north to Chicago, the Chicago-based Atchison, Topeka, and Santa Fe (AT&SF) had started to build south from Albuquerque toward Mexico. In company with Mexico-based entrepreneurs who had negotiated new railway concessions from the Mexican government, AT&SF elites organized new Mexican railroads to connect to their US-based network at the border. The first of these, the Sonoran Railway, was completed between the port of Guaymas and the Arizona border in 1882. The second, the Mexican Central Railway, was completed between El Paso and Mexico City two years later. Both would effectively extend the reach of US-centered infrastructures across the continent to the Pacific coast of Mexico and the terrestrial heart of Mexico, respectively.[41]

Myers had already been enmeshed in AT&SF circles. Through the US Army, he used their new rail system to begin supplying the nation's renewed war against Chihene and Chokonen bands of the Chiricahua Apaches. Under such leaders as Victorio and Geronimo, they had begun to evade government-enforced containment on Arizona's San Carlos Apache Reservation—fleeing into their former regions of refuge in the Sierra Madres of Sonora and Chihuahua. Myer's relationship with the AT&SF help me to make sense of the peculiarities of Hall's memoirs—why they were printed when they were; why it appeared that such care and expense was put into the printing (as evidenced by such things as their elaborately marbled fore-edges), but with a surprisingly small run (one that could hardly pretend to pay back expenses); and why it took so long for them to appear on the rare book market. Four of the surviving six copies, it turns out, were signed by AT&SF railroad magnates. One copy, at the Beinecke Library, contains Hall's map, traced in 1879 or 1880 by George Henckel—a civil engineer on Myers's Chicago staff. Henckel added a dotted line depicting the path of the Sonora Railway as it was projected in 1879. The line ran diagonally from eastern Arizona through the Mexican highlands that Hall called home, and which he described so carefully. A new cohort of adventurers had their eyes set on Mexico, and Hall's memoirs and map pointed the way.[42]

Chicago seemed to be at the center of it all, but most of the men interested in Hall's tale—and, indeed, those who invested in AT&SF development west and south across the continent—hailed from Boston, where the Hall family business had foundered so many years earlier. And these men, like Hall, approached the continent by way of Pacific worlds. Thomas Jefferson Coolidge, who was president of the AT&SF (and who owned a copy of Hall's memoirs that would later wind up in Harvard's Houghton Library), hailed

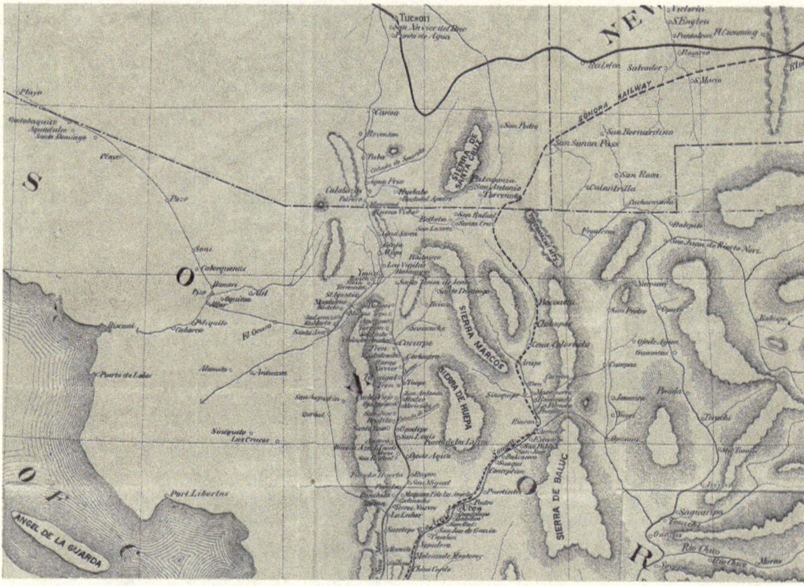

Figure 12.1. George Henckel's adaptation of John Denton Hall's map with the projected path of the Sonora Railway. John Denton Hall, *Sonora: Travels and Adventures in Sonora*, Beinecke Book and Manuscript Library, Yale University.

from a prominent opium trading family in Boston. Elijah Smith, the railroad magnate who owned the copies at the Huntington Library and the University of Western Ontario, was the son of a New Bedford whaling merchant and ship builder. The AT&SF owned much to the capital that the progenitors of its Boston-based directors accumulated in the China and Pacific trades. Perhaps these men saw something of their own family stories in Hall's tale. George Taber, from Smith's New Bedford cohort, saw the resonances this way when pitching Hall's mines to friends. Hall came to Mexico as a "sea-captain," he said, and then became "'capitan' of the district where he now resides." It was a path that the US sons of Pacific maritime entrepreneurs, turning maritime capital to new continental ends, might emulate in their own crossings south.[43]

At some point in the 1880s, after Myers printed *Travels and Adventures in Sonora* in Chicago, the original manuscript appears to have drifted west to San Francisco. Toward the end of the decade, the well-known California bibliophile Hubert Howe Bancroft referred to it twice in his multivolume works on Pacific North America. These are fleeting citations, but enough to tell us that the manuscript was organized differently than the printed version. How

Bancroft got his hands on the manuscript or where it went next, I do not yet (and may never) know. I assume, due to the attention that Myers gave to the account (he went so far as to take out a copyright in his own name), that he kept it. I cannot find any evidence that Hall himself ever went to Chicago—and I have no idea whether he even got a copy of the printed book. According to family lore, all of his belongings were burned in a fire years ago. In 1881, Myers was reassigned to the Department of Dakota. Not long after that, he fell ill—and by 1883, he had retired from active service. He died in 1887 in New York City.[44]

If he still had Hall's manuscript at that time, it probably went to his daughter, Sallie, who had married in San Francisco in 1878 and stayed there after her father relocated to Chicago. Her husband, Charlie Page, was a California lawyer with a specialty in admiralty law, but he was also a well-known book collector, a member of a cohort in San Francisco that included Bancroft. To get Hall's manuscript from Myers to Bancroft, one could hardly envision a better go-between than Sallie Myers Page. Charlie Page may have even heard of Hall's adventures before Myers did. He had co-founded the Yale Club of California, and one of the active members of that association was the mining engineer Louis Janin Jr. There is nothing in the Page family papers today to suggest any relationship with Hall or his tale. The San Francisco law office of Charlie Page burned in the fire that attended the great earthquake of 1906, an echo of the fire in that same city fifty-six years earlier that pushed Hall on the road with a villager named Don Chico. It may be that my trail ends there.[45]

If Hall's manuscript did wind up with Page, if even for a short time, he may have also found a great deal of resonance in its storyline. For Page knew something about washing up in a foreign land. As a young boy, he had sailed north to California from his childhood home in Valparaiso, Chile, with his parents: a New Jersey-born adventurer, Thomas Stokes Page, and Anna María Liljevalch, the latter born and raised in a household of US and Swedish immigrants to Valparaiso. Like Hall, Thomas Stokes Page had jumped ship to become a healer in the Spanish-speaking Pacific—parlaying his more formal US training in medicine to work as a doctor in Valparaiso. In 1849, he sailed north with other Chilean adventurers to try his luck in California. Where others went for gold, he went for land, acquiring the 17,234-acre Rancho Cotate near Sonoma that same year. He then returned to Valparaiso—operating his new investment as a cattle ranch from afar into the late 1860s, when he and his family bid farewell to Chile for the last time and moved north to San Francisco.[46]

I often wonder if Thomas Stokes Page and John Denton Hall boarded the same vessel in 1849 as they pursued new ventures in a foreign country. Once off the standard historical script, it becomes hard to know what to expect. When Hall later met Sir Henry Keppel on the beach in Guaymas, Keppel was astonished. "What are you doing here, Mr. Hall?" Keppel asked. Hall might have asked Keppel the same question. The English sea captain had come north from Valparaiso on an unexpected detour to pick up a load of silver. There he conducted business with Don Juan Robinson, a New York–born sailor who had jumped ship in Guaymas in 1823, married locally, and had a run of good luck. That luck later turned. The same battles that plagued Hall in the 1860s pushed Don Juan north as a political exile. He and his family settled into a new home at 59 South Park, San Francisco, to start over. Two years later, a new family moved next door to the Robinsons—a New Jersey–born doctor and his Chilean wife, said to be invested in land up north.[47]

"What are you doing here, Mr. Page?" I ask. But I let it go. Little about this amphibious, border-crossing realm surprises me anymore.

Notes

1. Karen Wulf, "Vast Early America: Three Simple Words for a Complex Reality," *Humanities* 40, no. 1 (Winter 2019): 26–47; Greg Dening, *Islands and Beaches: Discourse on a Silent Land: Marquesas 1774–1880* (University Press of Hawaii, 1980). Wulf gestures to a broader tapestry of Atlantic and continental histories, but individual studies contributing to this tapestry tend to be divided along older temporal and spatial divides, with work on Mexico and the Pacific at the margins, viewed separately.

2. Dr. J. [John Denton] Hall, *Sonora: Travels and Adventures in Sonora: A Description of its Mining and Agricultural Resources and a Narrative of a Residence of Fifteen Years* (Chicago: J. M. W. Jones Stationery and Printing Co., 1881), 9–11.

3. Most of this I draw from his memoirs, apart from the details about his family (which he does not discuss).

4. For Myers in Arizona, see "From Monday's Daily," *The Citizen* (Tucson), 2 May 1879; "Silver King Mine," *Daily Arizona Citizen*, 8 May 1879. For his larger trajectory, see George W. Cullum, *Biographical Register of the Officers and Graduates of the U.S. Military Academy at West Point, N.Y., From Its Establishment in 1802 to 1890*, 3rd ed. (Boston and New York, 1891), 2:502–3; for refugees, see *The Citizen* (Tucson), 28 March 1879. This revolt put Cucurpe at risk given the role of Francisco Altamirano y Altamirano, Don Chico's son. "Latest News from Sonora," *Arizona Weekly Citizen* (Tucson), 7 March 1879; "Sonora Revolution," *Arizona Weekly Citizen* (Tucson), 11 March 1879; and "Latest from Sonora," *Arizona Weekly Citizen* (Tucson), 28 March 1879.

5. Janin diary HM 64295 (1863–64), Papers of Louis Janin (Addenda), Huntington Library, San Marino, California; but also see his report in *Reports, Prospectus, and By-Laws of the Santa*

Teresa de Jesus Silver Mining Co. (Turnbull & Smith, 1864), 10–26. For early railroad visions, see Captain T. J. Cram, *Memoir Showing How to Bring the Lead, Copper, Silver, and Gold of Arizona into the Marts of the World* (R. V. Wilson, 1858). For context, see Sarah E. M. Grossman, *Mining the Borderlands: Industry, Capital, and the Emergence of Engineers in the Southwest Territories, 1855–1910* (University of Nevada Press, 1995), esp. 21–50.

6. For Simmel's figure of the "stranger," see *George Simmel on Individuality and Social Forms*, ed. Donald N. Levine (University of Chicago Press, 1971), 143–49.

7. Hall, *Travels and Adventures*, 23, 36, 48, 53.

8. W. E. Duellman and G. R. Zug, "amphibian," *Encyclopedia Britannica*; Lauren Benton, *Search for Sovereignty: Law and Geography in European Empires, 1400–1900* (Cambridge University Press, 2010); James C. Scott, *The Art of Not Being Governed: An Anarchist History of Upland Southeast Asia* (Yale University Press, 2009); and Eric Tagliacozzo, *Secret Trades, Porous Borders: Smuggling and States Along a Southeast Asian Frontier, 1865–1915* (Yale University Press, 2009).

9. John Denton Hall entry, 17 March 1837, in Children's Registers (1827–42), Christ's Hospital Ms. 12818/015, Guildhall Library, London, England; Gordon Lang and Ellen Paul Denker, *Pottery and Porcelain Marks* (Miller's, 1995), 300; Gillian Neale, *Encyclopedia of British Transfer-Printed Pottery Patterns, 1790–1930* (Miller's, 2005), 172; *London Gazette*, 10 July 1832; and *The Staffordshire Advertiser*, 14 July 1832.

10. Notes on Bankruptcy of John Hall & Sons, 814–15; Bankruptcy Enrolment Books; Certificates of Conformity; B 5/38, TNA; For Staffordshire's transatlantic trade, see Neil Ewins, *"Supplying the Present Wants of Our Yankee Cousins . . .": Staffordshire Ceramics and the American Market, 1775–1880* (City Museum & Art Gallery, 1997). For Bagster's moves, see directors' minute, Liverpool & Manchester (RAIL 371), Leicester & Swannington (RAIL 359), and London & Birmingham (RAIL 384), but also "Letters from Superintendent A. Bagster," RAIL 1008/98, TNA. For Stephenson, see Michael Bailey, ed., *Robert Stephenson: The Eminent Engineer* (Ashgate, 2003).

11. For Stephenson's work in South America, see Bailey, ed., *Eminent Engineer*, 14–21.

12. John Denton Hall entry, 17 March 1837. For Hannah's work as a teacher of "young gentlemen," see "Warwick-Place Small Heath," and "Warwick-Place," *Birmingham Gazette*, 11 June 1838. I suspect this shaped Hall's work, later, as a teacher in Cucurpe.

13. For Christ's Hospital and its Royal Mathematical School, see Ken Mansell, *Christ's Hospital in the Victorian Era* (Ashwater, 2011); for Hall's movements through the Royal Mathematical School, see Minutes of Committee of Almoners, 18 January and 7 June 1839, *Committee Minutes Book for 1836–1840*, 432 and 488–89; and 30 November 1841, 25 May, 29 November, and 6 December 1842, and 3 January 1843, *Committee Minutes Book for 1840–1845*, 131, 191, 256, 260, and 264 in Christ's Hospital Ms. 12811/019, Guildhall Library, London, England. For Cass, Roe, and Bushnan, start with Peter Bromley Maling, "Thomas Cass," *An Encyclopaedia of New Zealand*, ed. Alexander H. McLintock (R. E. Owen, 1966), I:318; Luciana Martins and Felix Driver, "John Septimus Roe and the Art of Navigation, c. 1815–30," in *Art and the British Empire*, ed. Tim Barringer et al. (Manchester University Press, 2007), 53–66; and "Lt. John Bushnan," *The New Monthly Magazine and Literary Journal* (1 November 1824), 520.

14. *Staffordshire Gazette* (Stafford), 16 May 1840; *Derby Mercury* (Derby), 10 July 1839. For Pirie, see "Alderman Sir John Pirie, Bart.," *The Gentleman's Magazine and Historical Review* (May 1851), 551–52. For Bagster and his business relationship to Moorsom, see A. Bagster, Superintendent of Coaching Department, to Capt. Moorsom, RAIL 384/282, TNA.

15. John C. Templer, ed., *The Private Letters of Sir James Brooke, K.C.B., Rajah of Sarawak, Narrating the Events of his Life from 1838 to the Present Time*, 3 vols. (Richard Bentley, 1853), esp. 1:1–278.

16. *A Selection from Papers Relating to Borneo and the Proceedings at Sarawak of James Brooke, Esq., Now Agent for the British Government in Borneo* (Robson, Levey & Franklyn, 1846); Henry Keppel, *The Expedition to Borneo of H.M.S. Dido for the Suppression of Piracy*, 2 vols. (Chapman and Hall, 1846), II:99–116; 305–6. For raiders from the Sulu Sea, begin with James Francis Warren, *The Sulu Zone, 1768–1898: The Dynamics of External Trade, Slavery and Ethnicity in the Transformation of a Southeast Asia Maritime State* (Singapore University Press, 1981); and J. H. Walker, *Power and Prowess: The Origins of Brooke Kingship in Sarawak* (University of Hawaii Press, 2002).

17. Capt. Edward Belcher to Rear Admiral Sir Thomas Cochrane, *H.M.S. Samarang*, Hong Kong, April 26, 1845, *Orderbook and Letterbook of H.M.S. Samarang*, National Maritime Museum, Greenwich, England (hereafter NMM). For the larger itinerary of HMS *Samarang*, see Edward Belcher, *Narrative of the Voyage of H.M.S. Samarang, During the Years 1843–46* (Reeve, Benham, and Reeve, 1848); and Frank S. Marryat, *Borneo and the Indian Archipelago* (Longman, Brown, Green, and Longmans, 1848).

18. *Royal Albert* schedules in the *Port of Registry: London Ships Names: RI–RZ*, 1848, Records of Board of Trade, Registry of Shipping and Seamen: Agreements & Crew Lists (BT 98/1680), TNA; "First Tea Sale in Ireland," *Morning Post* (London), 24 March 1835; and Hall, *Travels and Adventures*, 48, 53. My sense of the shadowy littoral comes from business records of Jardine, Matheson & Co., a China-trading company with whom Scanlan was associated, in the Jardine Matheson Archive, Cambridge University Library, UK.

19. Belcher, *Narrative*; Marryat, *Borneo and the Indian Archipelago*; and George F. McDougall, typescript journal kept in H.M.S. *Samarang* 1842, MSS 89/105.0, NMM; *Royal Albert* schedules; Templer, ed., *Private Letters of Sir James Brooke*, 286.

20. For Hall's arrival in Lima, see "Puerto del Callao," *El Comercio* (Lima, Peru), 16 August 1849 (where his name is misspelled as Juan Jonton Wale), and "Razon de los pasaportes presentados y expedidos," *El Comercio* (Lima, Peru), 22 August 1849.

21. Hall, *Travels and Adventures*, 16. A study of the ships that remained and rotted in San Francisco Bay is James P. Delgado's *Gold Rush Port: The Maritime Archaeology of San Francisco's Waterfront* (University of California Press, 2009). For a view of the connections between South America and California, see Jay Monaghan, *Chile, Peru, and the California Gold Rush of 1849* (University of California Press, 1973).

22. Frank Marryat, *Mountains and Molehills, or Recollections of a Burnt Journal* (Longman, Brown, Green, and Longman, 1855); Marryat, *Borneo and the Indian Archipelago*; Belcher, *Narrative*; Frederick Marryat, *Narrative of the Travels and Adventures of Monsieur Violet in California, Sonora, and Western Texas* (Longman, Brown, Green, and Longman, 1843). For Frederick Marryat and the Marryat family, see Florence Marryat, *The Life and Letters of Captain Marryat*, 2 vols. (Richard Bentley and Son, 1872).

23. Hall, *Travels and Adventures*, 9, 14.

24. For Hall's 1851–52 travels to California and return with Clarke, see Hall, *Travels and Adventures*, 47–98. For a more in-depth discussion of these families and my efforts to place them in village networks, see Samuel Truett, "Border-Crossing Microhistories: Reconnecting Small Worlds in the Global Nineteenth Century," in *To See the World: Microhistories of the Global United States*, ed. David Sim and Nora Lessersohn (Cornell University Press, forthcoming). For

intermarriage patterns in contexts of coastal and inland trade, start with Louise Pubols, *The Father of All: The de la Guerra Family, Power, and Patriarchy in Mexican California* (University of California Press, 2010), and Anne Hyde's *Empires, Nations, and Families: A History of the North American West, 1800–1860* (University of Nebraska Press, 2011). After visiting Hall and Palomino in 1864, Louis Janin sent "a case of ladies instruments for sewing" as a gift for her hospitality, indicating to me a sense of household parity, and possibly gesturing to the labor that Palomino gave to the village. Loose paper in Janin diary HM 64294 (1863), in Papers of Louis Janin (Addenda), Huntington Library, San Marino, California.

25. For changes following 1854, see Samuel Truett, *Fugitive Landscapes: The Forgotten History of the U.S.-Mexico Borderlands* (Yale University Press, 2006), 33–51.

26. For context, see Michel Gobat, *Empire by Invitation: William Walker and Manifest Destiny in Central America* (Harvard University Press, 2018); Joseph Allen Stout, Jr., *The Liberators: Filibustering Expeditions into Mexico, 1848–1862 and the Last Thrust of Manifest Destiny* (Westernlore Press, 1973); and my own, revisionist take on Crabb's filibuster in Samuel Truett, "Borderlands and Border Crossings," in *Cambridge History of America and the World*, ed. Mark Bradley, Vol. 2, *1812–1900*, ed. Kristin Hoganson and Jay Sexton (Cambridge University Press, 2022), 594–616. For the notion of the "Anglo-world," I draw on James Belich, *Replenishing the Earth: The Settler Revolution and the Rise of the Anglo-World, 1783–1939* (Oxford University Press, 2009).

27. Hall, *Travels and Adventures*, 174–96, 198–209. For a more in-depth discussion of the Sonoita Massacre and Hall's eyewitness account, see Truett, *Fugitive Landscapes*, 45–47.

28. For Indigenous borderlands, see David Yetman, *The Ópatas: In Search of a Sonoran People* (University of Arizona Press, 2010); Campbell W. Pennington, *The Pima Bajo of Central Sonora, Mexico*, 2 vols. (University of Utah Press, 1980); José Refugio de la Torre Curiel, *Twilight of the Mission Frontier: Shifting Interethnic Alliances and Social Organization in Sonora, 1768–1855* (Stanford University Press, 2012); and Lance R. Blyth, *Chiricahua and Janos: Communities of Violence in the Southwest Borderlands, 1680–1880* (University of Nebraska Press, 2012). For Hall's work as a scribe and teacher, see Janin diary HM 64295 (1863–64); his work as a healer runs throughout his memoirs, but I also draw on the testimony of descendants: email from Aída Bustamante to Samuel Truett, November 17, 2009, and Rafael Bustamante in Carlos Sánchez, "Cómo es vivir en Soledad en estos días de Lluvia," https://elmineral.com.mx/ver_anterior.php?artid=37368. Hall built a close relationship not only to the Sinohui family, but also to that of Don Chico's brother Atenogenes, who was connected to Ópata families by marriage and godparentage, if not also ethnic identity.

29. For the history and social relations of Jolo, see Warren, *Sulu Zone*; "birdcage-looking buildings" comes from Keppel, *A Sailor's Life*, 2:100–101. See also Belcher, *Narrative of the Voyage*, 1:114–18. For Marryat quote, see Marryat, *Borneo and Indian Archipelago*, 41.

30. For these spatial dynamics, see James Francis Warren, "Saltwater Slavers and Captives in the Sulu Zone, 1768–1878," *Slavery and Abolition* 31, no. 3 (September 2010): 429–49. For "kinetic power," I draw on Pekka Hämäläinen, "Empires in Motion," Kinetic Empire Workshop, Oxford Centre for Global History, November 6, 2013.

31. Blake gathered most of his information from informants in Manila. Captain J. J. Blake, "Information Regarding Illanoon Pirates," *Journal of the Asiatic Society of Bengal* 7, no. 83 (1838): 979–82. See also discussion of the Iranun inlands in Warren, *Sulu Zone*, 149–51.

32. Walker, *Power and Prowess*, 6–9. For a discussion of British-Malay intimate networks in Brooke's Sarawak, see Robert H. W. Reece, "European-Indigenous Miscegenation and Social

Status in Nineteenth-Century Borneo," in *Female and Male in Borneo: Contributions and Challenges to Gender Studies*, ed. Vinson Sutlive (Borneo Research Council, 1991), 455–88.

33. Spenser St. John, *Life in the Forests of the Far East* (London: Smith Elder & Co., 1862), II: 171–203; and Nicholas Loney, *A Britisher in the Philippines; or, the Letters of Nicholas Loney* (Manila: [National Library], 1964), 64.

34. For shape shifting, I draw from James Scott, who notes that in spaces of weak state power, a "mixed portfolio" of identities may offer "great protective value," especially in contested contexts where "a definite, fixed identity might prove fatal." Scott, *Art of Not Being Governed*, 256. For Conrad and the use of Wyndham as a basis for the Lingard character, see Laurence Davies et al., *The Collected Letters of Joseph Conrad*, Vol. 9, *Uncollected Letters and Indexes* (Cambridge University Press, 2007), 56–57.

35. For the longer history of these Apache groups and their political organization in the region, see Paul Conrad, *The Apache Diaspora: Four Centuries of Displacement and Survival* (University of Pennsylvania Press, 2021).

36. For these dynamics, see, for instance, Hall, *Travels and Adventures*, 38–40, 47–49, 64–70, and 92–96, but also see *Western Apache Raiding and Warfare, from the Notes of Grenville Goodwin*, ed. Keith H. Basso (University of Arizona Press, 1971); Morris E. Opler, "Chiricahua Apache," in *Southwest*, ed. Alfonso Ortiz; *Handbook of North American Indians*, ed. William C. Sturtevant (Smithsonian Institution, 1983), 401–18; and Morris Edward Opler, *An Apache Life-Way: The Economic, Social, and Religious Institutions of the Chiricahua Indians* (University of Nebraska Press, 1941).

37. Juan Nentvig, *Rudo Ensayo: A Description of Sonora and Arizona in 1764*, trans. and annot. Alberto Francisco Pradeau and Robert R. Rasmussen (University of Arizona Press, 1980), 70; Ignaz Pfefferkorn, *Sonora: A Description of the Province*, trans. and annot. Theodore E. Treutlein (University of New Mexico Press, 1949), 70–71; Asa Daklugie testimony in Eve Ball, *Indeh: An Apache Odyssey*, rev. ed. (University of Oklahoma Press, 1988), 57, 76.

38. For relations of borderlands mobility noted here, see John Russell Bartlett, *Personal Narrative of Explorations and Incidents in Texas, New Mexico, California, Sonora, and Chihuahua*, 2 vols. (D. Appleton, 1854), I:398–401, and Hall, *Travels and Adventures*, 44–45. For Apache control of the area around Cucurpe, begin with Edwin R. Sweeney, *Cochise: Chiricahua Apache Chief* (University of Oklahoma Press, 1991); see Hall, *Travels and Adventures*, 64–70, for the border-crossing story of the captive who guided Hall's companions into the land of his adopted kin.

39. Cooper was also known at the time for his sea novels.

40. I echo arguments in Jürgen Osterhammel, *The Transformation of the World: A Global History of the Nineteenth Century* (Princeton University Press, 2014); and Charles S. Maier, "Consigning the Twentieth Century to History: Alternative Narratives for the Modern Era," *American Historical Review* 105, no. 3 (June 2000): 805–31. For the state conquest of Apaches, see Conrad, *Apache Diaspora*, esp. 209–48; for the conquest of Samal and Iranun waters at this same time, see Warren, *Sulu Zone*, 104–43.

41. For Mexican railroads, see John H. Coatsworth's classic *Growth Against Development: The Economic Impact of Railroads in Porfirian Mexico* (Northern Illinois University Press, 1981); for larger contexts, see Richard White, *Railroaded: The Transcontinentals and the Making of Modern America* (W. W. Norton, 2011).

42. Copies at the Weldon (Western Ontario) and Huntington libraries are inscribed by Elijah Smith; a copy at the Houghton Library (Harvard) has the bookplate of T. Jefferson Coolidge and is inscribed by S. A. Kent; and a fourth, at the Beinecke Library (Yale), is inscribed by

H. C. Nutt. All were associated in one way or another with the Atchison, Topeka, and Santa Fe Railroad, though Smith, in particular, moved through several railroad circles based out of Massachusetts. I suspect these men gathered together less as railroad and army men, and more as brokers of inside knowledge for land speculation.

43. See the dedications and other bibliographic data for the various copies of Hall's *Travels and Adventures* in the libraries noted above for relevant clues to the larger cohort of railway engineers. For Taber quote, see "Notes," *Boston Daily Advertiser* (Boston), March 12, 1881. For a larger context for these railway magnates (and ties to the earlier China and Pacific trades), see Barry Supple, *Boston Capitalists and Western Railroads: A Study in the Nineteenth-Century Railroad Investment Processes* (Harvard University Press, 1967).

44. See Hubert Howe Bancroft, *Works*, Volume XIV, *History of Mexico*, Volume VI, *1861–1887* (The History Company, 1888), 329; and *Works*, Volume XVII, *History of Arizona and New Mexico, 1580–1888* (The History Company, 1889), 401, 499. For Myers's movements after Chicago, see Cullum, 2:502–3, and "Mustered Out," *The National Tribune* (Washington, DC), 24 November 1887.

45. "Death of Noted Admiralty Lawyer," *Railway and Marine News* (15 March 1912), 21–22, Prudence and Lloyd Draper, *Cotati* (Arcadia, 2004), and Flora H. Loughhead, *The Libraries of California; Containing Descriptions of the Principal Private and Public Libraries Throughout the State* (A. L. Bancroft and Co., 1878), 239; 24 March 2013 email with John Page for Page papers; for Page, Janin, and Yale Club of California, see *San Francisco Blue Book and Club Directory* (Hoag and Irving, 1891), 145–46.

46. For Thomas Stokes Page, Anna María Liljevalch, and their family (and its various peregrinations), begin with Draper and Draper's *Cotati*; but also see the family tree with documents at https://www.ancestry.com/family-tree/person/tree/21499153/person/110155710628/facts?_phsrc=wNz668&_phstart=successSource.

47. Hall, *Travels and Adventures*, 53; For Keppel's unexpected detour north, see Sir Henry Keppel, *A Sailor's Life Under Four Sovereigns*, 2 vols. (Macmillan and Co., 1899), II:181–90; Sir Henry Keppel's logbook-journal of HMS *Maeander* (HTN/69, NMM). For Page at 58 South Park and Robinson at 59 South Park Street, see 1870 US Federal Census for San Francisco Ward 9, image 197 of 274, accessed through Ancestry.com on March 9, 2025.

CONTRIBUTOR BIOGRAPHIES

Sarah Chute is a PhD candidate at the University of Toronto researching slavery and freedom in eighteenth- and nineteenth-century North America. Her research interests are in transnational free and forced migration, trade, and the connections that Nova Scotia, New Brunswick, and Prince Edward Island had with slavery in other parts of the Atlantic world, including the British Caribbean.

Edward G. Gray was Professor of History at Florida State University in Tallahassee. Gray passed away on December 22, 2023, as this volume was nearing publication. His early work focused on the history of ideas and especially the relationship between language, culture, and politics. Later work carried similar themes from linguistic history into the history of space and cartography with, shortly before his death, a monograph on the Mason-Dixon line. He was a generous and penetrating collaborator, mentor, and scholar. He is sorely missed by the editors and contributors to this volume. We wish to thank Stacey Rutledge for her assistance and permission to publish his essay.

Kim M. Gruenwald is Associate Professor of History at Kent State University. She first became interested in rivers as borders when her research showed how long it took the Northwest Ordinance to act as the boundary between North and South in the Ohio Valley during the early republic.

Rachel B. Herrmann is Senior Lecturer in Modern American History at Cardiff University. A specialist in the history of food, hunger, and cannibalism, she turns her attention now to how people constructed, navigated, and undermined borders in the Gulf South on rivers, bays, coasts, islands, and swamps.

Christian J. Koot is a Professor of History at Towson University. He is a historian of the British Atlantic world interested in the ways that that early modern peoples understood cartographic borders and how they alternatively manipulated them, transversed them, and gave them meaning.

Chad McCutchen is an Associate Professor of History at Minnesota State University, Mankato in Mankato, Minnesota. His research focuses on Spanish colonialism, the Andes, ethnohistory, and the Atlantic world. His current research analyzes the continuity and influence of Indigenous cultures within the development of colonial Latin American society.

Jennifer Monroe McCutchen is an Assistant Professor of History at the University of St. Thomas in St. Paul, Minnesota. She specializes in early American history and ethnohistory, with a focus on the themes of gender, power, exchange, and diplomacy. Her current scholarship explores the cultural and political impact of gunpowder among the Muscogee Creek in the late eighteenth-century Southeast.

John Morton has been a Visiting Assistant Professor at Boston College and at St. Joseph's University in Philadelphia. Currently he is Executive Director of the Paul Revere Museum of Discovery and Innovation in Canton, Massachusetts. He specializes in New England and Atlantic Canada, with a focus on church networks and border formation after the American Revolution.

Paul Musselwhite is Associate Professor of History at Dartmouth College. His research explores the conceptualization of settler colonial places in the early modern Atlantic world.

Charles Prior is Professor in History at the University of Birmingham (UK). He approaches treaties between Native nations and non-Native governments in North America as sites of negotiation over sovereignty, law, constitutionalism, resources, and wider questions of the status and rights of small nations.

Karen Rann is a practicing visual artist with an abiding interest in historical maps. Her scholarly work focuses on the emergence of contours, lines that traverse borders and enabled map-users to do likewise. She received her PhD from Queen's University Belfast in 2022. Her creative work has received funding from Arts Council England, Creative Scotland, and Arts Council Ireland.

Jessica Choppin Roney is Associate Professor of History at Temple University and Director of the Program in Early American Economy and Society (PEAES) at the Library Company of Philadelphia. Her current work focuses on the trans-Appalachian west and its impact on US constitutional development in the era of US independence.

Samuel Truett is a Professor of History at the University of New Mexico. He is a historian of borderlands and border crossings in North America and the world, with particular expertise in the US-Mexico border region and its Indigenous, cross-border, and environmental contexts.

Harvey Amani Whitfield is the Centennial Carnegie Chair in the History of Slavery in Canada at University of King's College and Professor of History at Dalhousie University. He is the author of several books and articles about Black enslavement in northeastern North America.

Alex Zukas, Professor of History, Emeritus at National University, is writing a book with the working title, *Herman Moll and the Entangled Cartography of the British Empire*. His interest in political borders juxtaposes assertions of authority and state power on maps with the malleability and contradictory nature of the reality on the ground.

INDEX

ACKNOWLEDGMENTS

In addition to thanking our authors for their hard work on this volume, the co-editors are grateful to the anonymous reviewers who commented on this book, and to the network of friends, colleagues, and interlocutors who gave life to this volume: S. Max Edelson, Travis Glasson, Edward Gray, Patrick Griffin, Pekka Hämäläinen, April Hatfield, Lawrence Hatter, Christian Koot, Andrew Lipman, Kris Lane, Michael McDonnell, Harvey Neptune, Joshua Piker, Nancy Shoemaker, Alec Reichardt, Daniel K. Richter, Patrick Spero, Samuel Truett, Molly Warsh, Lissa Wadewitz, Mark Williams, and Laurie Wood. Our thanks to Ryan Langton for his careful work creating our index. We are grateful to Bob Lockhart and all the people at UPenn Press who helped make this volume a reality.

Jess would like to thank her family, Adam, Lexi, and Caitlin, who inspire her questions and enable her research. She thanks Rachel for her drive to get this volume done and Marc for the time he got up at dawn to escort her to the Bristol train station even though she said she knew the way. (She did not.) Rachel would like to thank her other half, Marc, for his constant support, and in this specific instance for supporting her unanticipated turn to tariffs. She depends on Jess to write some of the only emails she looks forward to reading, and appreciates Adam, Lexi, Caitlin, and the cats for opening their home to her in Philly. Jess and Rachel thank the McNeil Center fellowship committee that gave them the extraordinary good fortune to be fellows at the same time. Finally, they give thanks for one another: for the many years' project of producing this volume, friendship and knitwear, the excellent food, the many laughs, occasional cries, thought-provoking conversations, and (always) the unwavering support.

Kolibri 360 GmbH
Ferdinand-Jühlke-Straße 7
99095 Erfurt, Deutschland
produktsicherheit@kolibri360.de